Cyberattack Strategies Against Cyber-Physical Systems: From a Control Perspective

Sheng Gao · Huaicheng Yan · Hao Zhang ·
Yunkai Lv · Zhichen Li · Meng Wang

Cyberattack Strategies Against Cyber-Physical Systems: From a Control Perspective

Springer

Authors
See next page

ISBN 978-981-95-9644-7 ISBN 978-981-95-9645-4 (eBook)
https://doi.org/10.1007/978-981-95-9645-4

This Springer imprint is published by the registered company Springer Nature Singapore Pte Ltd.
The registered company address is: 152 Beach Road, #21-01/04 Gateway East, Singapore 189721, Singapore

Sheng Gao
Department of Control Science
and Engineering, College of Electronic
and Information Engineering
Tongi University
Shanghai, China

Key Laboratory of Smart Manufacturing
in Energy Chemical Process of the Ministry
of Education, School of Information
Science and Engineering
East China University of Science
and Technology
Shanghai, China

Hao Zhang
Shanghai Key Laboratory of Wearable
Robotics and Human-Machine Interaction,
Department of Control Science
and Engineering, College of Electronic
and Information Engineering
Tongji University
Shanghai, China

National Key Laboratory of Autonomous
Intelligent Unmanned Systems, Frontiers
Science Center for Intelligent Autonomous
Systems of the Ministry of Education
Shanghai Institute of Intelligent Science
and Technology
Shanghai, China

Zhichen Li
Key Laboratory of Smart Manufacturing
in Energy Chemical Process of the Ministry
of Education, School of Information
Science and Engineering
East China University of Science
and Technology
Shanghai, China

Huaicheng Yan
Faculty of Artificial Intelligence
Shanghai University of Electric Power
Shanghai, China

Key Laboratory of Smart Manufacturing
in Energy Chemical Process of the Ministry
of Education, School of Information
Science and Engineering
East China University of Science
and Technology
Shanghai, China

Yunkai Lv
Key Laboratory of Smart Manufacturing
in Energy Chemical Process of the Ministry
of Education, School of Information
Science and Engineering
East China University of Science
and Technology
Shanghai, China

Meng Wang
Key Laboratory of Smart Manufacturing
in Energy Chemical Process of the Ministry
of Education, School of Information
Science and Engineering
East China University of Science
and Technology
Shanghai, China

Upon the publication of this book, the author would especially like to thank the China Postdoctoral Science Foundation, and the National Natural Science Foundation of China for their long-term support. The author is also grateful to colleagues in the academic and industrial communities at home and abroad, whose valuable exchanges have deepened the author's understanding of cyber-physical systems and provided inspiration. Special thanks are also extended to the author's team leaders and fellow graduate students for their strong support and assistance.

Preface

In this era of rapid technological advancement, cyber-physical systems (CPSs) are fundamentally transforming the way we live, work, and operate as a society. With continuous breakthroughs in computing hardware and information technology, networks have expanded into diverse domains, fostering the deep integration of cyberspace and the physical world and thereby giving rise to the concept of CPSs.

The CPSs are multidimensional and complex systems that integrate computational, network, and physical environments through 3C technologies, i.e., computation, communication, and control. They enable seamless integration and deep collaboration among system components, supporting real-time sensing, dynamic control, and information services in large-scale engineering applications. By incorporating multidisciplinary theories from communication technology, sensor technology, computer science, and control science, CPSs have become increasingly vital in key domains such as smart manufacturing, intelligent transportation, and smart grid systems. Consequently, they have become a central focus for achieving the modernization of nations worldwide in advancing digitalization, networking, and intelligence.

In 2017, China's State Council issued the "New Generation Artificial Intelligence Development Plan," explicitly elevating intelligent system cybersecurity to a national strategic level and identifying it as a key priority. That same year, the Ministry of Industry and Information Technology released the "Cyber-Physical Systems White Paper," further emphasizing its critical importance. In 2018, the United States enacted the "Cybersecurity and Infrastructure Security Agency Act," establishing the protection of national networks and critical infrastructure as a core mission. In 2020, the European Union introduced the "Directive on Measures for a High Common Level of Network and Information Security across the Union," also prioritizing the security of CPSs. Singapore implemented its "Cybersecurity Act" in 2024, focusing on strengthening the cyber resilience of such systems. In 2025, the Chinese Academy of Engineering's report "China's Engineering Science and Technology 2040 Development Strategy: Information and Electronic Fields" once again listed cybersecurity in

CPSs as a critical technology for future development. Thus, it is evident that cybersecurity in CPSs has become a core research direction in the global development of intelligent systems.

In practical scenarios, the plants and information centers within CPSs are typically distributed across different geographical locations. They rely on wireless communication networks to facilitate information exchange among components, which brings various conveniences to accomplishing overall tasks. However, the open network transmission environment is vulnerable and susceptible to intrusion, making it prone to external interference and threats. Consequently, security issues in CPSs have garnered extensive attention from both academia and industry. Yet cyberattack techniques continue to evolve, with new tactics and strategies emerging constantly. Studying the design of the attack strategies from a control perspective aims to provide defenders with insights into attack behaviors. This enables defenders to understand and predict potential future threats, fostering the development of more robust defense mechanisms to ensure the security of CPSs. To this end, this book adopts an attacker's viewpoint, integrating control theory, optimization theory, and game theory to investigate the design of DoS attack and FDI strategies.

This book primarily draws upon the author's systematic research on key technologies and critical issues related to the design of the attack strategies for cybersecurity in CPSs, carried out since the beginning of the author's Ph.D. studies. Chapter 1 introduces the concept of cyber-physical systems, the significance of their cybersecurity, and research progress both domestically and internationally. Chapters 2–8 present research results on DoS attacks and FDI attacks under model-based, data-driven, time-varying delay, and game-based scenarios. The book validates the effectiveness and superiority of the theoretical methods through simulation and physical examples.

This book is mainly intended for scholars in fields such as cybersecurity, intelligent control, cyber-physical systems, and unmanned systems, as well as Ph.D. and Master's students. It is also useful for control engineers working in industries such as robotics and aviation.

Due to the author's limited abilities, there are inevitably some oversights in the book. Readers are kindly requested to provide criticism and suggestions.

Shanghai, China Sheng Gao
September 2025

Acknowledgements This book has been supported by the National Natural Science Foundation of China (62503174, 62433014, 62333005, 62403200, 62473133, 62533002), the China University Industry-Research Innovation Fund (2024IT032), the State Key Laboratory of Autonomous Intelligent Unmanned Systems (ZZKF2025-1-24), the Shanghai International Science and Technology Cooperation Project (24510714000), the Shanghai Natural Science Foundation (24ZR1416200), and the Fundamental Research Funds for the Central Universities.

Competing Interests The authors have no competing interests to declare that are relevant to the content of this manuscript.

Ethics Approval The chapter does not include studies with humans, and no ethics approval is required.

Contents

Acronyms

$\mathbb{N}$	Integer field
$\mathbb{N}^+$	Positive integer field
$\mathbb{R}$	Real number field
$\mathbb{R}^n$	n-dimensional Euclidean space of real vectors
$\mathbb{R}^{n\times n}$	$n \times n$-dimensional real matrix space
$\mathbb{S}_+^n$	$n \times n$ positive semi-definite matrix space
$\mathbb{S}_{++}^n$	$n \times n$ positive definite matrix space
$\mathbf{A} \succ 0$	Positive definite matrix **A**
$\mathbf{A} \prec 0$	Negative definite matrix **A**
$\mathbf{A} \succeq 0$	Positive semi-definite matrix **A**
$\mathbf{A} \preceq 0$	Negative semi-definite matrix **A**
$\mathbf{A}^T$	Transpose matrix of the matrix **A**
$\mathbf{A}^{-1}$	Inverse matrix of the matrix **A**
$tr(\mathbf{A})$	Trace of matrix **A**
$\mathbb{E}[\mathbf{A}]$ or $\mathbb{E}\{\mathbf{A}\}$	Expectation of matrix **A**
$\|\mathbf{A}\|$	2-norm of the matrix **A**
$\|\mathbf{A}\|_\infty$	Infinity norm of the matrix **A**
$\prod_{j=1}^n \mathbf{A}(j)$	Product of the n matrices $\mathbf{A}(j)$
$\lambda \cdot \mathbf{A}$ or $M \cdot \lambda$	The product of number λ and matrix **A**
$\mathbf{A} \times \mathbf{B}$	The product of matrix **A** and matrix **B**
$\mathbf{A}/\lambda$	The division of matrix **A** and number λ
$\mathrm{diag}\{\cdot\}$	Diagonal matrix
$\mathrm{diag}\{\mathrm{d}_1, ..., \mathrm{d}_\mathrm{n}\}$	Diagonal matrix with the diagonal elements $d_1, ..., d_n$
$Sign$	Sign function
Sat	Saturation function
$\frac{\partial f}{\partial M}$	The first-order partial derivative of the function f with respect to the matrix M
$\frac{\partial^2 f}{\partial M^2}$	The second-order partial derivative of the function f with respect to the matrix M

Chapter 1
Introduction

Abstract This chapter presents an introduction of the foundational concepts, background, and security challenges in cyber-physical systems (CPSs). By analyzing documented security incidents and demonstrating representative case studies, the critical importance of CPSs security is highlighted. Furthermore, this study emphasizes the necessity of analyzing denial-of-service (DoS) and false data injection (FDI) attack strategies from a control-theoretic perspective, particularly considering whether attackers possess knowledge of the system model. It also underscores the value of modeling such attack behaviors within a game-theoretic framework to capture adversarial interactions. The chapter provides the comprehensive overviews of DoS and FDI attack strategies against CPSs, covering model-based, data-driven, and game-based approaches to the design of the attack, along with the discussions of the current research landscape in these areas.

1.1 Cyber-Physical Systems Concept

1.1.1 Introduction to Cyber-Physical Systems

With the continuous advancement of computer data processing capabilities and the rapid development of wireless communication technologies, CPSs, which integrate computing, communication, control, and physical processes, have experienced accelerated growth. CPSs can typically be characterized as complex networked control systems that seamlessly bridge physical entities in the real world with computational resources in the virtual cyberspace through embedded systems [1–5]. Different from the conventional embedded systems, a mature CPS is generally architected as a network of interconnected components featuring physical inputs and outputs, rather than functioning as isolated devices. This enables the coordinated operation and deep integration of physical components and software elements across multidimensional spatiotemporal scales. This concept also is closely associated with robotics and sensor networks, where intelligent computational mechanisms guide system behavior. Continued progress in science and engineering enhances the integration between computational and physical elements through the implementation of smart strate-

S. Gao et al., *Cyberattack Strategies Against Cyber-Physical Systems: From a Control Perspective*, https://doi.org/10.1007/978-981-95-9645-4_1

Fig. 1.1 The practical applications of cyber-physical systems [9–16]

gies, thereby boosting the adaptability, autonomy, efficiency, functionality, reliability, safety, and usability of CPSs [6–8]. These advantages have made CPSs widely applicable across multiple cutting-edge technologies and critical areas of social development, including but not limited to intelligent manufacturing, aerospace, intelligent warfare, intelligent marine, UAV swarms, smart cities, intelligent transportation, and smart healthcare, as detailed in Fig. 1.1.

1.1.2 Background and Significance of Cyber-Physical Systems

Cyber-physical systems possess strong self-organizing, learning, and reasoning capabilities, which facilitate the completion of overall tasks. However, open wireless communication networks also provide malicious attackers with avenues to disrupt these tasks [17, 18]. In recent years, as illustrated by global statistics on information technology (IT) security vulnerabilities and exposures alongside global hacker/IT incident counts which can be observed from Fig. 1.2, the frequency of cybersecurity incidents has escalated significantly. To name a few, which can be seen from Fig. 1.3, in March 2021, a cyberattack on the U.S. oil pipeline operator forced pipeline shutdowns, triggering a national emergency [21]. In March 2022, European satellite communications provider Viasat suffered a cyberattack that caused 5,800 Enercon wind turbines in Germany to malfunction [22]. In August 2023, attacks on Danish hosting providers CloudNordic and AzeroCloud resulted in the near-total loss of customer data [23]. In May 2024, Japan's solar grid infrastructure was attacked, resulting in the hijacking of approximately 800 remote monitoring devices (SolarView Compact) at solar power facilities [24]. Fundamentally, these attacks represent disruptions to cyber-physical systems at both the information and physical layers, ultimately causing massive economic losses and severe societal impacts. This has brought cybersecurity issues to the forefront of academic and industrial attention. Moreover, these

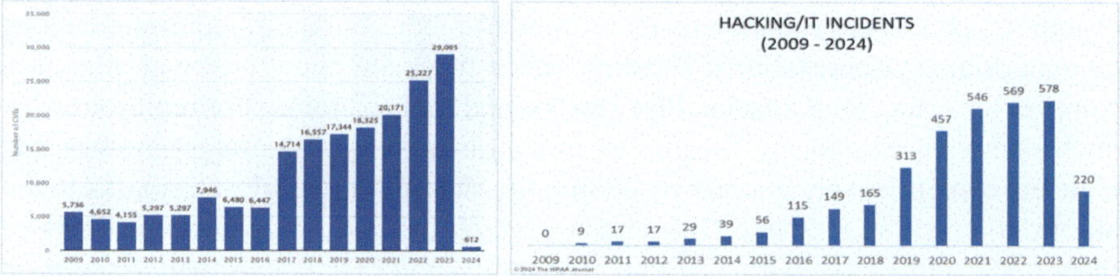

Fig. 1.2 Global IT security vulnerabilities with exposures and global hacking/IT incident count statistics chart [19, 20]

Fig. 1.3 Typical cybersecurity incidents [21–24]

incidents demonstrate that national security is inseparable from cybersecurity, making the cybersecurity of cyber-physical systems a critical challenge.

In the cybersecurity domain of cyber-physical systems, the core objective centers on building robust defense mechanisms designed to comprehensively resist and protect systems from various cyberattacks. To achieve this goal, gaining a deep understanding of attackers' motivations, strategies, and behavioral patterns is crucial. By analyzing the response characteristics of the attackers during cyberattacks, potential threats can be accurately identified and effectively mitigated, thereby enhancing the security resilience of the systems. Therefore, designing and executing attack strategies from the perspective of the attackers not only provides defenders with insights into attacker behavior but also offers a theoretical basis for optimizing defense strategy design. In practical engineering applications, attackers possess limited energy resources, preventing sustained high-frequency attack signal transmission over extended periods. Additionally, the complex and dynamic network communication environment of cyber-physical systems necessitates attackers to consider both energy constraints and the selection of communication channels for their attacks. This dual constraint transforms attack strategy formulation into a complex optimization problem. When confronting complex network environments with inherent resource constraints, attackers primarily focus on how to precisely disrupt communications and undermine system stability. Existing attack types are numerous, with common examples including DoS attacks, FDI attacks, and replay attacks. For replay attacks, which involve continuously feeding historical data back to the target system, the defenders can effectively counter by adding timestamps or digital watermarks to the data. In contrast, attacks like DoS attacks, which disrupt communication links or render nodes inoperable with a certain probability, and FDI attacks, which focus on tampering with signals undetected, pose significant challenges for defenders. Consequently, the core research focus of this book lies in exploring DoS and FDI attack strategies.

It is noteworthy that for the attacked systems with known models, the attackers design various attack strategies using known system model parameters. However, in real-world attack scenarios, sensitive information is often difficult to steal. Thus, designing effective attack strategies under completely unknown model parameters has emerged as a critical research challenge. In the design of attack strategies without models, where the attackers possess no prior knowledge of the system model, strategies can only be constructed based on wirelessly transmitted data from the attacked system. For instance, Kim et al. [25] proposed a data framing attack strategy that operates without prior knowledge of network parameters. Research on cyberattacks where system parameter models are entirely unknown remains scarce. Consequently, the design of attack strategies under completely unknown models warrants further exploration.

Moreover, with rapid advances in computing technologies and the widespread adoption of intelligent algorithms, the proactive countermeasures employed by defenders must also be taken into account. Consequently, designing attack strategies under adversarial scenarios involving both attackers and defenders has become an important research frontier. Experts worldwide have conducted studies on active defenses against DoS attacks and FDI attacks [26–30], primarily focusing on filtering and resilient control. However, these approaches are based on designing one party's strategy after the other's strategy is determined. Given the increasing frequency and sophistication of cyberattacks in recent years, the development of attack strategies under adversarial conditions has gained significant research attention. This enables defenders to more comprehensively grasp the latest dynamics and strategic shifts of attackers.

1.2 Overview of Research on DoS Attack

Over the past two decades, DoS attacks against cyber-physical systems have garnered significant attention from both academia and industry, particularly in analyzing the design of DoS attack strategies from a control perspective. This subsection provides a detailed analysis of the current state of research worldwide regarding the modeling of DoS attacks and the design of game-theoretic DoS attack strategies.

1.2.1 DoS Attack Strategies via Attack Model

Designing DoS attack strategies from the side of the attacker helps defenders develop reasonable countermeasures against new forms of DoS attacks. Due to their unpredictable nature and diverse methods, there is no unified approach to modeling DoS attacks. Experts both domestically and internationally have achieved some preliminary results in this area. Table 1.1 summarizes these findings.

Initially, Professor Ling Shi's team at the Hong Kong University of Science and Technology proposed an optimal periodic DoS attack strategy under energy constraints [31], building upon their earlier work on optimal periodic scheduling of sensors under limited resources [32, 33]. They also presented an optimal DoS attack strategy when the receiver employs an intrusion detection system (IDS). For modeling the DoS attack itself, the paper employs a discrete sequence described by binary values 0 and 1. Subsequently, they extended this work to designing optimal attack strategies for networked control systems [34], focusing on the optimal disturbance attack that maximizes the cost function of linear quadratic gaussian (LQG) control under energy constraints. After analyzing the characteristics of the cost function under arbitrary attack plans, the optimal interference attack plan and corresponding cost function were derived, and the stability of the control system under this optimal

Table 1.1 Summary on the modeling methods of the DoS Attack

Methods index	Modeling approaches for DoS attacks	Discrete/Continuous	Periodic/Aperiodic
Shi et al. [31]	Periodic discrete sequence described by 0,1	Discrete	Periodic
Martinez et al. [35]	Pulse-width-modulated periodic signal	Continuous	Periodic
Peng et al. [38]	Signal-to-interference-and-noise ratio model	Discrete	Aperiodic
Li et al. [39]	Signal-to-interference-and-noise ratio model, limited attack times	Discrete	Aperiodic
Alsaadi et al. [40]	Gaussian white noise satisfying the Bernoulli binomial distribution	Discrete	Aperiodic
Peng et al. [41]	Define two time intervals to describe the existence of DoS attacks	Continuous	Aperiodic
Ye et al. [42]	Average dwell time	Discrete	Aperiodic

attack strategy was analyzed. Subsequently, Martinez et al. [35] described the periodic behavior of the DoS attacker using a pulse-width-modulated continuous signal. Whenever the attack signal transitions to a high level, the communication channel between the sensor and controller becomes instantaneously congested, causing the control signal input to drop to zero and rapidly destabilizing the system. Inspired by the signal-to-noise ratio (SNR) and signal-to-interference-and-noise ratio (SINR) in communication networks [36, 37], Peng et al. [38] proposed a SINR-based DoS attack model. This model reveals the interactions among attacker energy, sensor transmission energy, and transmission channels during network defense. By transforming the problem model into a markov decision process (MDP) and applying bellman's optimality theorem, they derived an optimal attack strategy. Li et al. [39] similarly employed SINR for DoS attack modeling. By incorporating inequality constraints on terminal state error covariance and sufficiency conditions regarding attack frequency's impact on terminal error, they derived a closed-form solution for optimal attack energy allocation. Alsaadi et al. [40] proposed a binomial distribution model, characterizing occasional DoS attacks using Gaussian white noise following a Bernoulli distribution. Unlike the aforementioned models based on probability distributions or periodic properties. Peng et al. [41] introduced a non-periodic DoS attack strategy, defining two time intervals to describe the presence of DoS attacks in the communication channel. Compared to other models, the non-periodic DoS attack strategy proposed in this paper requires only a constant M to characterize the maximum consecutive packet loss count, making it formally simpler. An event-triggered sliding mode control method based on an observer for nonlinear unmanned surface vehicle systems under DoS attack is proposed in [42]. It employs the average dwell time to model the DoS attack, characterizing its frequency and the timing of its initiation and termination.

1.2.2 DoS Attack Strategies via Game Theory

Cyberattacks almost invariably involve interactions between two or more participants with conflicting and competing interests. These participants may be attackers or defenders. The probability of a successful attack depends on the strategic interactions among these participants. Therefore, whenever strategic interactions exist between two or more participants, a game emerges. Game theory serves as a suitable mathematical tool for studying the behavior of attackers and defenders in cybersecurity [43–45]. Basar et al. [46] pioneered the application of game theory to DoS attacks, examining the transmission of sequences of independent Gaussian random variables with identical probability distributions through a Gaussian memoryless channel with given input power constraints, under the presence of an intelligent jammer. Since then, research on DoS attacks from a game-theoretic perspective has proliferated. Table 1.2 summarizes several typical game-theoretic approaches to the design of DoS attack strategies.

Table 1.2 Summary on several typical game-theoretic approaches to the design of DoS attack strategies

Methods index	Number of attackers and defenders	Types of game	Limitations or shortcomings
He et al. [47]	One versus one		Although analytical solutions exist, the method lacks portability and is only applicable to specific game scenarios
Shi et al. [49, 50]	One versus one		
Li et al. [51]	One versus one	Static games under Nash equilibrium theory	1. These models are based on perfect information, and become invalid when information asymmetry exists between attackers and defenders
Guan, Hao, Lv et al. [52–55]	One versus one/One versus multiple		2. They lack analysis of multi-party games involving multiple attackers and defenders
Basar et al. [56]	One versus one		3. Some studies focus primarily on resilience control, with attack strategies being merely a byproduct
Zhou et al. [57]	One versus one		
Liu et al. [60, 61]	One versus one/Multiple versus one		Only the optimal static strategies for attackers and defenders have been obtained, lacking analysis of their dynamic strategies
Yang et al. [62]	One versus one	Starkberg game	Although analytical solutions exist, the method lacks portability and is only applicable to specific game scenarios
Xu et al. [63]	Multiple versus one		The starting point of the research is the design of cooperative defense strategies rather than offensive ones; the outcomes of offensive strategies are merely byproducts

In the beginning, researchers treated the DoS attack-defense process as a two-player static game [46], where the decision set of the attacker was simply defined as attack or no attack, and the decision set of the defender was defined as defend or no defend. The objective functions for both sides were constructed based on attack costs, defense costs, and the damage inflicted on the system. Subsequently, Nash equilibrium (NE) theory [47, 48] was applied to compute the mixed-strategy Nash equilibrium, recommending that both parties make decisions according to the probability values within the mixed strategy. However, such game models did not incorporate the specific behavioral patterns of DoS attacks. Subsequently, Professor Shi Ling's team constructed game models based on periodic DoS attack models [49, 50] and SINR-based DoS attack models [51]. The former demonstrated that the optimal strategies of attackers and defenders form a Nash equilibrium in a zero-sum game, yielding a closed-form solution for the strategies. The latter established a game framework based on Markov decision processes and employed reinforcement learning methods to obtain optimal solutions for the game optimization problem. These works were subsequently extensively expanded [52–55], but most of the results utilizing Nash equilibrium theory addressed games concerning energy allocation problems at the information layer in physical information systems, with little focus on games at the control layer. When addressing the control layer, numerous researchers have approached game problems from a security control perspective. Initially, the concept of cross-level games and the proposed "game-in-game principle" was introduced in [56]. It employed hybrid dynamic game theory methods, integrating discrete-time Markov models with continuous-time dynamical models to describe the controlled physical processes. Thus, security policies formulated at the information layer influence the extent to which physical systems are vulnerable to attacks, while robust control designs employing zero-sum differential games at the physical layer also impact strategy formulation at the information layer. Zhou et al. [57] simultaneously address optimal strategies for controllers, triggers, and external attackers within a game-theoretic framework, utilizing fully informed zero-sum static games to investigate optimal strategies for both offensive and defensive parties. The aforementioned results all apply Nash equilibrium theory to derive equilibrium decisions when attackers and defenders reach a steady state in their offensive–defensive interactions. However, in real-world scenarios, defenders are often unaware when malicious hackers launch cyberattacks, becoming passive victims. By the time protective measures are taken, it is frequently too late. Therefore, a leader-follower game framework is needed to describe this type of cyberattack and defense.

The Stackelberg game, first proposed by von Stackelberg in the 1930s, simulates output leadership phenomena reflecting asymmetric competition among commercial firms [58]. Applying the Stackelberg game model to cybersecurity analysis effectively captures the unequal relationship between attackers and defenders. Research based on Stackelberg game theory has gradually emerged in recent years. Tang et al. [59] investigated power control for signal sources and relay nodes in cooperative wireless networks to resist interference. They employed the Stackelberg game to model the behavior of transmitters and jammers, analyzing optimal strategies for both parties to derive the Stackelberg equilibrium (SE) of the game. Additionally, the Nash equilibrium for the anti-jamming game was derived. Simulation results indicate that jammers employing the SE strategy exhibit greater intelligence than those using the NE strategy. Li et al. [94] proposed an optimal FDI attack strategy under energy constraints based on Stackhouse game theory. Building upon this approach, Liu et al. [60, 61] developed optimal DoS attack strategies grounded in Stackhouse theory. Simulation results indicate that the probability of a DoS attacker existing in the environment does not significantly impact the final gains for either the attacker or defender. Yang et al. [62] employ the Stackhouse game framework to analyze interactions between an intelligent attacker and a defender, utilizing variable sampling to address packet loss between sensors and controllers. The work [63] considers a multi-leader, single-follower Stackelberg game model, designing an anti-interference mechanism based on the "no pains, no gains" principle. The proposed mechanism sacrifices some defenders' interests to trap the attacker, while the remaining defenders achieve better performance. This approach improves system throughput. However, most of the above research focuses on the defender side, analyzing the optimal defense strategy for defenders. The attacker's strategy is merely a byproduct of the research. Furthermore, although the works [60, 61] obtained the optimal DoS attack strategy under the Stackberg game framework, it is a static equilibrium solution. It lacks analysis of the dynamic decision-making process between attackers and defenders and requires further research.

1.3 Overview of Research on FDI Attack

Similar to the design of DoS attack strategies discussed in the previous section, FDI attacks that degrade system performance by injecting false data have also garnered significant attention and research from scholars. This section provides a detailed analysis of the current state of research on FDI attack strategies based on system models, data-driven approaches, and game theory.

1.3.1 FDI Attack Strategies via System Model

In recent years, the in-depth exploration of injection attacks in cyber-physical systems from a control theory perspective has become a research hotspot attracting widespread attention in both academia and industry. Numerous researchers have focused on how injection attacks can degrade state estimation accuracy and disrupt control performance, particularly targeting attacks on the controller-to-actuator (C-A) and sensor-to-controller (S-C) channels, namely, actuator attacks and sensor attacks. Tables 1.3 summarizes several typical FDI attack strategy designs based on system models.

For the degradation of estimation performance, under the assumption of complete information, where attackers possess all parameter matrices and data information

Table 1.3 Summary on several typical model-based FDI attack strategies

Methods index	Types of systems	Types of attack	Limitations or shortcomings
Mo et al. [64]	Linear discrete-Time cyber-Physical systems	Sensor attacks	Relying on the precision of the system parameter matries, without considering the energy consumption of the attack
Yang et al. [65]			
Huo et al. [66]			
Huo et al. [67]		Actuator attacks	
Ye et al. [68]		Sensor attacks	
Lu et al. [69]		Actuator attacks and sensor attacks	
Moura et al. [70, 71]			
Sun et al. [72, 73]	Linear continuous-Time cyber-Physical systems	Actuator attacks	Relying on the precision of the system parameter matries, without considering the stealthiness of attacks
Wu et al. [74]			
Gao et al. [75]			
Wu et al. [76]		Sensor attacks	
Murguia et al. [79]	Nonlinear discrete-Time cyber-Physical systems	Actuator attacks and sensor attacks	Without considering the energy consumption of the attack
Mikhaylenko et al. [77]	Nonlinear continuous-Time cyber-Physical systems		
Keliris et al. [78]			

of the attacked system, Mo et al. [64] proposed a false data injection attack strategy for linear discrete cyber-physical systems aimed at degrading state estimation performance. However, this strategy disregarded energy consumption constraints, limiting its practical applicability. Subsequently, Yang et al. [65] further investigated injection attack strategies similar to those degrading state estimation performance based on polling scheduling algorithms, again without addressing energy consumption considerations. Huo et al. [66, 67] proposed switching-based injection attack strategies for sensors and actuators under the same assumptions, as well as noise injection strategies for actuators. These studies similarly failed to incorporate attack energy consumption factors into their designs. Under partial information assumptions, where attackers possess only partial system knowledge, Ye et al. [68] designed a self-generating injection attack strategy targeting sensors to degrade the state estimation performance of cyber-physical systems, again disregarding attack energy constraints. Differing from the above studies, the work [69] designed simultaneous injection attack strategies targeting both actuators and sensors without assuming attacker knowledge of the estimator, demonstrating a more comprehensive attack perspective.

For the degradation of control performance, Moura et al. [70, 71] introduced quadratic control objectives and designed an injection attack strategy for linear discrete cyber-physical systems, though without addressing attack energy consumption. Subsequently, Sun et al. [72, 73] designed non-switching and switching injection attack strategies for actuators in linear continuous time-invariant systems, considering attack energy constraints. Further, for quadratic control objective functions involving single-target and multi-target summation, Wu et al. [74] proposed an optimal partial feedback injection attack strategy. Additionally, Gao et al. investigated optimal injection strategies for hybrid attack signals and their derived signals under the same system [75], though specific constraints must be satisfied to ensure strategy optimality. Notably, Wu et al. [76] extend to sensor channels, designing injection attack strategies in both non-switching and switching forms. However, these strategies were not fully designed with stealth requirements in mind, limiting their practical applicability. Mikhaylenko et al. [77] investigated injection attack strategies targeting simultaneous attacks on selected actuators and sensors within continuous-time cyber-physical systems, disregarding energy constraints. For continuous-time nonlinear cyber-physical systems containing unknown nonlinear components, Keliris et al. [78] designed injection attack strategies aimed at maximizing control performance degradation, assuming the attacker possesses information about the linear modeling portion, yet similarly neglected attacker energy consumption. For fully nonlinear discrete-time cyber-physical systems, Murguia et al. [79] proposed injection attack strategies based on node reconstruction methods to degrade system control performance. However, this approach also failed to incorporate attack energy consumption considerations. In summary, designing stealth injection attack strategies for cyber-physical systems while accounting for attack energy constraints remains an urgent and highly valuable research problem.

1.3.2 FDI Attack Strategies via Data-Driven Methods

As the data security of cyber-physical systems continues to strengthen, the ability of the attackers to obtain internal information from the attacked systems is gradually being constrained. Compared to FDI attack strategies based on system models, confronting scenarios where the system model is unknown, the so-called "black-box" environment, has become a more common challenge for the attackers. Against this backdrop, data-driven system identification techniques have been thoroughly validated in numerous references [80–83] as a key approach to solving the problem of unknown system parameters. Over the past decade, the exceptional performance of data-driven methods in handling complex and unmodeled dynamics has attracted not only widespread academic interest but also significant industrial attention. For nonlinear affine systems, a series of adaptive control strategies based on input/output data have emerged, as described in [82, 84–87], significantly advancing research in this field. However, it is only in recent years, with the enhancement of computational capabilities and the widespread adoption of intelligent learning algorithms, that the design of FDI attack strategies using data-driven methods under model-unknown conditions has gained increasing attention. Despite the immense research potential in this domain, design complexity and broad applicability remain significant challenges. Table 1.4 summarizes the current status of several typical data-driven FDI attack strategy designs.

In the study of discrete-time linear cyber-physical systems, researchers proposed an FDI attack strategy designed using historical data to degrade system state estimation performance [88]. However, this strategy still relies on the assumption that the attacker knows part of the system parameter matrix. Building upon this, Konstantinou et al. [89] designed a data-driven injection attack strategy for power grid-type cyber-physical systems using robust linear regression (RLR) techniques. Similarly,

Table 1.4 Summary on several typical data-driven FDI attack strategies

<table>
<tr><th>Methods index</th><th>Types of systems</th><th>Types of attack</th><th>Limitations or shortcomings</th></tr>
<tr><td>Yang et al. [88]</td><td rowspan="5">Linear discrete-Time cyber-Physical systems</td><td rowspan="5">Sensor attacks</td><td>Relying on the precision of the system parameter matries</td></tr>
<tr><td>Zhao et al. [91]</td><td>Without considering the complex of the network environment</td></tr>
<tr><td>Lakshminarayana et al. [90]</td><td rowspan="4">Without considering the energy consumption of the attack and the complex of the network environment</td></tr>
<tr><td>Konstantinou et al. [89]</td></tr>
<tr><td>Mukherjee et al. [92]</td></tr>
<tr><td>Santoso et al. [93]</td><td>Nonlinear continuous-Time cyber-Physical systems</td><td>Actuator attacks and sensor attacks</td></tr>
</table>

Lakshminarayana et al. [90] employed random matrix methods to design injection attack strategies based on power grid measurement data collected within a finite time window. Both of these injection attack strategies neglected the factor of attack energy consumption. Notably, in practical applications, factors such as communication delays and device failures may cause measurement data loss, thereby limiting the effectiveness of data-driven attack strategies within finite measurement windows. Subsequently, Zhao et al. [91] proposed a data-driven false data injection attack scheme for linear discrete-time invariant systems using subspace identification techniques, aiming to degrade system estimation performance. However, this strategy may not be optimal under suboptimal network conditions or latency. For instance, Mukherjee et al. [92] optimized the data processing workflow using singular value decomposition to reduce computational overhead, designing a data-driven injection attack strategy on the low-rank subspace of the mapping matrix, yet this approach similarly relies on data integrity and timeliness. For continuous-time nonlinear cyber-physical systems, Santoso et al. [93] proposed an injection attack strategy by discretizing data and integrating it with convolutional neural networks from deep learning. However, this approach also neglects attack energy consumption considerations. In summary, under factors such as attack energy consumption, complex network environments, and active defense by controllers, research on data-driven injection attacks has become a highly significant and challenging topic.

1.3.3 FDI Attack Strategies via Game Theory

With the continuous improvement of computer hardware performance and the widespread application of artificial intelligence technology, the importance of proactive defense strategies for controllers has become increasingly prominent, driving research on the adversarial relationship between attackers and controllers to the forefront. In this subsection, the attackers are considered that they can train more efficient FDI attack strategies through interactive learning and iterative optimization under game theory. Table 1.5 systematically summarizes several typical game-theoretic-based FDI attack strategy designs.

Specifically, Li et al. [94] constructed a leader-follower game framework, the Stackelberg game model, assuming attackers possess complete information about the targeted system. In this model, attackers dynamically adjust their strategies based on defenders' behavioral patterns. However, this model excludes energy consumption factors for both sides. Tian et al. [95] designed injection attack strategies under imperfect information using a dynamic Bayesian game framework. Their approach determines the payoffs for both sides based on a proposed two-layer optimization model, continuously updates the attacker's prior belief based on historical profiles and measurement relationships, and ultimately achieves a Bayesian Nash equilibrium for defense and attack strategies. Furthermore, based on the attack-defense graph game, which integrates attack-defense graphs with game theory, Liu et al. [96] modeled and analyzed cyberattacks and defenses in compromised systems under per-

Table 1.5 Summary on several typical game-based FDI attack strategies

Methods index	Types of systems	Types of game	Limitations or shortcomings
Li et al. [94]		Stackelberg game	
Tian et al. [95]		Bayesian game	
Liu et al. [96]		attack-defense graph game	
Esmalifalak et al. [97]	Linear discrete-Time cyber-Physical systems		Relying on the precision of the system parameter matries, without considering the energy consumption of the attack
Sanjab et al. [98]			
Yi et al. [99]		Zero-sum game	
wang et al. [100]	Nonlinear discrete-Time cyber-Physical systems		
Venkitasubramaniam et al. [101]			

fect information, proposing injection attack strategies grounded in this framework. It is important to note that the game-based attack-defense graph models the attacker's steps toward achieving their ultimate goal as an attack path, where attack steps are represented as edges connecting adjacent system states, while the defender's strategy is to block these edges. This provides a new perspective for designing injection attack strategies. Finally, unlike the above modeling approaches for attackers and defenders, the most common framework models the relationship between attackers and defenders as a zero-sum game. According to game theory, zero-sum games characterize adversarial behavior by maintaining a global performance metric sum of zero in the Nash equilibrium sense. Esmalifalak et al. [97] designed zero-sum game-based injection attack strategies for power grid cyber-physical systems, yet neglected attack energy consumption considerations. Subsequently, Sanjab et al. [98] developed a game-based injection attack strategy under multi-attacker scenarios. Additionally, a multi-stage zero-sum game framework incorporating data tampering, strategy adaptation, and multi-path attacks was designed for injection strategies under perfect information [99], though energy consumption for both sides remained unaccounted. Subsequently, Wang et al. [100] innovatively introduced a two-layer adversarial game structure. Specifically, the upper layer involves methods for identifying malicious data and constructing false data, while the lower layer employs attack-defense strategies based on a two-person zero-sum game. Under full information, they designed injection attack strategies for nonlinear discrete-time cyber-physical systems aimed at degrading state estimation performance. Notably, Venkitasubramaniam et al. [101] further enriched the field by designing injection attack strategies for linear discrete-time cyber-physical systems within a zero-sum game framework. This work focused solely on controller energy consumption and assumed attackers possessed complete knowledge of the targeted system. In summary, current game-theoretic injection attack strategy designs predominantly focus on fully informed scenarios with limited consideration of attack energy consumption. Therefore, conducting in-depth research on game-theoretic injection attack strategies under conditions of attack energy consumption and information uncertainty holds significant importance for enhancing the security of cyber-physical systems.

References

1. Wolf, W.: Cyber-physical systems. Computer **42**(03), 88–89 (2009)
2. Baheti, R., Gill, H.: Cyber-physical systems. Impact Control. Technol. **12**(1), 161–166 (2011)
3. Wu, G., Sun, J., Chen, J.: A survey on the security of cyber-physical systems. Control Theory Technol. **14**(1), 2–10 (2016)
4. Liu, Y., Peng, Y., Wang, B., Yao, S., Liu, Z.: Review on cyber-physical systems. IEEE/CAA J. Automatica Sin. **4**(1), 27–40 (2017)
5. Alguliyev, R., Imamverdiyev, Y., Sukhostat, L.: Cyber-physical systems and their security issues. Comput. Ind. **100**, 212–223 (2018)
6. Alippi, C.: Intelligence for Embedded Systems. Springer, Springer Verlag, Berlin (2014)
7. Zhou, Y., Yu, F.R., Chen, J., Kuo, Y.: Cyber-physical-social systems: a state-of-the-art survey, challenges and opportunities. IEEE Commun. Surv. Tutor. **22**(1), 389–425 (2020)
8. Wang, Z., Xie, W., Wang, B., Tao, J., Wang, E.: A survey on recent advanced research of CPS security. Appl. Sci. **11**(9), 3751 (2021)
9. Ram, M.: Advances in Mathematics for Industry 4.0. Academic Press (2021)
10. BeiDou navigation system construction completes successfully! satellite-based solar power generation "black technology". (2020). https://mp.ofweek.com/solar/a645693221926
11. Lin, B., Zhang, Z., Han, X., Liu, L., Liu, Z., Che, Y.: Key technologies for space-air-ground-ocean integrated networks towards maritime emergency: an overview. Mob. Commun **44**(9), 19–26 (2020)
12. In search of fishing, from "digital marine" to "smart marine". (2022). https://www.sohu.com/a/562039565_120104192
13. Britain's royal air force chief says drone swarms are ready to overpower enemy defenses. (2022). https://interestingengineering.com/innovation/britain-air-force-drone-swarms-ready
14. Huzhou and lenovo explore new opportunities for smart cities, jointly promoting green development. (2025). https://baijiahao.baidu.com/s?id=1842031694888501971&wfr=spider&for=pc
15. Quick introduction to blockchain in automobile industry. (2025). https://noorhantrdg.com/quick-introduction-to-blockchain-in-automobile-industry/
16. Innovative data infrastructure breaks through data bottlenecks, enabling the development of smart campuses. (2021). https://t.cj.sina.com.cn/articles/view/1730695047/67284f8700100s0tw
17. Gao, S., Zhang, H., Wang, Z., Huang, C., Yan, H.: Data-driven injection attack strategy for linear cyber-physical systems: an input-output data-based approach. IEEE Trans. Netw. Sci. Eng. **10**(6), 4082–4095 (2023)

18. Wang, Z., Gao, S., Zhang, H., Yan, H., Li, T.: Optimal DoS attack strategy for cyber-physical systems via energy allocation. IEEE Trans. Control Netw. Syst. **11**(4), 2022–2032 (2024)
19. Number of common IT security vulnerabilities and exposures (CVEs) worldwide from 2009 to 2024. (2024). https://www.statista.com/statistics/500755/worldwide-common-vulnerabilities-and-exposures/
20. Alder, S.: Healthcare data breach statistics. HTPAA J. (2024). https://www.hipaajournal.com/healthcare-data-breach-statistics/
21. Colonial pipeline, the ransomware task force, and your business. (2021). https://sase.checkpoint.com/blog/news/colonial-pipeline-the-ransomware-task-force-and-your-business
22. Greig, J.: German wind turbine maker shut down after cyberattack. (2022). https://therecord.media/german-wind-turbine-maker-shut-down-after-cyberattack
23. Swinhoe, D.: Danish hosting firms lose all customer data in ransomware attack, cloudnordic and azerocloud have lost everything. (2023). https://www.datacenterdynamics.com/en/news/danish-hosting-firms-lose-all-customer-data-in-ransomware-attack/
24. Brumfield, C.: Hijack of monitoring devices highlights cyber threat to solar power infrastructure. (2024). https://www.csoonline.com/article/2119281/hijack-of-monitoring-devices-highlights-cyber-threat-to-solar-power-infrastructure.html
25. Kim, J., Tong, L., Thomas, R. J.: Data framing attack on state estimation with unknown network parameters. In: Proceedings of the 2013 Asilomar Conference on Signals, Systems and Computers, pp. 1388–1392 (2013)
26. Zheng, X., Zhang, H., Wang, L., Yan, H.: Dynamic event-based distributed piecewise filtering over sensor networks under periodic DoS attacks. IEEE Trans. Control Netw. Syst. **10**(2), 853–864 (2023)
27. Gao, S., Zhang, L., He, L., Deng, X., Yin, H., Zhang, H.: Attack detection for intelligent vehicles via CAN-bus: a lightweight image network approach. IEEE Trans. Veh. Technol. **72**(12), 16624–16636 (2023)
28. Tang, Q., Deng, C., Wang, Y., Guo, F., Fan, S.: Iterative observer-based resilient control for energy storage systems in microgrids under FDI attacks. IEEE Trans. Smart Grid **15**(5), 4744–4753 (2024)
29. Zhang, H., He, L., Wang, Z., Gao, S., Yan, H.: The detection mechanism for false data injection attack via the ellipsoidal set-membership approach. Asian J. Control **25**(6), 4853–4863 (2023)
30. Zhu, P., Jin, S., Bu, X., Hou, Z.: Distributed data-driven control for a connected autonomous vehicle platoon subjected to false data injection attacks. IEEE Trans. Autom. Sci. Eng. **21**(4), 7527–7538 (2024)
31. Zhang, H., Cheng, P., Shi, L., Chen, J.: Optimal denial-of-service attack scheduling with energy constraint. IEEE Trans. Autom. Control **60**(11), 3023–3028 (2015)
32. Shi, L., Cheng, P., Chen, J.: Optimal periodic sensor scheduling with limited resources. IEEE Trans. Autom. Control **56**(9), 2190–2195 (2011)
33. Shi, L., Cheng, P., Chen, J.: Sensor data scheduling for optimal state estimation with communication energy constraint. Automatica **47**(8), 1693–1698 (2011)
34. Zhang, H., Cheng, P., Shi, L., Chen, J.: Optimal DoS attack scheduling in wireless networked control system. IEEE Trans. Control Syst. Technol. **24**(3), 843–852 (2015)
35. Shisheh Foroush, H., Martínez, S.: On triggering control of single-input linear systems under pulse-width modulated DoS signals. SIAM J. Control. Optim. **54**(6), 3084–3105 (2016)
36. Tse, D., Viswanath, P.: Fundamentals of Wireless Communication. Cambridge university press (2005)
37. Poisel, R.: Modern Communications Jamming Principles and Techniques. Artech house (2011)
38. Peng, L., Shi, L., Cao, X., Sun, C.: Optimal attack energy allocation against remote state estimation. IEEE Trans. Autom. Control **63**(7), 2199–2205 (2017)
39. Qin, J., Li, M., Shi, L., Kang, Y.: Optimal denial-of-service attack energy management over an SINR-based network. arXiv:1810.02558 (2018)

40. Ding, D., Wang, Z., Wei, G., Alsaadi, F.E.: Event-based security control for discrete-time stochastic systems. IET Control Theory Appl. **10**(15), 1808–1815 (2016)
41. Peng, C., Sun, H.: Switching-like event-triggered control for networked control systems under malicious denial of service attacks. IEEE Trans. Autom. Control **65**(9), 3943–3949 (2020)
42. Ye, Z., Zhang, D., Cheng, J., Wu, Z.G.: Event-triggering and quantized sliding mode control of UMV systems under DoS attack. IEEE Trans. Veh. Technol. **71**(8), 8199–8211 (2022)
43. Fudenberg, D., Tirole, J.: Game Theory. MIT press (1991)
44. Owen, G.: Game Theory. Emerald Group Publishing (2013)
45. Kumar, B., Bhuyan, B.: Using game theory to model DoS attack and defence. Sādhanā **44**(12), 1–12 (2019)
46. Basar, T.: The Gaussian test channel with an intelligent jammer. IEEE Trans. Inf. Theory **29**(1), 152–157 (1983)
47. He, W., Xia, C., Wang, H., Zhang, C., Ji, Y.: A game theoretical attack-defense model oriented to network security risk assessment. In: Proceedings of the 2008 International Conference on Computer Science and Software Engineering, vol. 6. pp. 498–504 (2008)
48. Nash, J.F.: Non-cooperative games. Ann. Math. **54**, 286–295 (1951)
49. Li, Y., Shi, L., Cheng, P., Chen, J., Quevedo, D.E.: Jamming attacks on remote state estimation in cyber-physical systems: a game-theoretic approach. IEEE Trans. Autom. Control **60**(10), 2831–2836 (2015)
50. Li, Y., Quevedo, D.E., Dey, S., Shi, L.: A game-theoretic approach to fake-acknowledgment attack on cyber-physical systems. IEEE Trans. Signal Inf. Proc. Netw. **3**(1), 1–11 (2016)
51. Li, Y., Quevedo, D.E., Dey, S., Shi, L.: SINR-based DoS attack on remote state estimation: a game-theoretic approach. IEEE Trans. Control Netw. Syst. **4**(3), 632–642 (2016)
52. He, J., Chen, C., Zhu, S., Yang, B., Guan, X.: Antijamming game framework for secure state estimation in power systems. IEEE Trans. Ind. Inform. **15**(5), 2628–2637 (2018)
53. Li, Y., Zhu, S., Chen, C., Guan, X.: Optimal denial-of-service attack strategy on state estimation over infinite-time horizon. IEEE Trans. Circuits Syst. II-Express Briefs **68**(8), 2860–2864 (2021)
54. Liu, R., Hao, F., Yu, H.: Optimal SINR-based DoS attack scheduling for remote state estimation via adaptive dynamic programming approach. IEEE Trans. Syst. Man Cybern. -Syst. **51**(12), 7622–7632 (2020)
55. Dai, P., Yu, W., Wang, H., Wen, G., Lv, Y.: Distributed reinforcement learning for cyber-physical system with multiple remote state estimation under DoS attacker. IEEE Trans. Netw. Sci. Eng. **7**(4), 3212–3222 (2020)
56. Zhu, Q., Basar, T.: Game-theoretic methods for robustness, security, and resilience of cyber-physical control systems: games-in-games principle for optimal cross-layer resilient control systems. IEEE Control Syst. Mag. **35**(1), 46–65 (2015)
57. Zhao, Y., He, X., Zhou, D.: Optimal joint control and triggering strategies against denial of service attacks: a zero-sum game. IET Control Theory Appl. **11**(14), 2352–2360 (2017)
58. Fujiwara-Greve, T.: Non-Cooperative Game Theory. Springer Japan, Tokio, Japan (2015)
59. Li, Y., Xiao, L., Liu, J., Tang, Y.: Power control Stackelberg game in cooperative anti-jamming communications. In: Proceedings of the 2014 5th International Conference on Game Theory for Networks, pp. 1–6 (2014)
60. Liu, H.: SINR-based multi-channel power schedule under DoS attacks: a Stackelberg game approach with incomplete information. Automatica **100**, 274–280 (2019)
61. Liu, H., Wang, S.: SINR-based multi-channel power control under DoS attacks: a Stackelberg game approach. In: Proceedings of the 2020 39th Chinese Control Conference, pp. 4359–4364 (2020)
62. Yang, H., Shi, M., Xia, Y., Zhang, P.: Security research on wireless networked control systems subject to jamming attacks. IEEE Trans. Cybern. **49**(6), 2022–2031 (2018)
63. Zhang, Y., Xu, Y., Xu, Y., Yang, Y., Luo, Y., Wu, Q., Liu, X.: A multi-leader one-follower Stackelberg game approach for cooperative anti-jamming: no pains, no gains. IEEE Commun. Lett. **22**(8), 1680–1683 (2018)

64. Mo Y., Sinopoli B.: False data injection attacks in control systems. In: Proceedings of the Preprints of the 1st workshop on Secure Control Systems, pp. 1–7 (2010)
65. Zhang, X.G., Yang, G.H.: Optimal sensor attacks in cyber-physical systems with round-robin protocol. Inf. Sci. **548**, 85–100 (2021)
66. Huo, J.R., Li, X.J.: Stealthy switching attacks on sensors against state estimation in cyber-physical systems. Int. J. Robust Nonlinear Control **33**(2), 1169–1183 (2023)
67. Huo, J.R., Li, X.J.: Stealthy noise injection attacks against state estimation in interconnected systems. IEEE Trans. Circuits Syst. II-Express Briefs **70**(6), 2042–2046 (2023)
68. Zhang, T.Y., Ye, D.: False data injection attacks with complete stealthiness in cyber-physical systems: a self-generated approach. Automatica **120**, 109117 (2020)
69. Lu, A.Y., Yang, G.H.: False data injection attacks against state estimation without knowledge of estimators. IEEE Trans. Autom. Control **67**(9), 4529–4540 (2022)
70. Chen, Y., Kar, S., Moura, J.M.: Cyber-physical attacks with control objectives. IEEE Trans. Autom. Control **63**(5), 1418–1425 (2018)
71. Chen, Y., Kar, S., Moura, J.M.: Cyber physical attacks with control objectives and detection constraints. In: Proceedings of the IEEE 55th Conference on Decision and Control, pp. 1125–1130 (2016)
72. Wu, G., Sun, J., Chen, J.: Optimal data injection attacks in cyber-physical systems. IEEE Trans. Cybern. **48**(12), 3302–3312 (2018)
73. Wu, G., Sun, J.: Optimal data integrity attack on actuators in cyber-physical systems. In: Proceedings of the American Control Conference, pp. 1160–1164 (2016)
74. Wu, G., Wang, G., Sun, J., Chen, J.: Optimal partial feedback attacks in cyber-physical power systems. IEEE Trans. Autom. Control **65**(9), 3919–3926 (2020)
75. Gao, S., Zhang, H., Wang, Z., Huang, C.: A class of optimal switching mixed data injection attack in cyber-physical systems. IEEE Robot. Autom. Lett. **6**(2), 1598–1605 (2021)
76. Wu, G., Wang, G., Sun, J., Xiong, L.: Optimal switching attacks and countermeasures in cyber-physical systems. IEEE Trans. Syst. Man Cybern. Syst. **51**(8), 4825–4835 (2021)
77. Mikhaylenko, D., Zhang, P.: Stealthy local covert attacks on cyber-physical systems. IEEE Trans. Autom. Control **67**(12), 6778–6785 (2022)
78. Zhang, K., Keliris, C., Parisini, T., Polycarpou, M.M.: Stealthy integrity attacks for a class of nonlinear cyber-physical systems. IEEE Trans. Autom. Control **67**(12), 6723–6730 (2022)
79. Yang, T., Murguia, C., Lv, C., Nešić, D., Huang, C.: On joint reconstruction of state and input-output injection attacks for nonlinear systems. IEEE Control. Syst. Lett. **6**, 554–559 (2022)
80. Tanaskovic, M., Fagiano, L., Novara, C., Morari, M.: Data-driven control of nonlinear systems: an on-line direct approach. Automatica **75**, 1–10 (2017)
81. Hou, Z., Chi, R., Gao, H.: An overview of dynamic-linearization-based data-driven control and applications. IEEE Trans. Ind. Electron. **64**(5), 4076–4090 (2016)
82. Yu, W., Bu, X., Hou, Z.: Security data-driven control for nonlinear systems subject to deception and false data injection attacks. IEEE Trans. Netw. Sci. Eng. **9**(4), 2910–2921 (2022)
83. Bazanella, A.S., Campestrini, L., Eckhard, D.: Data-driven controller design: the H2 approach. Springer Sci. Bus. Media (2011)
84. Lee, J.M., Lee, J.H.: Approximate dynamic programming-based approaches for input–output data-driven control of nonlinear processes. Automatica **41**(7), 1281–1288 (2005)
85. Hou, Z., Xiong, S.: On model-free adaptive control and its stability analysis. IEEE Trans. Autom. Control **64**(11), 4555–4569 (2019)
86. Hou, Z., Jin, S.: Model Free Adaptive Control: Theory and Applications. CRC, Boca Raton, FL, USA (2019)
87. Chi, R., Lv, Y., Hou, Z.: Compensation-based data-driven ILC with input and output package dropouts. Int. J. Robust Nonlinear Control **30**(3), 950–965 (2020)
88. Li, Y.G., Yang, G.H.: Optimal stealthy innovation-based attacks with historical data in cyber-physical systems. IEEE Trans. Syst. Man Cybern. Syst. **51**(6), 3401–3411 (2021)
89. Tian, J., Wang, B., Li, J., Konstantinou, C.: Datadriven false data injection attacks against cyber-physical power systems. Comput. Secur. **121**, 102836 (2022)

90. Lakshminarayana, S., Kammoun, A., Debbah, M., Poor, H.V.: Data-driven false data injection attacks against power grids: a random matrix approach. IEEE Trans. Smart Grid **12**(1), 635–646 (2021)
91. Zhao, Z., Huang, Y., Zhen, Z., Li, Y.: Data-driven false data-injection attack design and detection in cyber-physical systems. IEEE Trans. Cybern. **51**(12), 6179–6187 (2021)
92. Mukherjee, D.: Data-driven false data injection attack: a low-rank approach. IEEE Trans. Smart Grid **13**(3), 2479–2482 (2022)
93. Santoso, F., Finn, A.: A data-driven cyber-physical system using deep-learning convolutional neural networks: study on false-data injection attacks in an unmanned ground vehicle under fault-tolerant conditions. IEEE Trans. Syst. Man Cybern. Syst. **53**(1), 346–356 (2023)
94. Li, Y., Shi, D., Chen, T.: False data injection attacks on networked control systems: a Stackelberg game analysis. IEEE Trans. Autom. Control **63**(10), 3503–3509 (2018)
95. Tian, M., Dong, Z., Wang, X.: Analysis of false data injection attacks in power systems: a dynamic Bayesian game-theoretic approach. ISA Trans. **115**, 108–123 (2021)
96. Liu, S.Z., Shao, C.W., Li, Y.F., Yang, Z.: Game attack–defense graph approach for modeling and analysis of cyberattacks and defenses in local metering system. IEEE Trans. Autom. Sci. Eng. **19**(3), 2607–2619 (2022)
97. Esmalifalak, M., Shi, G., Han, Z., Song, L.: Bad data injection attack and defense in electricity market using game theory study. IEEE Trans. Smart Grid **4**(1), 160–169 (2013)
98. Sanjab, A., Saad, W.: Data injection attacks on smart grids with multiple adversaries: a game-theoretic perspective. IEEE Trans. Smart Grid **7**(4), 2038–2049 (2016)
99. Yi, N., Wang, Q., Yan, L., Tang, Y., Xu, J.: A multi-stage game model for the false data injection attack from attacker's perspective. Sustain. Energy, Grids Netw. **28**, 100541 (2021)
100. Wang, Q., Tai, W., Tang, Y., Ni, M., You, S.: A two-layer game theoretical attack-defense model for a false data injection attack against power systems. Int. J. Elect. Power Energy Syst. **104**, 169–177 (2019)
101. Zhang, R., Venkitasubramaniam, P.: False data injection and detection in LQG systems: a game theoretic approach. IEEE Trans. Control Netw. Syst. **7**(1), 338–348 (2020)

Chapter 2
Model-Based DoS Attack Strategy Against CPSs via Energy Allocation

Abstract The energy allocation problem of malicious denial-of-service (DoS) attacks for state estimation in cyber-physical systems (CPSs) is investigated in this chapter. An optimal energy allocation strategy for DoS attacks in CPSs under multi-sensor network fusion estimation is proposed. To describe the channel packet loss characteristic, a signal-to-interference-plus-noise ratio (SINR) model is introduced. Compared with the existing DoS attack strategies, the increase of the channel packet loss rate with the increase in energy injected by the attacker is considered. Meanwhile, the behavioral tendency of the attacker concerning the sensors with different parameters is taken into account, leveraging the additional sensing accuracy matrix. Moreover, the quantified relationship between the attacker and the remote estimator error covariance is derived and the optimal energy allocation strategy problem is transformed into a convex optimization problem to be solved. Finally, a simulation example is presented to verify the effectiveness of the proposed mechanism.

2.1 Introduction

One of the issues in securing CPSs is the investigation of attack strategies, as understanding the various avenues of attackers is critical to improve the design of defense mechanisms. There are various categories of attacks, just to name a few, integrity attacks and DoS attacks. Integrity attacks can jeopardize data integrity by modifying the content of the sensor or control data to make it more versatile and covert. Extensive study findings on integrity attack scheme design, detection, and security control have been documented in [1–5] and references therein. DoS attack is that an attacker sends attack signals in network to block the network communication channel, thus preventing the receiving end from receiving valid information. In the study of DoS attacks on network security, the resources of the attacker are usually restricted, such as the limited energy [6] and limited range of attacks at the moment [7], etc. Zhang et al. [8] designed the optimal attack sequence for energy-constrained DoS attackers, i.e., to cause as many consecutive packet losses of the channel as possible. Different from the former way of modeling the channel subject to cyberattacks, a binary model

S. Gao et al., *Cyberattack Strategies Against Cyber-Physical Systems: From a Control Perspective*, https://doi.org/10.1007/978-981-95-9645-4_2

that considers only two cases of channel packet loss or no packet loss under DoS attacks is developed in [9]. Furthermore, under the channel with random characteristics, the energy cost of each time for the attacker to block the channel is calculated in [10]. By introducing the SINR channel model, the injected energy of the attacker is translated into the corresponding packet loss probability [11, 12]. Thus, one of the motivations for the research is to adopt the SINR model to describe the channel packet loss generated by DoS attacks in more practical situations.

The single-sensor system is the primary focus of the network security model in the aforementioned attack strategy design problems. Since multi-sensor can gather a wealth of data and have strong fault tolerance for objects of interest, this framework has more widespread applications in CPSs. In a multi-sensor framework, it is necessary to limit the performance deterioration of system state estimates brought by cyberattacks. Multi-sensor security in several contexts, including smart grid and power market operations, is considered in [13, 14]. When compared to single-sensor systems, the complex coupling across channels and the rise in computing complexity provide certain challenges in the design of optimal attack strategies. Yang et al. export analytical results for the optimal DoS attack strategy against the single-sensor system and reduce it to a convex optimization problem, which was generalized to multiple systems [7]. A SINR-based network was introduced in [11], which enables DoS attackers and sensors to choose their energy level. Based on SINR-based network model, Peng et al. [12] investigated multi-sensor estimation systems and obtained an optimal attack strategy by modeling it as a Markov decision process. The primary research goal is to develop DoS attack strategies for multi-sensor networks.

Based on the above considerations, deterministic straightforward attack models in remote state estimation have been extensively studied. It is important to study more flexible DoS attacks, which can provide worse results by injecting enough energy into the channel to increase the packet loss rate. However, most of the previously released works on optimal DoS attack strategies have focused on single sensor systems. Multi-sensor frameworks have a greater scope of application than single-sensor frameworks due to the advantages of comprehensive data collection for better data fusion and robustness to sensor corruption. Optimal DoS attack strategy design is a more difficult task due to the difficulties of dimensionality enhancement and additional limitations in multi-channel networks. This inspires us to look into the issue of determining the optimal DoS attack scheme in a multi-sensor system while considering energy restrictions.

The main technical difficulties of this chapter are as follows:

(1) How to construct the relationship with the successful transmission rate ϵ_m in the presence of DoS attacks is one of the main difficulties. In the digital communication theory, the symbol error rate $\text{SER} = 2Q[\sqrt{\alpha \text{SNR}}]$ is frequently used. However, the use of SNR, which solely considers background noise, can result in significant errors when calculating the symbol error rate under the DoS attack.
(2) How to transform the energy allocation problem into a positive semi-definite convex optimization problem is another difficulty. Since there is a nonlinear

relationship between the energy consumed by the DoS attacker and the successful transmission rate, it is very difficult to directly calculate the energy of the attacker.

Motivated by the above, a novel DoS attack strategy against CPSs is proposed. The main contributions of this chapter are as follows:

(1) A DoS attack energy allocation problem in multi-sensor framework is proposed, in which the energy of an attacker can be continuously changed. Compared with the existing DoS attack strategies, the relationship between the energy utilized by the attacker and the probability of transmission packet loss is considered.
(2) The quantitative relationship between system state estimation and error covariance is derived. The proposed optimization problem is formulated into a positive semi-definite programming problem and solved.
(3) In order to evaluate the propensity of the attacker to allocate energy for different sensors when implementing an optimal DoS attack strategy, the sensing accuracy matrix is introduced.

The remainder of this chapter is organized as follows. The problem formulation about the cyber-physical systems with multi-sensor is presented in Sect. 2.2. Section 2.3 derives the evolution of error covariance and analyzes the effect of the DoS attack on Kalman filter. The designed optimal energy allocation strategy for the DoS attack is also described in Sect. 2.3. In Sect. 2.4, the efficiency of the proposed strategy is illustrated by numerical simulations. Finally, this chapter is concluded in Sect. 2.5.

2.2 Problem Formulation

Consider the cyber-physical system structured by M wireless sensors and a remote state estimator under cyber attacks as depicted in Fig. 2.1. Smart sensors utilize wireless networks to deliver data packets to a remote estimator after estimating the

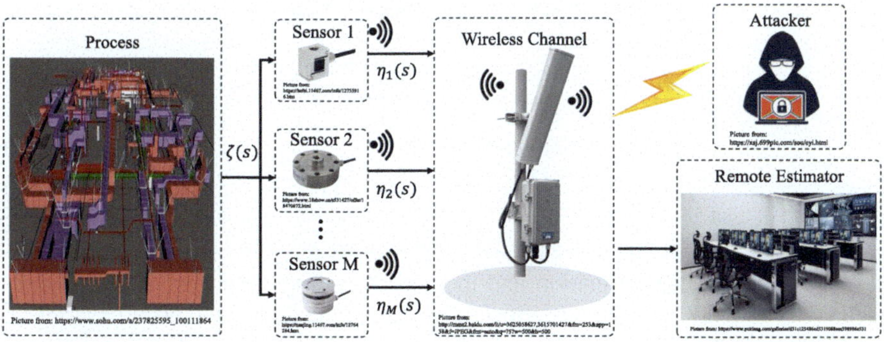

Fig. 2.1 The schematic of the attack on the cyber-physical system with multi-sensor state estimation

physical process. It should be noted that the DoS attacker is limited in its ability to block packets in the wireless channel due to power limitations. The following is a description of the detailed model and the problem formulation.

2.2.1 System Model

The cyber-physical system modeled as a linear time-invariant system based on M sensors is described as

$$\begin{aligned} \zeta(s+1) &= \mathcal{A}\zeta(s) + w(s), \\ \eta_m(s) &= C_m\zeta(s) + v_m(s), \quad m = 1, 2, \ldots, M, \end{aligned} \tag{2.1}$$

where $\zeta(s)$ and $\eta_m(s)$ represent the system process state and the measurement output of the m-th sensor, $\zeta(s) \in \mathbb{R}^n$ and $\eta_m(s) \in \mathbb{R}^{q_m}$. The process noise $w(s)$ and the measurement noise $v_m(s)$ of the m-th sensor are mutually independent zero-mean Gaussian white noises, where $w(s)$ and $v_m(s)$ satisfy $N(0, Q)$ and $N(0, \mathcal{R}_m)$, respectively. For the above system, the pair $(\mathcal{A}, C_m)$ is detectable and $(\mathcal{A}, \sqrt{Q})$ is stabilizable.

2.2.2 Kalman Filter

The cyber-physical system employs smart sensors to acquire the measured output values of the physical process and innovatively transmits this data to the remote estimator through the wireless channel. Then, the remote estimator fuses the received data and estimates the process state of the system by utilizing the standard Kalman filter which is as follows.

$$\hat{\zeta}(s|s-1) = \mathcal{A}\hat{\zeta}(s-1\, s-1), \tag{2.2}$$

$$\mathcal{P}(s|s-1) = \mathcal{A}\mathcal{P}(s-1\, s-1)\mathcal{A}^T + Q, \tag{2.3}$$

$$\mathcal{P}(s|s) = (\mathcal{P}^{-1}(s|s-1) + \sum_{m=1}^{M} C_m^T \mathcal{R}_m^{-1} C_m)^{-1}, \tag{2.4}$$

$$\mathcal{K}(s) = \mathcal{P}(s|s)\widetilde{C}^T \mathcal{R}^{-1}, \tag{2.5}$$

$$\hat{\zeta}(s|s) = \hat{\zeta}(s|s-1) + \mathcal{K}(s)\mathcal{Z}(s), \tag{2.6}$$

where $\hat{\zeta}(s|s-1)$ and $\mathcal{P}(s|s-1)$ stand for the priori minimum-mean-square-error-based (MMSE-based) estimation and the corresponding error covariance, respectively. $\hat{\zeta}(s)$ and $\mathcal{P}(s)$ are the posteriori estimate and the associated error covariance. The Kalman gain is denoted by $\mathcal{K}(s) \triangleq [k_1(s), \ldots, k_M(s)]$. $\widetilde{C} \triangleq [C_1^T, \ldots, C_M^T]^T$ represents the measurement matrix. $\mathcal{R} \triangleq \text{diag}\{\mathcal{R}_1, \ldots, \mathcal{R}_\text{M}\}$ indicates the noise covariance. The overall measurement at time s is $\eta(s) \triangleq [\eta_1^T(s), \ldots, \eta_M^T(s)]^T$. Further, $\mathcal{Z}_m(s) = \eta_m(s) - C_m\hat{\zeta}(s|s-1)$ is denoted as the local innovation $\mathcal{Z}(s) \triangleq [\mathcal{Z}_1^T(s), \ldots, \mathcal{Z}_M^T(s)]^T$, which is calculated by smart sensors and transmitted to a remote estimator.

2.2.3 Communication Model

The innovation $\mathcal{Z}(s)$ will be transmitted as a data packet to the remote estimator. Random data drops occur in the case of fading and network invader. Thus, an additive white gaussian noise (AWGN) network utilizing quadrature amplitude modulation (QAM) is assumed to model the communication between the sensors and the remote state estimator. Then, the relationship between the symbol error rate (SER) and the signal to noise ratio (SNR) is revealed by digital communication theory as the following (similar to [15, 16]),

$$\text{SER} = 2Q[\sqrt{\alpha \text{SNR}}], \tag{2.7}$$

where α is a positive parameter factor. The function $Q[\cdot]$ is defined as $Q[x] = \frac{1}{\sqrt{2\pi}}\int_x^{\infty} exp(-t^2/2)dt$.

In the presence of the DoS attacker, the SNR for the communication network can be replaced with $\text{SINR} = \frac{p}{(E_a+\varpi^2)}$, where p is the transmission power used by the sensor, E_a is the injected power from the attacker, and ϖ^2 is the power from the noise. Then, the symbol error rate (2.7) needs to be rewritten as

$$\text{SER} = 2Q[\sqrt{\alpha \text{SINR}}]. \tag{2.8}$$

Remark 2.1 It is worth noting that this chapter focuses on the design of DoS attacks. Consequently, the communication process within this scenario includes not only the background noise but also the interference generated by the DoS attack. As a result, in the SER, the SINR is applied to make it more accurate, rather than solely relying on the SNR. Expressing the SER solely in terms of the SNR may introduce significant bias.

Consequently, the successful transmission rate ϵ_m of data packet from the m-th sensor can be described as

$$\begin{aligned} \epsilon_m &= 1 - 2Q[\sqrt{\alpha \text{SINR}}], \\ &= 1 - 2Q\left[\sqrt{\alpha \frac{p_m}{(E_{a_m} + \varpi_m^2)}}\right]. \end{aligned} \tag{2.9}$$

Then, the relationship between the successful transmission of the channel and the energy injected by the attacker is displayed as Fig. 2.2. It can be seen that the probability of successful channel transmission gradually decreases with the increase of the

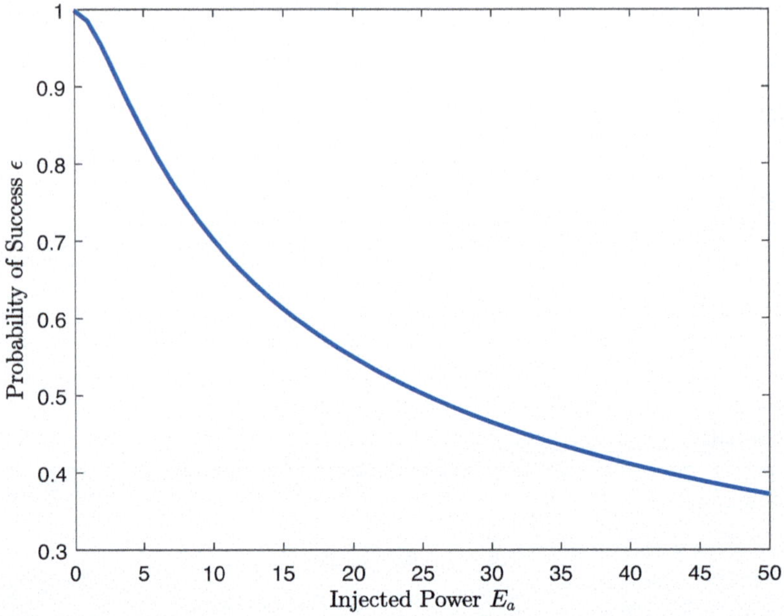

Fig. 2.2 The relationship between the injected power E_a and the successfully transmission rate ϵ

energy injected by the attacker. The SINR model depends not only on the transmit power used by the sensor, but also on the injected power of the DoS attacker. Different SINR models lead to different loss rates and long-range estimation performance.

2.2.4 *Attack Model*

In this subsection, the DoS attacker jams the wireless transmission channel of the m-th sensor by injecting the signal with energy E_{a_m}. From (2.9), it can be known that the attacker injecting signals with different extent of energy will result in different probabilities of packet loss in the wireless transmission channel, i.e., different successful transmission rates of the channel. γ_m is defined as the jamming variable whether the wireless transmission channel of the m-th sensor is jammed, as follows:

$$\gamma_m = \begin{cases} 1, \text{ channel of the } m\text{-th sensor is jammed,} \\ 0, \text{ channel of the } m\text{-th sensor is free.} \end{cases}$$

Then, the relationship between the jamming variable γ_m and the successfully transmission rate ϵ_m of the m-th sensor can be expressed as

$$\epsilon_m = \mathbb{P}[\gamma_m = 0]. \tag{2.10}$$

Combining the jamming variable γ_m and the innovation $\mathcal{Z}_m(s)$ received by the remote estimator, the attack model can be described as

$$\tilde{\mathcal{Z}}_m(s) = (1 - \gamma_m)\mathcal{Z}_m(s) + \vartheta_m(s), \tag{2.11}$$

where $\tilde{\mathcal{Z}}_m(s)$ stands for the attacked innovation that is transmitted from the m-th sensor to the remote estimator. $\vartheta_m(s)$ denotes the additional independent noise of the channel generated by the attacker after injecting energy signals. $\vartheta_m(s)$ satisfies $E[\vartheta_m(s)(\vartheta_m(s))^T] = \mu_m\Theta_m$, μ is a positive real number, Θ_m is the q_m-dimensional identity matrix, $m = 1, 2, \ldots, M$.

Remark 2.2 It is a quite general situation that the DoS attacker injects the energy signal into the wireless channel utilized by the m-th sensor and jams it. In addition to a certain probability that it will lead to the packet loss of innovation $\mathcal{Z}_m(s)$ delivered to the remote estimator, noise may also be generated. Thus, the attack model is established as (2.11). When the channel noise generated by DoS attackers is not considered, it is only necessary to set $\vartheta_m(s)$ to 0. It needs to be pointed out that μ_m is chosen by the attacker and the dimension of Θ_m is determined by the system's physical property.

The DoS attacker has the ability to inject energy signals whose power can vary continuously. It should be noted that there is an upper bound on the energy consumed by the attacker. To this end, the energy consumed by the DoS attacker at each moment satisfies

$$\sum_{m=1}^{M} E_{a_m} \leq \bar{E}_a, \tag{2.12}$$

where $\bar{E}_a$ is the upper bound of the energy consumption for the attacker.

2.2.5 *Optimal DoS Attack Problem*

The purpose of the DoS attacker in this chapter is to destroy the estimation performance of the remote estimator to the greatest extent when the total energy of the injected signals at each time instant meets the condition of (2.12). Thus, the objective function of the attacker can be described as

$$\mathcal{J}(E_a) \triangleq \lim_{S\to\infty} \sup \frac{1}{S}\sum_{s=1}^{S} tr[\mathcal{P}(s|s)], \tag{2.13}$$

where $\mathcal{P}(s|s)$ is the error covariance of the remote estimation under the DoS attack. $E_a \triangleq [E_{a_1}^T, E_{a_2}^T, \ldots, E_{a_M}^T]$ is the energy consumption vector of the attacker.

According to [17], the gain and error covariance of the Kalman filter are converged to the steady-state values with exponential speed. Therefore, the steady-state value $\bar{\mathcal{P}}$ of the priori estimation error covariance can be characterized as

$$\bar{\mathcal{P}} \triangleq \lim_{s \to \infty} \mathcal{P}(s|s), \tag{2.14}$$

where $\bar{\mathcal{P}}$ is the unique positive semi-definite solution of (2.4).

Integrating (2.13) and (2.14), the objective function of the attacker can be rewritten as

$$\mathcal{J}(E_a) = tr(\bar{\mathcal{P}}). \tag{2.15}$$

Since the objective of the attacker is to reduce the estimation performance of the remote estimator, the optimal energy allocation problem for the DoS attacker can be described as Problem 2.1 below, which is a static decision problem.

Problem 2.1

$$\max_{E_a} \quad \mathcal{J}(E_a)$$

$$s.t. \quad \sum_{m=1}^{M} E_{a_m} \leq \bar{E}_a.$$

Remark 2.3 It should be indicated that the DoS attacker injects the signals with the energy E_{a_m} into the wireless transmission channel of the m-th sensor to jam the channel. Combining the successful transmission rate of the data packet and the injected energy E_{a_m}, it can be seen that the relationship between the jamming variable γ_m and the energy injected by DoS attackers meets $\mathbb{P}[\gamma_m = 0] = 1 - 2Q\left[\sqrt{\alpha \frac{p_m}{(E_{a_m} + (\varpi_m)^2)}}\right]$. For this reason, the innovation of the remote estimator under attack calculated by the jamming variable is an equivalent description of the data transmitted from the sensor to the remote estimator after DoS attackers injected the signal of the energy E_{a_m}.

2.3 Main Results

In this section, the variation of the performance on the remote estimator is analyzed in depth for the cyber-physical systems under the designed DoS attack strategy from a probabilistic point of view. The sensing accuracy matrix is defined to analyze the trend of DoS attacks on different sensors. Meanwhile, the optimal energy allocation strategy for the DoS attack with limited energy is given.

2.3.1 Performance Analysis of Remote Estimator

The performance variation of the remote estimator is reflected by the variation of its system error covariance. Therefore, in this subsection, the evolution of the remote state estimator error covariance under the DoS attack is given as follows.

Theorem 2.1 *Under the designed DoS attack (2.11), the error covariance of the remote estimator can be rewritten as*

$$\mathcal{P}(s|s) = \left(\mathcal{P}^{-1}(s|s-1) + \sum_{m=1}^{M}(1-\gamma_m)C_m^T(\mathcal{R}_m + \Theta_m)^{-1}C_m\right)^{-1}. \tag{2.16}$$

Proof The innovation of the m-th sensor under the DoS attack is re-expressed as

$$\begin{aligned}\tilde{\mathcal{Z}}_m(s) &= (1-\gamma_m)\mathcal{Z}_m(s) + \vartheta_m(s),\\ &= (1-\gamma_m)(\eta_m(s) - C_m\hat{\zeta}(s|s-1)) + \vartheta_m(s).\end{aligned}$$

All the innovations $\tilde{\mathcal{Z}}_m(s)$ is superimposed in the form of a vector as follows:

$$\tilde{\mathcal{Z}}(s) = \tilde{\eta}(s) - \bar{\tilde{C}}\hat{\zeta}(s|s-1) + \vartheta(s),$$

where $\tilde{\eta}(s)$, $\bar{\tilde{C}}$, $\tilde{\mathcal{Z}}(s)$, and $\vartheta(s)$ are defined as

$$\begin{aligned}\tilde{\eta}(s) &\triangleq \Phi\eta(s), \quad \bar{\tilde{C}} \triangleq \Phi\tilde{C},\\ \tilde{\mathcal{Z}}(s) &\triangleq [\tilde{\mathcal{Z}}_1^T(s), \tilde{\mathcal{Z}}_2^T(s), \ldots, \tilde{\mathcal{Z}}_M^T(s)]^T,\\ \vartheta(s) &\triangleq [\vartheta_1^T(s), \vartheta_2^T(s), \ldots, \vartheta_M^T(s)]^T,\\ \Phi &\triangleq \mathrm{diag}\{(1-\gamma_1)I_{q_1}, (1-\gamma_2)I_{q_2}, \ldots, (1-\gamma_M)I_{q_M}\},\end{aligned}$$

and I_{q_m} is the q_m-dimensional identity matrix.

Then, $\hat{\zeta}(s|s)$ in (2.6) is changed into

$$\begin{aligned}\hat{\zeta}(s|s) &= \hat{\zeta}(s|s-1) + \mathcal{K}(s)(\tilde{\eta}(s) - \bar{\tilde{C}}\hat{\zeta}(s|s-1) + \vartheta(s)),\\ &= \hat{\zeta}(s|s-1) + \mathcal{K}(s)(\bar{\tilde{C}}\zeta(s) + \tilde{v}(s) - \bar{\tilde{C}}\hat{\zeta}(s|s-1) + \vartheta(s)),\end{aligned}$$

where $\tilde{v}(s)$ is $[(1-\gamma_1)v_1^T(s), (1-\gamma_2)v_2^T(s), \ldots, (1-\gamma_M)v_M^T(s)]^T$.

Therefore, the error between $\zeta(s)$ and $\hat{\zeta}(s|s)$ can be described as

$$\begin{aligned}\zeta(s) - \hat{\zeta}(s|s) &= \zeta(s) - \hat{\zeta}(s|s-1) - \mathcal{K}(s)(\bar{\tilde{C}}\zeta(s) + \tilde{v}(s) - \bar{\tilde{C}}\hat{\zeta}(s|s-1) + \vartheta(s)),\\ &= (I - \mathcal{K}(s)\bar{\tilde{C}})(\zeta(s) - \hat{\zeta}(s|s-1)) - \mathcal{K}(s)\tilde{v}(s) - \mathcal{K}(s)\vartheta(s),\end{aligned}$$

where I is the identity matrix with the appropriate dimension.

Combining the error $(\zeta(s) - \hat{\zeta}(s|s))$, the error covariance $\mathcal{P}(s|s)$ under the DoS attack can be calculated as

$$\begin{aligned}\mathcal{P}(s|s) &= E[(\zeta(s) - \hat{\zeta}(s|s))(\zeta(s) - \hat{\zeta}(s|s))^T], \\ &= (I - \mathcal{K}(s)\bar{\tilde{C}})\mathcal{P}(s|s-1)(I - \mathcal{K}(s)\bar{\tilde{C}})^T + \mathcal{K}(s)(\tilde{\mathcal{R}} + \Theta)\mathcal{K}^T(s),\end{aligned} \tag{2.17}$$

where $\tilde{\mathcal{R}}$ and Θ are denoted by $\tilde{\mathcal{R}} = E[\tilde{\upsilon}(s)\tilde{\upsilon}^T(s)] = \text{diag}\{(1-\gamma_1)\mathcal{R}_1, (1-\gamma_2)\mathcal{R}_2, \ldots, (1-\gamma_{\text{M}})\mathcal{R}_{\text{M}}\}$, and $\Theta = E[\vartheta(s)\vartheta^T(s)]$, respectively.

The optimal Kalman gain is acquired as follows by $\frac{\partial tr(\mathcal{P}(s|s))}{\partial \mathcal{K}(s)} = 0$.

$$\mathcal{K}(s) = \mathcal{P}(s|s)\bar{\tilde{C}}^T(\tilde{\mathcal{R}} + \Theta)^{-1}, \tag{2.18}$$

where $\mathcal{P}(s|s)$ can be obtained by solving the equation (2.17). Bringing (2.18) into (2.17), $\mathcal{P}^{-1}(s|s)$ can be written as

$$\begin{aligned}\mathcal{P}^{-1}(s|s) &= \mathcal{P}^{-1}(s|s-1) + \bar{\tilde{C}}^T(\tilde{\mathcal{R}} + \Theta)^{-1}\bar{\tilde{C}}, \\ &= \mathcal{P}^{-1}(s|s-1) + [(1-\gamma_1)C_1^T, \ldots, (1-\gamma_M)C_M^T]\text{diag}\{[(1-\gamma_1)\mathcal{R}_1 + \Theta_1]^{-1}, \\ &\quad \ldots, [(1-\gamma_M)\mathcal{R}_M + \Theta_M]^{-1}\}[(1-\gamma_1)C_1^T, \ldots, (1-\gamma_M)C_M^T]^T, \\ &= \mathcal{P}^{-1}(s|s-1) + \sum_{m=1}^{M}(1-\gamma_m)C_m^T((1-\gamma_m)\mathcal{R}_m + \Theta_m)^{-1}C_m.\end{aligned}$$

For the term $\sum_{m=1}^{M}(1-\gamma_m)C_m^T((1-\gamma_m)\mathcal{R}_m + \Theta_m)^{-1}C_m$ in the error covariance $\mathcal{P}(s|s)$, when $\gamma_m = 1$, the term $\sum_{m=1}^{M}(1-\gamma_m)C_m^T((1-\gamma_m)\mathcal{R}_m + \Theta_m)^{-1}C_m = \mathbf{0}$. In addition, when $\gamma_m = 0$, the term $\sum_{m=1}^{M}(1-\gamma_m)C_m^T((1-\gamma_m)\mathcal{R}_m + \Theta_m)^{-1}C_m = \sum_{m=1}^{M}C_m^T(\mathcal{R}_m + \Theta_m)^{-1}C_m$. Thus, through the above two situations, (2.16) can be obtained. This is the end of the proof. □

Integrating the error covariance of the remote estimator under the DoS attack in Theorem 2.1, the sensing accuracy matrix Λ_m of the m-th sensor can be defined as

$$\Lambda_m \triangleq C_m^T(\mathcal{R}_m + \Theta_m)^{-1}C_m. \tag{2.19}$$

The objective of the DoS attacker in this chapter is to deteriorate the performance of the remote estimator, and the attacker wants the error covariance of the remote estimator to be as large as possible. Therefore, the DoS attacker expects that the larger the sensor accuracy matrix of the sensor is, the greater the probability of packet loss is. That is, the DoS attacker prefers to attack the sensor which has a larger sensing accuracy matrix by using a high energy signal, this can make the overall error covariance larger. In practice, the DoS attacker prefers to attack the sensor containing a larger measurement matrix C_m and a smaller observation error matrix $\mathcal{R}_m$, $m = 1, 2, \ldots, M$.

2.3.2 Optimal DoS Attack Strategy

In this subsection, in order to obtain the optimal DoS attack strategy, the optimization problem is transformed into a semi-positive definite programming problem to be solved.

Problem 2.1 is transformed into an equivalent optimization problem given in the following lemma.

Lemma 2.1 *The optimal energy allocation problem for the DoS attacker described as Problem 2.1 is equivalent to the following problem:*

Problem 2.2

$$
\begin{aligned}
\max_{\Re, E_a} \quad & tr(\Re) \\
s.t. \quad & \Re^{-1} \succeq [\hbar(\Re)]^{-1} + \mathcal{H}^T \bar{\Phi} \mathcal{H}, \\
& \sum_{m=1}^{M} E_{a_m} \leq \bar{E}_a,
\end{aligned}
$$

where $\Re = \mathcal{P}(s|s)$, the operator $\hbar$ is defined as $\hbar(\Re) \triangleq \mathcal{A}\Re\mathcal{A}^T + Q$, $\mathcal{H} = [\mathcal{H}_1, \mathcal{H}_2, \ldots, \mathcal{H}_M]^T$, $\mathcal{H}_m$ is the Cholesky factorization of the sensing accuracy matrix Λ_m of the m-th sensor, $\bar{\Phi}$ is the matrix $\text{diag}\{\bar{\gamma}_1 I_{q_1}, \bar{\gamma}_2 I_{q_2}, \ldots, \bar{\gamma}_M I_{q_M}\}$, and $\bar{\gamma}_m = 1 - \gamma_m$.

Proof Using Cholesky factorization, the sensing accuracy matrix Λ_m can be factorized as $\mathcal{H}_m^T \mathcal{H}_m = C_m^T (\mathcal{R}_m + \Theta_m)^{-1} C_m$. Then, one can get

$$
\widetilde{C}^T (\mathcal{R} + \Theta)^{-1} \widetilde{C} = \mathcal{H}^T \mathcal{H},
$$

where $\mathcal{H}$ is defined in Lemma 2.1. Therefore, the iterative relationship for the error covariance of the remote estimator can be re-expressed as

$$
\begin{aligned}
\mathcal{P}^{-1}(s|s) &= \mathcal{P}^{-1}(s|s-1) + \mathcal{H}^T \bar{\Phi} \mathcal{H}, \\
&= [\hbar(\mathcal{P}(s-1\, s-1))]^{-1} + \mathcal{H}^T \bar{\Phi} \mathcal{H}, \\
&\succeq [\hbar(\mathcal{P}(s|s))]^{-1} + \mathcal{H}^T \bar{\Phi} \mathcal{H},
\end{aligned}
$$

thus, $\Re^{-1} \succeq [\hbar(\Re)]^{-1} + \mathcal{H}^T \bar{\Phi} \mathcal{H}$ can be obtained. This is the end of the proof. □

Remark 2.4 Lemma 2.1 transforms the original energy constrained optimization problem into the optimization problem with error covariance and energy constraints by combining Theorem 2.1, namely the error covariance of the remote estimator under the designed DoS attack strategy, which makes the description of the optimization problem more complete.

Theorem 2.2 *For the cyber-physical systems (2.1) with multi-sensor utilizing the Kalman filter (2.2)–(2.6), the optimal DoS attack strategy of Problem 2.2 can be solved by the following positive semi-definite programming problem:*

Problem 2.3

$$\min_{\tilde{\Re},\delta_m} \; tr(\tilde{\Re})$$
$$s.t. \quad \begin{bmatrix} \tilde{\Re} - \{\hbar(E(\Re))\}^{-1} & \mathcal{H}^T \\ \mathcal{H} & \Psi^{-1} \end{bmatrix} \succeq 0,$$
$$\sum_{m=1}^{M} \delta_m \leq \Gamma, \quad 1 \leq \delta_m \leq \frac{1}{\underline{\epsilon}},$$

where

$$\tilde{\Re} = E[\mathcal{P}^{-1}(s|s)], \qquad \delta_m = \frac{1}{\epsilon_m},$$
$$\Gamma = M\underline{\epsilon}, \qquad \Psi = E[\bar{\Phi}],$$
$$\underline{\epsilon} = \arg\min_{m\in(1,2,\ldots,M)} \{1 - 2Q[\sqrt{\alpha\frac{p_m}{(\bar{E}_a+\varpi_m^2)}}]\}.$$

Proof When the expectation is calculated on both sides of $\mathcal{P}^{-1}(s|s) = \mathcal{P}^{-1}(s|s-1) + \mathcal{H}^T\bar{\Phi}\mathcal{H}$, one can obtain

$$\begin{aligned} E[\mathcal{P}^{-1}(s|s)] &= E[\mathcal{P}^{-1}(s|s-1)] + \mathcal{H}^T\Psi\mathcal{H}, \\ &= E[(\hbar(\mathcal{P}(s-1s-1)))^{-1}] + \mathcal{H}^T\Psi\mathcal{H}, \end{aligned}$$

where Ψ is defined in Theorem 2.2. According to probability theory and (2.10), $\Psi = \mathrm{diag}\{\epsilon_1 I_{q_1}, \epsilon_2 I_{q_2}, \ldots, \epsilon_M I_{q_M}\}$, where the successfully transmission rate of the m-th sensor is $\epsilon_m \in (0, 1)$.

Through the Jensen inequality [18], one has

$$\begin{aligned} E[\mathcal{P}^{-1}(s|s)] &= E[(\hbar(\mathcal{P}(s-1\,s-1)))^{-1}] + \mathcal{H}^T\Psi\mathcal{H}, \\ &\geq E[(\hbar(\mathcal{P}(s|s)))^{-1}] + \mathcal{H}^T\Psi\mathcal{H}, \\ &\geq \{E[\hbar(\mathcal{P}(s|s))]\}^{-1} + \mathcal{H}^T\Psi\mathcal{H}. \end{aligned} \tag{2.20}$$

For the defined operator $\hbar(\cdot)$, it has no effect on the expectation of the error covariance, that is,

$$\begin{aligned} E[\hbar(\cdot)] &= E[\mathcal{A}(\cdot)\mathcal{A}^T + Q], \\ &= \mathcal{A}E[(\cdot)]\mathcal{A}^T + Q, \\ &= \hbar(E[(\cdot)]). \end{aligned}$$

Therefore, the inequality (2.20) can be rewritten as

$$\begin{aligned} E[\mathcal{P}^{-1}(s|s)] &\geq \{E[(\hbar(\mathcal{P}(s|s)))]\}^{-1} + \mathcal{H}^T\Psi\mathcal{H}, \\ &= \{\hbar(E[\mathcal{P}(s|s)])\}^{-1} + \mathcal{H}^T\Psi\mathcal{H}. \end{aligned} \tag{2.21}$$

Then, the Schur complement [19] is applied to convert (2.21) to the first condition of Problem 2.3.

The attacker is subject to the following energy constraints, $\sum_{m=1}^{M} E_{a_m} \leq \bar{E}_a$. Since the purpose of the attacker is to maximize the trace of estimate error covariance $tr(\bar{\mathcal{P}})$, they will use up all their energy.

Figure 2.2 indicates that the successful transmission rate ϵ_m gradually decreases with the increase of the energy E_{a_m} used by the attacker. Consequently, the value of ϵ_m is bounded when the energy of the attacker is finite, i.e., the data transmitted by the sensor network will not lose all packets. Thus, there is a threshold $\underline{\epsilon}$, and the successful transfer rate ϵ_m is subject to the following constraints,

$$0 < \underline{\epsilon} \leq \epsilon_m \leq 1,$$

where $\underline{\epsilon}$ can be calculated through (2.9).

Further, the above inequality can be reformulated as

$$1 \leq \frac{1}{\epsilon_m} \leq \frac{1}{\underline{\epsilon}}.$$

Hence, the second and third conditions of Problem 2.3 can be obtained.

The purpose of the attacker is to maximize $tr(\bar{\mathcal{P}})$, which is equivalent to maximize $E[\mathcal{P}(s|s)]$, that is, minimize $E[\mathcal{P}^{-1}(s|s)]$. This is the end of the proof. □

Utilizing the probability theory and (2.9)–(2.10), Theorem 2.2 transforms the energy allocation problem under the optimal DoS attacks containing the jamming variable γ_m into the positive semi-definite programming problem based on the expectation for the inverse of the error covariance by employing the successful transmission rate ϵ_m. Problem 2.3 in Theorem 2.2 can be solved by adopting the CVX toolkit in MATLAB to obtain the optimal δ_m^*. The successful transmission rate ϵ_m can be calculated through $\epsilon_m = \frac{1}{\delta_m^*}$. Subsequently, the optimal allocated energy E_{a_m} employed by the DoS attacker at the m-th sensor is solved by $\epsilon_m = 1 - 2Q\left[\sqrt{\alpha\frac{p_m}{(E_{a_m}+\varpi_m^2)}}\right]$. It should be noted that the function $Q[\cdot]$ is a nonlinear function such that the optimal energy cannot be solved directly. Therefore, vpasolve function in the MATLAB toolkit is applied to obtain the optimal allocated energy $E_{a_m}^*$ employed by the DoS attacker at the m-th sensor.

2.4 Numerical Simulation

In this section, both qualitative and quantitative analyzes are presented on the impact that the optimal DoS attack energy allocation strategy has on the performance of fusion estimation systems with multi-sensor networks. Two different physical processes of state stable and unstable are observed by the multi-sensor networks. The optimality of the proposed strategy is verified by implementing it on two physical processes and quantitatively calculating the error covariance of the fusion estimation. In addition, the tendency of the attacker to attack different sensors is validated by two sets of measurement matrix parameters. The parameter settings for stable and unstable physical processes are detailed below.

For the stable physical process, $\mathcal{A}_1 = 0.6I_2$, while for the unstable physical process, $\mathcal{A}_2 = 1.02I_2$. And two sets of measurement matrices are as follows.
CASE 1:

$$C_1 = \begin{bmatrix} 1.2 & 0 \\ 0 & 1.2 \end{bmatrix}, C_2 = \begin{bmatrix} 0.1 & 0 \\ 0 & 0.1 \end{bmatrix}, C_3 = \begin{bmatrix} 0.3 & 0 \\ 0 & 0.3 \end{bmatrix},$$

CASE 2:

$$C_1 = \begin{bmatrix} 1.8 & 0 \\ 0 & 1.2 \end{bmatrix}, C_2 = \begin{bmatrix} 0.1 & 0 \\ 0 & 1 \end{bmatrix}, C_3 = \begin{bmatrix} 0.3 & 0 \\ 0 & 0.1 \end{bmatrix},$$

with $Q = \mathcal{R}_m = \Theta_m = I_2, m = 1, 2, 3$, the upper bound of the energy consumption for the attacker $\bar{E}_a = 100$. To avoid the influence of random numbers on the experimental results, it has been solved 10,000 times and averaged. The simulation results are shown in Tables 2.1, 2.2, 2.3, and 2.4 and Figs. 2.3, 2.4, 2.5, 2.6, 2.7, 2.8, 2.9, 2.10, 2.11, 2.12, 2.13, and 2.14.

2.4.1 The Tendency of The Attacker

This subsection is to verify the preference trend of the DoS attacker for the sensors with different parameters under the optimal attack strategy. On the one hand, the energy allocated by the attacker to two sets of sensors with different measurement matrices C_m during stable and unstable physical processes is compared for the given observation error matrices $\mathcal{R}_m$, as shown in Tables 2.1 and 2.2. It can be seen from the tables that the attacker tends to allocate more energy to the sensor containing a larger measurement matrix C_m, thereby reducing the probability of successfully transmitting its information.

On the other hand, under the given measurement matrices C_m, compare the energy allocated by the attacker to sensors with different observation error matrices $\mathcal{R}_m$ in stable and unstable physical processes, as shown in Tables 2.3 and 2.4. It is noteworthy that setting the same measurement matrix C_m causes divergent results when

Table 2.1 The tendency of the attacker for stable physical processes $\mathcal{A}_1 = 0.6I_2$ when $\mathcal{R}_m = I_2$

Measurement matrix	Case 1		Case 2	
Allocated energy and successfully transmission rate	E_{a_m}	ϵ_m	E_{a_m}	ϵ_m
Sensor 1	88.5296	0.2857	85.7041	0.2901
Sensor 2	0.0191	0.9986	14.2770	0.6245
Sensor 3	11.4513	0.6737	0.0189	0.9994

Table 2.2 The tendency of the attacker for unstable physical processes $\mathcal{A}_2 = 1.02I_2$ when $\mathcal{R}_m = I_2$

Measurement matrix	Case 1		Case 2	
Allocated energy and successfully transmission rate	E_{a_m}	ϵ_m	E_{a_m}	ϵ_m
Sensor 1	63.0682	0.3348	62.1841	0.3368
Sensor 2	17.0478	0.5852	23.7027	0.5142
Sensor 3	19.8840	0.5516	14.1132	0.6270

Table 2.3 The tendency of the attacker for stable physical processes $\mathcal{A}_1 = 0.6I_2$ when $\mathcal{R}_m$ is set to different values

Settings	Variables	Sensor 1	Sensor 2	Sensor 3
$C_m = 3I_2 m = 1, 2, 3$	$\mathcal{R}_m$	$1I_2$	$2I_2$	$3I_2$
	E_{a_m}	48.1929	30.3739	21.4331
	ϵ_m	0.3786	0.4637	0.5355
Case 1	$\mathcal{R}_m$	$300I_2$	$1I_2$	$30I_2$
	E_{a_m}	38.4061	40.3871	21.2068
	ϵ_m	0.4189	0.4097	0.5377
Case 2	$\mathcal{R}_m$	$80I_2$	$50I_2$	$0.5I_2$
	E_{a_m}	41.1975	10.4644	48.3381
	ϵ_m	0.4062	0.6937	0.3781

implementing simulations of attacker tendencies for different $\mathcal{R}_m$ in unstable physical processes. For this reason, we use two different sets of observation error matrices $\mathcal{R}_m$ under the same measurement matrix C_m for comparison experiments. It can be seen from Tables 2.3 and 2.4 that the attacker tends to allocate more energy to the sensor containing a smaller observation error matrix $\mathcal{R}_m$, thereby the probability of successfully transmitting its information is decreased, which verifies the previous analysis.

Table 2.4 The tendency of the attacker for unstable physical processes $\mathcal{A}_2 = 1.02I_2$ when $\mathcal{R}_m$ is set to different values

Settings	Variables	Sensor 1	Sensor 2	Sensor 3
Case 1	$\mathcal{R}_m$	$1I_2$	$1I_2$	$0.03I_2$
	E_{a_m}	42.4803	6.2841	51.2356
	ϵ_m	0.4007	0.8007	0.3683
Case 1	$\mathcal{R}_m$	$1.2I_2$	$0.01I_2$	$300I_2$
	E_{a_m}	62.5036	19.2893	18.2071
	ϵ_m	0.3362	0.5581	0.5707
Case 2	$\mathcal{R}_m$	$0.4I_2$	$20I_2$	$0.1I_2$
	E_{a_m}	60.4212	19.7096	19.8692
	ϵ_m	0.3415	0.5535	0.5517
Case 2	$\mathcal{R}_m$	$0.15I_2$	$0.4I_2$	$30I_2$
	E_{a_m}	59.5301	24.3481	16.1218
	ϵ_m	0.3439	0.5086	0.5975

2.4.2 *Effectiveness of The DoS Attack Strategy*

Two sets of state estimation tracking curves for the stable system are shown in Figs. 2.3 and 2.4. As the Kalman filter iterates, the estimator converges and the

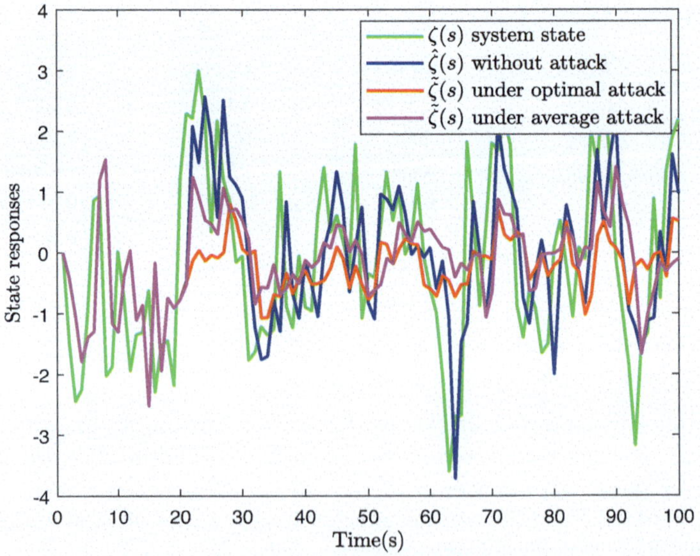

Fig. 2.3 Health state, estimation without attack, estimation under optimal attack, and estimation under average attack of stable physical processes $\mathcal{A}_1 = 0.6I_2$ in CASE 1

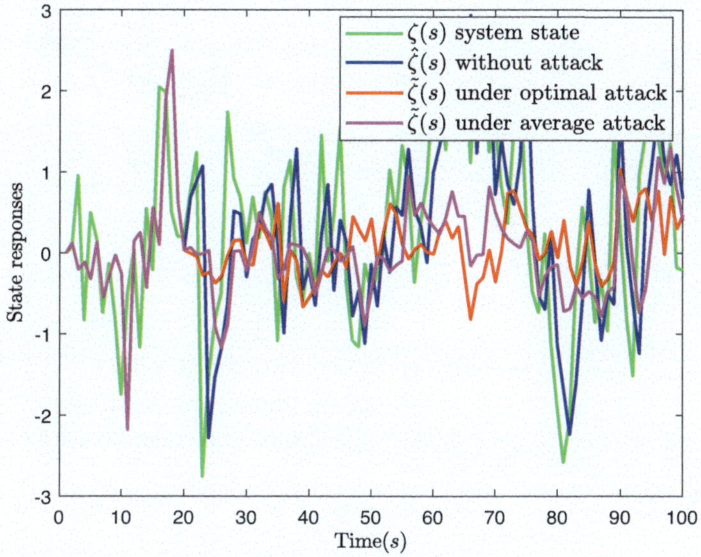

Fig. 2.4 Health state, estimation without attack, estimation under optimal attack, and estimation under average attack of stable physical processes $\mathcal{A}_1 = 0.6I_2$ in CASE 2

simulation implements the DoS attack from the 20th step. The green curve represents the local actual system state, reflecting the real physical process. The blue curve stands for the healthy remote fusion state estimate. The blue curve basically fits the green curve, indicating that the applied Kalman filter does a good job of following the physical process. The magenta curve is the state estimation under the average attack. The attacker evenly distributes his own energy and injects it into each channel, causing channel packet loss, thus affecting the system tracking state. Furthermore, the red curve is denoted as the state estimation under the optimal attack, where the attacker implements the optimal DoS attack strategy obtained by solving Problem 3 for energy allocation. From the norm of estimation error for the stable systems as exhibited in Figs. 2.5 and 2.6, it can be seen that the tracking performance of the remote fusion estimation system deteriorates more severely under optimal attacks than under no attack and average attack. The traces of error covariance expectation for stable systems are presented in Figs. 2.7 and 2.8, which implies that the traces of error covariance expectation under the optimal attack is significantly higher than the average attack strategy. That is, the optimal attack strategy will deviate the estimated value of the sensor network from the true value to the greatest extent.

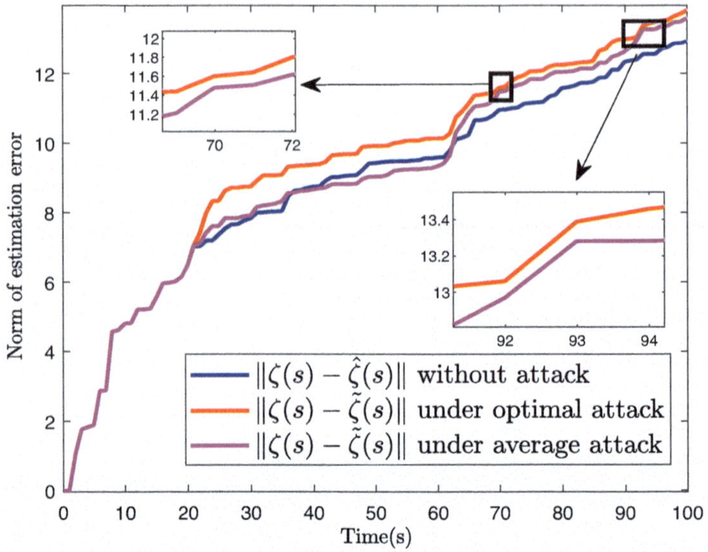

Fig. 2.5 Norm of estimation error without attack, norm of estimation error under optimal attack, and norm of estimation error under average attack of stable physical processes $\mathscr{A}_1 = 0.6I_2$ in CASE 1

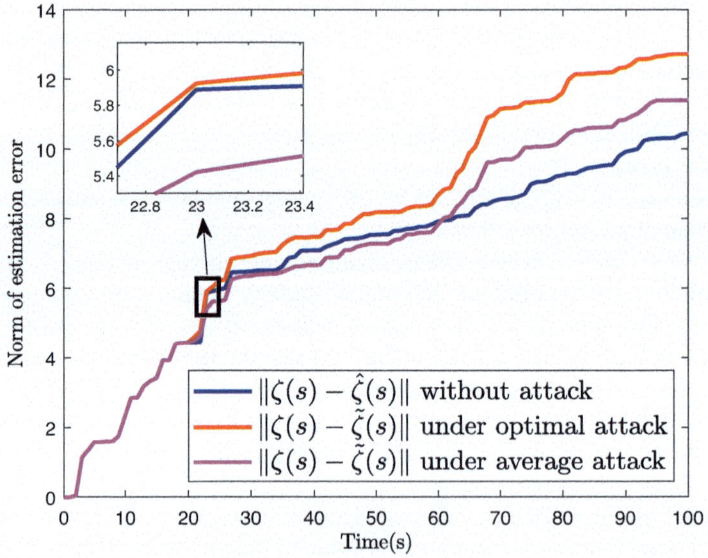

Fig. 2.6 Norm of estimation error without attack, norm of estimation error under optimal attack, and norm of estimation error under average attack of stable physical processes $\mathscr{A}_1 = 0.6I_2$ in CASE 2

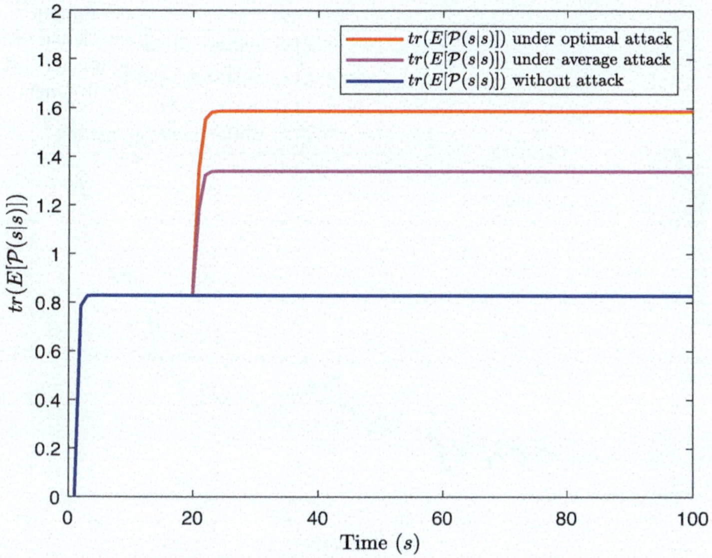

Fig. 2.7 Traces of error covariance expectation without attack, traces of error covariance expectation under optimal attack, and traces of error covariance expectation under average attack for the stable physical processes $\mathscr{A}_1 = 0.6I_2$ in CASE 1

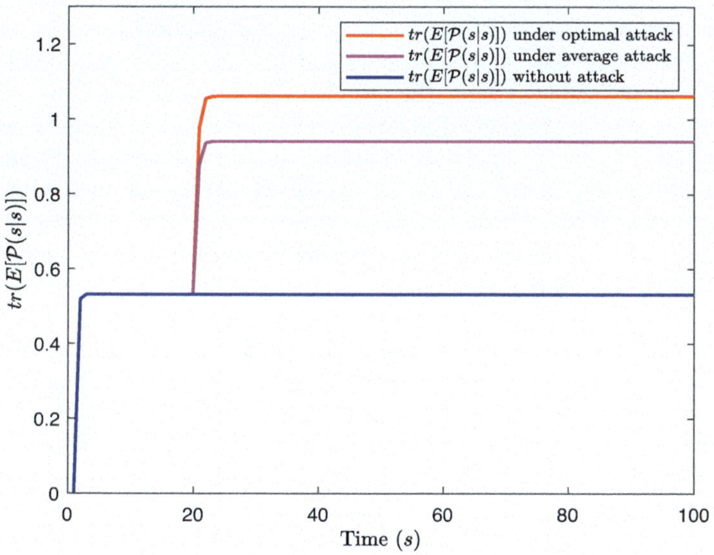

Fig. 2.8 Traces of error covariance expectation without attack, traces of error covariance expectation under optimal attack, and traces of error covariance expectation under average attack for the stable physical processes $\mathscr{A}_1 = 0.6I_2$ in CASE 2

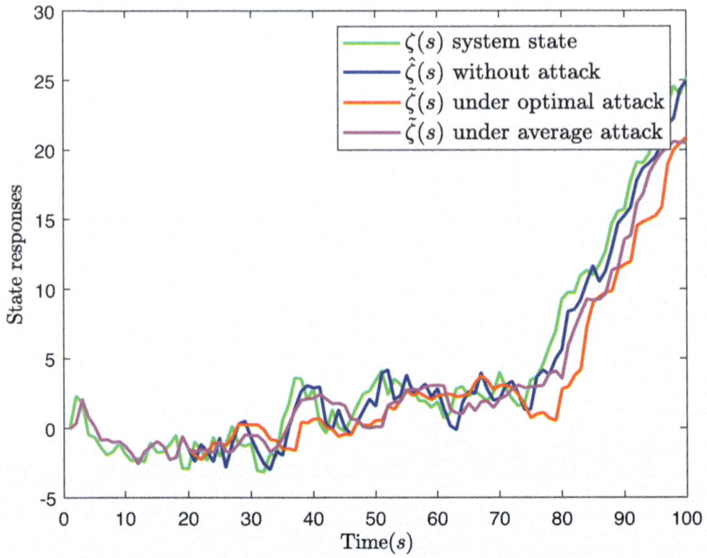

Fig. 2.9 Health state, estimation without attack, estimation under optimal attack, and estimation under average attack of stable physical processes $\mathscr{A}_2 = 1.02 I_2$ in CASE 1

For the unstable systems, the two sets of state estimation curves are shown in Figs. 2.9 and 2.10, the starting time of attack and the meaning of each color curve are consistent with the stable system. The unstable system accumulates over time, and the state of the system gradually deviates from the initial value. It can be seen from the figure that with the accumulation of errors caused by the attack, the attack will lead to that the fusion estimate deviates from the actual physical state of the system to a greater extent. The norm of the estimation error for unstable systems, as shown in Figs. 2.11 and 2.12, reveals that the tracking performance of the remote fusion estimation system deteriorates is more severely under the optimal attack compared to no attack and the average attack. From the traces of error covariance expectation for the unstable systems as shown in Figs. 2.13 and 2.14, it can be verified again that the traces of error covariance expectation under the optimal attack strategy are significantly higher than those of the average attack strategy. It can be seen that the cumulative effect of the unstable system on the error is more obvious, and the network attack is more likely to cause the unstable system's state to diverge.

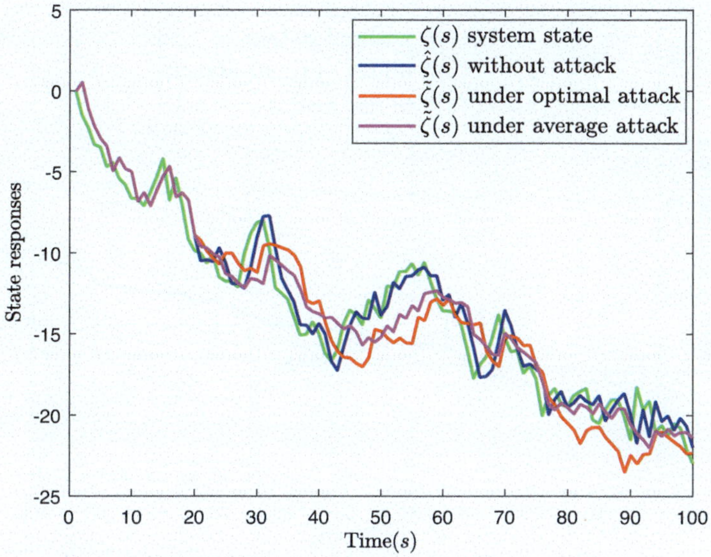

Fig. 2.10 Health state, estimation without attack, estimation under optimal attack, and estimation under average attack of stable physical processes $\mathcal{A}_2 = 1.02I_2$ in CASE 2

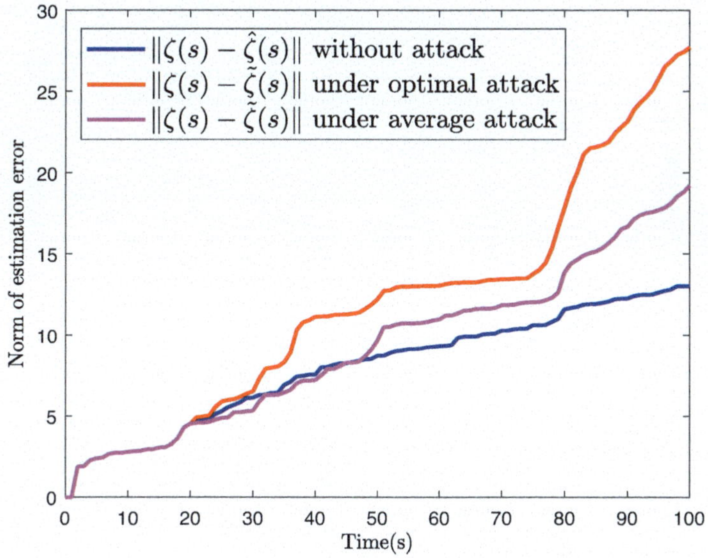

Fig. 2.11 Norm of estimation error without attack, norm of estimation error under optimal attack, and norm of estimation error under average attack of stable physical processes $\mathcal{A}_2 = 1.02I_2$ in CASE 1

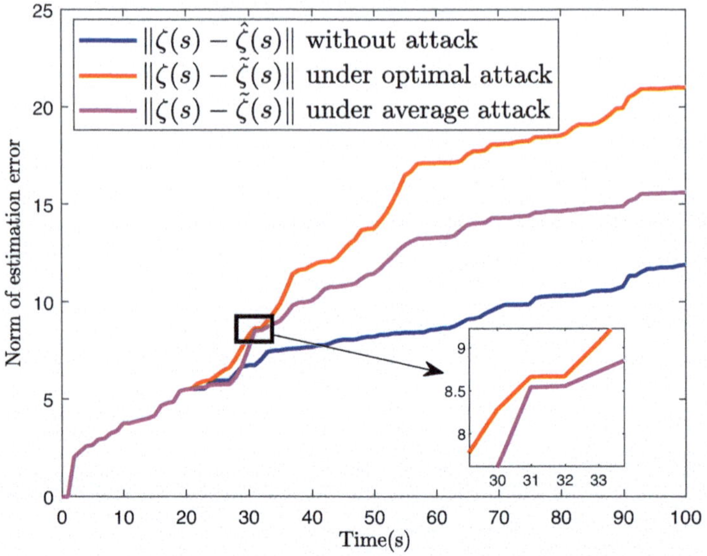

Fig. 2.12 Norm of estimation error without attack, norm of estimation error under optimal attack, and norm of estimation error under average attack of stable physical processes $\mathcal{A}_2 = 1.02I_2$ in CASE 2

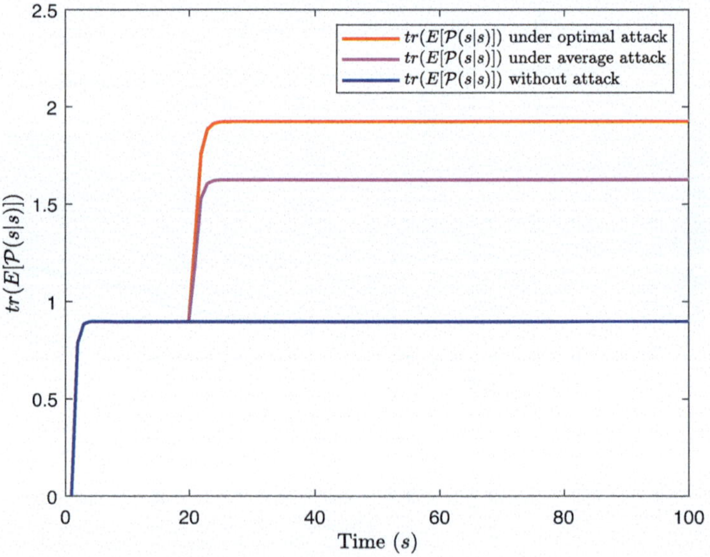

Fig. 2.13 Traces of error covariance expectation without attack, traces of error covariance expectation under optimal attack, and traces of error covariance expectation under average attack for the stable physical processes $\mathcal{A}_2 = 1.02I_2$ in CASE 1

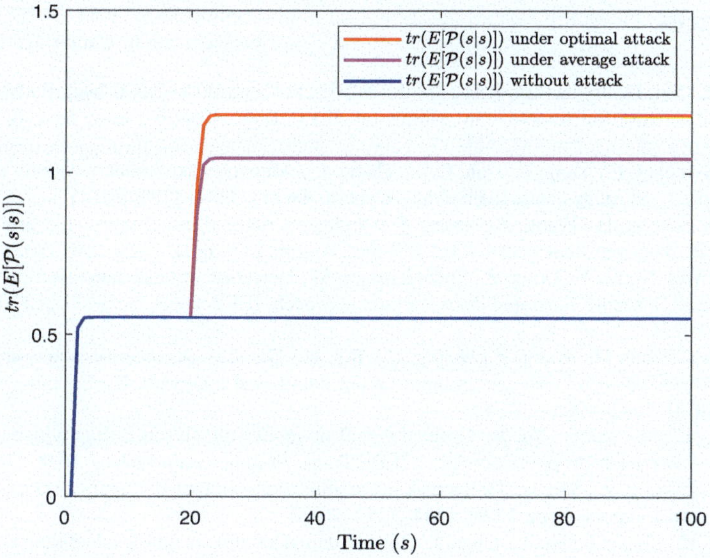

Fig. 2.14 Traces of error covariance expectation without attack, traces of error covariance expectation under optimal attack, and traces of error covariance expectation under average attack for the stable physical processes $\mathcal{A}_2 = 1.02I_2$ in CASE 2

2.5 Conclusion

For the cyber-physical systems with multi-sensor remote state estimation, this chapter investigates the optimal DoS attack problem in which the attacker attempts to interrupt data delivery in wireless channels by combining the probability model between the successful transmission rate and the injected energy of the attacker. The expression of error covariance under the condition of the limited energy consumption by the attacker is derived, and the tendency of the DoS attacker to hit the sensors with different parameters is analyzed. Furthermore, the original optimal attack problem is converted into a solvable positive semi-definite programming problem, and the trace of the expectation matrix for the inverse to the covariance is minimized, that is, the trace of the error covariance in the original problem is maximized under probabilistic conditions. In future research work, the focus will be on the problem of designing attack strategies under conditions of model-free and delay caused by environments, as well as the DoS attacks on the control inputs transmitted over a wireless channel.

References

1. Guo, Z., Shi, D., Johansson, K.H., Shi, L.: Worst-case innovation-based integrity attacks with side information on remote state estimation. IEEE Trans. Control Netw. Syst. **6**(1), 48–59 (2018)
2. Mousavinejad, E., Yang, F., Han, Q.L., Vlacic, L.: A novel cyber attack detection method in networked control systems. IEEE Trans. Cybern. **48**(11), 3254–3264 (2018)
3. Gao, S., Zhang, H., Wang, Z., Huang, C.: A class of optimal switching mixed data injection attack in cyber-physical systems. IEEE Robot. Autom. Lett. **6**(2), 1598–1605 (2021)
4. Li, Y., Wei, X., Li, Y., Dong, Z., Shahidehpour, M.: Detection of false data injection attacks in smart grid: a secure federated deep learning approach. IEEE Trans. Smart Grid **13**(6), 4862–4872 (2022)
5. Gao, S., Zhang, H., Wang, Z., Huang, C., Yan, H.: Data-driven injection attack strategy for linear cyber-physical systems: an input-output data-based approach. IEEE Trans. Netw. Sci. Eng. **10**(6), 4082–4095 (2023)
6. Qin, J., Li, M., Shi, L., Yu, X.: Optimal denial-of-service attack scheduling with energy constraint over packet-dropping networks. IEEE Trans. Autom. Control **63**(6), 1648–1663 (2017)
7. Yang, C., Yang, W., Shi, H.: DoS attack in centralised sensor network against state estimation. IET Control Theory Appl. **12**(9), 1244–1253 (2018)
8. Zhang, H., Cheng, P., Shi, L., Chen, J.: Optimal denial-of-service attack scheduling with energy constraint. IEEE Trans. Autom. Control **60**(11), 3023–3028 (2015)
9. Li, Y., Shi, L., Cheng, P., Chen, J., Quevedo, D.E.: Jamming attacks on remote state estimation in cyber-physical systems: a game-theoretic approach. IEEE Trans. Autom. Control **60**(10), 2831–2836 (2015)
10. Zhang, H., Qi, Y., Wu, J., Fu, L., He, L.: DoS attack energy management against remote state estimation. IEEE Trans. Control Netw. Syst. **5**(1), 383–394 (2016)
11. Li, Y., Quevedo, D.E., Dey, S., Shi, L.: SINR-based DoS attack on remote state estimation: a game-theoretic approach. IEEE Trans. Control Netw. Syst. **4**(3), 632–642 (2016)
12. Peng, L., Shi, L., Cao, X., Sun, C.: Optimal attack energy allocation against remote state estimation. IEEE Trans. Autom. Control **63**(7), 2199–2205 (2017)
13. Liu, Y., Ning, P., Reiter, M.K.: False data injection attacks against state estimation in electric power grids. ACM Trans. Inf. Syst. Secur. **14**(1), 1–33 (2011)
14. Xie, L., Mo, Y., Sinopoli, B.: Integrity data attacks in power market operations. IEEE Trans. Smart Grid **2**(4), 659–666 (2011)
15. Li, Y., Quevedo, D.E., Lau, V., Shi, L.: Optimal periodic transmission power schedules for remote estimation of ARMA processes. IEEE Trans. Signal Process. **61**(24), 6164–6174 (2013)
16. Li, Y., Quevedo, D.E., Lau, V., Shi, L.: Online sensor transmission power schedule for remote state estimation. In: Proceedings of the 52nd IEEE Conference on Decision and Control, pp. 4000–4005 (2013)
17. Anderson, B.D., Moore, J.B.: Optimal Filtering. Prentice-Hall, Englewood Cliffs, NJ, USA (2005)
18. Dragomir, S.S.: Some reverses of the Jensen inequality with applications. Bull. Australian Math. Soc. **87**(2), 177–194 (2013)
19. Zhang, F.: The Schur Complement and Its Applications. Springer, Berlin, Germany (2006)

Chapter 3
Model-Based FDI Attack Strategy Against CPSs via Moment Matrices

Abstract This chapter investigates injection attacks incorporating stochastic noise against cyber-physical systems, which is more general but also more challenging to defend than deterministic injection attacks. Based on the optimization theory and novel key defining matrices, an optimal stochastic injection attack strategy is proposed. Unlike existing strategies, this strategy is stochastic, making it harder for defenders to predict. Therefore, the newly designed attack is expected to be widely used to disrupt system performance. By leveraging estimated data from the observer and the attack input, a virtual residual system is established, which more accurately reflects changes in the system error before and after the attack than the traditional error system. Using the state and output residuals along with the stochastic attack input, two performances are defined to ensure the stealthiness and effectiveness of the attack, respectively. Subsequently, an optimal attack problem with non-convex objective function and constraint is formulated. The key to acquiring the designed optimal stochastic injection attack strategy is to apply a semi-definite relaxation involving moment matrices for transforming this non-convex optimization problem into a convex optimization problem and solving it. Finally, the effectiveness of the proposed attack strategy is validated through numerical simulation of a networked mass-spring-damper system and a V-formation experiment involving three quadrotors.

3.1 Introduction

Modern networks face threats from a variety of attack types, among which denial-of-service (DoS) attacks, injection attacks, and replay attacks are particularly common. In any case, the characteristic of the injection attacks is that the attacker attempts to tamper with the attacked signal without being noticed, which poses a significant challenge to defenders. Unfortunately, according to current research, apart from deploying a highly secure network architecture and relying on resilient defense strategies, there is no proven and effective method to fully defend against such attacks. Therefore, in-depth understanding of the behavior of injection attackers has become

S. Gao et al., *Cyberattack Strategies Against Cyber-Physical Systems: From a Control Perspective*, https://doi.org/10.1007/978-981-95-9645-4_3

the focus of this chapter. Numerous scholars have conducted in-depth explorations in the design of injection attack strategies, making heavy use of control theory and optimization theory. An optimal injection attack strategy for linear discrete time-invariant system was proposed, which was optimized under a convex quadratic control objective function [1]. Subsequently, an injection attack mechanism for partial information was proposed [2], taking into account the peak value of the attack signal to ensure stealthiness. For the problem of undermining the system state estimation performance, an injection attack strategy via the estimation error was proposed, which considered convex value functions and convex constraints of infinite-norm performance [3]. Further, an injection attack mechanism based on error learning design was proposed [4]. In order to reflect the changes in estimation errors before and after the attack, a self-generated injection attack strategy considering the residual of the estimation error was designed [5]. For linear continuous time-invariant systems, an optimal partial feedback injection attack strategy was designed for a quadratic control objective function that simultaneously considers single and the sum of multiple objectives, and a position switching attack form was constructed in [6]. Furthermore, the optimal injection attack of the mixed attack signal and its derivative signal was given [7], but it was necessary to satisfy constraint conditions to ensure the optimality of the attack strategy. Compared with time-invariant systems, time-varying systems are more common in practical applications. However, injection attack strategies for time-varying systems have only been explored recently. A data injection attack that can evade bad data detection (BDD) was designed for linear discrete-time time-varying smart grid systems [8]. Under the above background, one of the motivations of this chapter is to design an injection attack strategy for time-varying cyber-physical systems. Meanwhile, it is worth noting that most existing injection attack strategies are based on deterministic methods, and there are relatively few studies on stochastic injection attacks. In recent years, although Zhang et al. [9] designed an injection attack strategy containing arbitrary signals for linear time-invariant systems containing nonlinear parts, this strategy did not fully consider the energy consumption of the attack signal. Therefore, another research motivation of this chapter is to design the optimal injection attack strategy containing stochastic signals while considering attack energy consumption. In addition, the current research on the optimal injection attack problem mainly focuses on convex objective functions with convex constraints [1, 3] and non-convex objective function with convex constraints [6, 7, 10] scenarios. The optimal injection attack problem for non-convex objective function with non-convex constraints is rarely addressed, which makes in-depth research on such problems particularly important.

Motivated by the above, a novel model-based stochastic injection attack strategy against linear discrete time-varying cyber-physical system is designed in this chapter, with the output residuals constrained, which can maximize the bias between the state estimation errors before and after the attack using minimum attack energy, i.e., deteriorate the estimation performance of the observer. The main contributions of this chapter are summarized as follows:

(1) The attacked linear discrete time-varying virtual residual system is constructed to reflect the changes in system state and output residuals before and after the attack. Compared to the existing injection attack strategies that consider estimation errors [3, 8], the performance indexes defined based on the established residual system can ensure the destructiveness of the attack while maintaining its stealthiness, particularly for residual-based detectors. Furthermore, in contrast to the existing injection attack strategies that consider residuals for linear time-invariant cyber-physical systems [5, 11], this chapter addresses a more general time-varying cyber-physical system.
(2) Moment matrices are utilized for transforming the non-convex optimal attack problem into a convex optimization problem. Then, the proposed strategy can be solved using the semi-definite programming toolkit in MATLAB. In comparison to the existing optimal injection attack strategies [1, 3, 6, 7, 10], a more general optimal attack problem is considered in this chapter, specifically, its objective function and constraint are non-convex.
(3) An optimal stochastic injection attack strategy is proposed. Compared to the existing deterministic injection attack strategies [1, 3, 6, 7, 10], the proposed strategy is stochastic, making it more difficult for defenders to predict and reconstruct attack signals. Additionally, unlike the existing injection attack strategy containing arbitrary signals [9], the proposed strategy takes the energy consumption of the attacker into consideration.

The remaining part of the chapter is organized as follows. Section 3.2 introduces the problem formulation of the stochastic injection attack against the linear discrete time-varying cyber-physical system. The transformation of the original non-convex attack optimization problem into a convex optimization problem using moment matrices is described in Sect. 3.3. Section 3.4 presents the optimal stochastic injection attack strategy and algorithm. Subsequently, the effectiveness of the proposed strategy is verified through numerical simulation and experiment in Sect. 3.5. Finally, Sect. 3.6 concludes the chapter.

3.2 Problem Formulation

3.2.1 System Model

Consider a discrete time-varying cyber-physical system expressed as

$$\begin{cases} x_{k+1} = A_k x_k + B_k u_k + w_k, \\ y_k = C_k x_k + v_k, \end{cases} \tag{3.1}$$

where $x_k \in \mathbb{R}^{n_x}$, $u_k \in \mathbb{R}^{n_u}$, and $y_k \in \mathbb{R}^{n_y}$ are the unknown system state, the control input, and the measurable system output, respectively. $w_k \sim \mathbb{P}_k^w$ and $v_k \sim \mathbb{P}_k^v$ are the independent external process and measurement disturbances. $A_k \in \mathbb{R}^{n_x \times n_x}$, $B_k \in \mathbb{R}^{n_x \times n_u}$, and $C_k \in \mathbb{R}^{n_y \times n_x}$ are the known time-varying matrices at time k.

An observer based on the Luenberger form for estimating the unknown state x_k of the system (3.1) can be constructed as

$$\begin{cases} \hat{x}_{k+1} = A_k \hat{x}_k + B_k u_k + L_k (y_k - \hat{y}_k), \\ \hat{y}_k = C_k \hat{x}_k, \end{cases} \tag{3.2}$$

where $\hat{x}_k$ is the estimation of the unknown state x_k and $\hat{y}_k$ stands for the estimation of the system output y_k. The observer gain L_k is chosen to ensure that all eigenvalues of the matrix $(A_k - L_k C_k)$ are within the unit circle.

3.2.2 *Attacked Model*

First of all, the following assumptions are made for the attacker.

Assumption 3.1 The attacker can utilize stolen traffic data to obtain the parameter matrices of the attacked system, containing A_k, B_k, C_k, and L_k.

Assumption 3.2 The attacker possesses the capability to tamper with the system output sent by the sensor. Specifically, the attacker can inject the designed false data and the system output into the remote observer, simultaneously.

Remark 3.1 Assumption 3.2 describes the ability of the attacker to invade the attacked system. In fact, due to the openness and fragility of the wireless network communication channel, the assumption of the attacker ability to invade the system is reasonable. And the cover attack technique can be used [2]. In this case, the system output signal and the attack input signal are exactly transmitted to the remote observer synchronously. For the asynchronous transmission of the system output and the attack input signal, a method studied in [12] can be adopted to deal with asynchronous clocks by introducing a time label.

The attacked measurable system output can be described as

$$\tilde{y}_k = y_k + y_k^a, \tag{3.3}$$

where $\tilde{y}_k \in \mathbb{R}^{n_y}$ is the attacked measurable system output. y_k^a represents a randomly distributed attack input.

Combining the system (3.1) and the attacked system output (3.3), the attacked system model can be formulated as

$$\begin{cases} \tilde{x}_{k+1} = A_k \tilde{x}_k + B_k \tilde{u}_k + w_k, \\ \tilde{y}_k = C_k \tilde{x}_k + y_k^a + v_k, \end{cases} \tag{3.4}$$

where $\tilde{x}_k$, $\tilde{u}_k$, and $\tilde{y}_k$ are the unknown attacked system state, the attacked control input, and the attacked system output, respectively.

Together with the observer (3.2) and the attacked system output (3.3), the attacked observer can be expressed as

$$\begin{cases} \hat{\tilde{x}}_{k+1} = A_k \hat{\tilde{x}}_k + B_k \tilde{u}_k + L_k(\tilde{y}_k - \hat{\tilde{y}}_k), \\ \hat{\tilde{y}}_k = C_k \hat{\tilde{x}}_k, \end{cases} \tag{3.5}$$

where $\hat{\tilde{x}}_k$ and $\hat{\tilde{y}}_k$ are the attacked estimation of the unknown state x_k and the attacked estimation of the system output, respectively.

Remark 3.2 The reason for the attack being carried out only on the system output, i.e., the communication channel from the sensor to the remote observer, and not on the controller to actuator channel, is that the attack strategy of this chapter is based on the residual. Using the virtual residual system (3.6) described in Sect. 3.2.3, it can be seen that if the attacker performs the attack on the wireless channel from the controller to the actuator, it is not able to affect the estimator performance. So the attacker will not waste energy on the wireless channel from the controller to the actuator in the scenario studied in this chapter.

3.2.3 *Optimal FDI Attack Problem*

Before the presentation of the attack problem, it is imperative to first introduce the virtual residual system. This system characterises the residual of the error between the real signal and the estimation of the observer before and after the attack, which is described as

$$\begin{cases} e^a_{k+1} = (A_k - L_k C_k) e^a_k - L_k y^a_k, \\ z^a_k \;\; = C_k e^a_k + y^a_k, \end{cases} \tag{3.6}$$

where

$$e_k = x_k - \hat{x}_k, \tilde{e}_k = \tilde{x}_k - \hat{\tilde{x}}_k, e^a_k = \tilde{e}_k - e_k, e^a_0 \sim \mathbb{P}^{ea}_0,$$
$$z_k = y_k - \hat{y}_k, \tilde{z}_k = \tilde{y}_k - \hat{\tilde{y}}_k, z^a_k = \tilde{z}_k - z_k.$$

The specific process is detailed as follows. Combined with the systems (3.1) and (3.2), the error e_{k+1} and z_k can be described as

$$\begin{aligned} e_{k+1} &= x_{k+1} - \hat{x}_{k+1}, \\ &= A_k x_k + B_k u_k + w_k - A_k \hat{x}_k - B_k u_k - L_k(y_k - \hat{y}_k), \\ &= (A_k - L_k C_k) e_k + w_k - L_k v_k, \end{aligned} \tag{3.7}$$

and,

$$\begin{aligned} z_k = y_k - \hat{y}_k &= C_k x_k + v_k - C_k \hat{x}_k, \\ &= C_k e_k + v_k. \end{aligned} \tag{3.8}$$

Subsequently, together with the attacked systems (3.4) and (3.5), the error $\tilde{e}_{k+1}$ and $\tilde{z}_k$ can be expressed as

$$\begin{aligned} \tilde{e}_{k+1} &= (A_k - L_k C_k)\tilde{e}_k + w_k - L_k v_k - L_k y_k^a, \\ \tilde{z}_k &= C_k \tilde{e}_k + v_k + y_k^a. \end{aligned} \tag{3.9}$$

Finally, in combination with (3.7), (3.8), and (3.9), the residual e_{k+1}^a and z_k^a can be obtained.

For the above virtual residual system, the following assumptions are made about the attacker.

Assumption 3.3 The attacker knows the initial value of the virtual residual system, that is, $e_0^a \sim \mathbb{P}_0^{ea}$.

Remark 3.3 In the actual operation process, the attacker can measure the initial state and its estimation. It should be noted that due to the existence of process and measurement noise, the subsequent state is still unknown for the attacker. Then, the initial state error is calculated based on the measurement results before and after the attack, and finally the initial residual value is obtained by the error. As there is random noise in the actual measurement, the initial residual is also random, that is, $e_0^a \sim \mathbb{P}_0^{ea}$.

The objective of the attacker is to maximize the degradation of estimation performance of the observer using minimal energy consumption without being detected by anomaly detector. Specifically, the attacker aims to use minimal energy consumption to maximize the state residual, denoted as e_k^a, and minimize the output residual, denoted as z_k^a, simultaneously. Then, the objective function of the attacker in this chapter uses $e_k^{a^T} R_{3,k} e_k^a$ and $z_k^{a^T} R_{2,k} z_k^a$ to denote the offsets of the state and output residuals, respectively, and defines $y_k^{a^T} R_{1,k} y_k^a$ as the energy consumption of the attack at each moment. For these two residuals, given two prescribed scalars $\eta_1 > 0$ and $\eta_2 > 0$, the attacker wants the following two constraints to be satisfied,

$$\mathbb{E}\{\|z_k^a\|^2\} \leq \eta_1^2 \mathbb{E}\{\|y_k^a\|^2\}, \tag{3.10a}$$

$$\mathbb{E}\{\|e_k^a\|^2\} \geq \eta_2^2 \mathbb{E}\{\|y_k^a\|^2\},\ k = 1, \ldots, \bar{N}, \tag{3.10b}$$

where y_k^a is a randomly distributed attack input satisfying $y_k^a \sim \mathbb{P}_k^a(e_k^a)$. The physical meaning of (3.10a) is to ensure the upper bound on the sum of residuals of the output estimation error before and after the attack. For a given scalar η_1 that is sufficiently small, employing the minimum attack energy consumption ensures that the output residual z_k^a remains bounded to a level that evades detection by the anomaly detector.

Conversely, the physical meaning of (3.10b) is to establish the lower bound on the sum of the residuals of the state estimation error before and after the attack. In other words, its role is to ensure that there is a minimum lower bound on the degree of degradation of the observer's performance. With a sufficiently large scalar η_2, it is still possible for the observer's performance to degrade significantly even when using the minimum attack energy consumption.

The attack problem studied in this chapter is written as follows.

Problem 3.1

$$\begin{aligned}
\min_{y_k^a} \quad & \mathbb{E}\left\{\sum_{k=1}^{\bar{N}} (y_k^{a^T} R_{1,k} y_k^a + z_k^{a^T} R_{2,k} z_k^a - e_k^{a^T} R_{3,k} e_k^a)\right\} \\
s.t. \quad & e_{k+1}^a = (A_k - L_k C_k) e_k^a - L_k y_k^a, \\
& z_k^a \quad = C_k e_k^a + y_k^a, \\
& \mathbb{E}\{\|z_k^a\|^2\} \leq \eta_1^2 \mathbb{E}\{\|y_k^a\|^2\}, \\
& \mathbb{E}\{\|e_k^a\|^2\} \geq \eta_2^2 \mathbb{E}\{\|y_k^a\|^2\}, \\
& y_k^a \sim \mathbb{P}_k^a(e_k^a), \\
& e_0^a \sim \mathbb{P}_0^{ea}, \\
& R_{i,k} \succeq 0, \ i = 1, 2, 3, \ k = 1, ..., \bar{N}.
\end{aligned}$$

Remark 3.4 There are two points to be noted about Problem 3.1. On the one hand, Problem 3.1 is a non-convex optimization problem as both the cost function and the inequality constraints contain non-convex terms, which makes the attack problem difficult to be solved. On the other hand, compared with the existing attack strategies, the attack strategy designed in Problem 3.1 is a stochastic strategy, which makes it difficult for the defender to predict and reproduce the attack inputs.

Remark 3.5 The constraint similar to (3.10) is defined in [13]. The parameters η_1 and η_2 of the constraint (3.10) are determined by the attacker. Theoretically, the smaller the η_1 and the larger the η_2, the better. However, in actual applications, considering the solvability of optimization problem and the maximum power consumption of controllers, actuators and other devices, the values of the parameters η_1 and η_2 may not be infinite. The attacker can adjust the parameters η_1 and η_2 in real time by stealing the alarm message and communication traffic data of the attacked system. If the alarm message is not triggered, increase parameter η_2 until the alarm is just not triggered. If the alarm message is triggered, turn parameter η_1 down until the alarm is not triggered.

3.3 Attack Problem Transformation via Moment Matrices

In order to facilitate the subsequent transformation of Problem 3.1 using moment matrices, the equivalent form of Problem 3.2 is first introduced. The detailed process of converting Problem 3.1 into an equivalent form of Problem 3.2 is described as follows. Plugging (3.6) into the cost function of Problem 3.1, the summation term of the cost function can be rewritten as

$$\begin{aligned}
& y_k^{a^T} R_{1,k} y_k^a + z_k^{a^T} R_{2,k} z_k^a - e_k^{a^T} R_{3,k} e_k^a \\
&\qquad = y_k^{a^T} R_{1,k} y_k^a + (C_k e_k^a + y_k^a)^T R_{2,k} (C_k e_k^a + y_k^a) - e_k^{a^T} R_{3,k} e_k^a, \\
&\qquad = y_k^{a^T} (R_{1,k} + R_{2,k}) y_k^a + e_k^{a^T} (C_k^T R_{2,k} C_k - R_{3,k}) e_k^a \\
&\qquad\quad + e_k^{a^T} C_k^T R_{2,k} y_k^a + y_k^{a^T} R_{2,k} C_k e_k^a, \\
&\qquad = \begin{bmatrix} e_k^a \\ y_k^a \end{bmatrix}^T \begin{bmatrix} C_k^T R_{2,k} C_k - R_{3,k} & C_k^T R_{2,k} \\ R_{2,k} C_k & R_{1,k} + R_{2,k} \end{bmatrix} \begin{bmatrix} e_k^a \\ y_k^a \end{bmatrix},
\end{aligned}$$

hence, the above equation can be further expressed as

$$\begin{bmatrix} e_k^a \\ y_k^a \end{bmatrix}^T \begin{bmatrix} C_k^T R_{2,k} C_k - R_{3,k} & C_k^T R_{2,k} \\ R_{2,k} C_k & R_{1,k} + R_{2,k} \end{bmatrix} \begin{bmatrix} e_k^a \\ y_k^a \end{bmatrix} = \begin{bmatrix} 1 \\ e_k^a \\ y_k^a \end{bmatrix}^T R_k \begin{bmatrix} 1 \\ e_k^a \\ y_k^a \end{bmatrix}$$

where

$$R_k = \begin{bmatrix} 0 & \mathbf{0} & \mathbf{0} \\ \mathbf{0} & C_k^T R_{2,k} C_k - R_{3,k} & C_k^T R_{2,k} \\ \mathbf{0} & R_{2,k} C_k & R_{1,k} + R_{2,k} \end{bmatrix}.$$

The inequality constraint for Problem 3.1 can be equivalently described as follows,

$$\begin{aligned}
\mathbb{E}\{\|z_k^a\|^2\} - \eta_1^2 \mathbb{E}\{\|y_k^a\|^2\} &= \mathbb{E}\{(z_k^{a^T} z_k^a - \eta_1^2 y_k^{a^T} y_k^a)\}, \\
&= \mathbb{E}\{(C_k e_k^a + y_k^a)^T (C_k e_k^a + y_k^a) - \eta_1^2 y_k^{a^T} y_k^a\}, \\
&= \mathbb{E}\{y_k^{a^T} (C_k^T C_k - \eta_1^2 I) y_k^a + e_k^{a^T} C_k^T C_k e_k^a + e_k^{a^T} C_k^T y_k^a + y_k^{a^T} C_k e_k^a\}, \\
&= \mathbb{E}\left\{ \begin{bmatrix} 1 \\ e_k^a \\ y_k^a \end{bmatrix}^T \Phi_{1,k} \begin{bmatrix} 1 \\ e_k^a \\ y_k^a \end{bmatrix} \right\} \le 0,
\end{aligned}$$

and

$$\eta_2^2\mathbb{E}\{\|y_k^a\|^2\} - \mathbb{E}\{\|e_k^a\|^2\} = \mathbb{E}\{\eta_2^2 y_k^{a^T} y_k^a - e_k^{a^T} e_k^a\},$$
$$= \mathbb{E}\left\{\begin{bmatrix} 1 \\ e_k^a \\ y_k^a \end{bmatrix}^T \Phi_{2,k} \begin{bmatrix} 1 \\ e_k^a \\ y_k^a \end{bmatrix}\right\} \le 0,$$

where

$$\Phi_{1,k} = \begin{bmatrix} 0 & \mathbf{0} & \mathbf{0} \\ \mathbf{0} & C_k^T C_k & C_k^T \\ \mathbf{0} & C_k & (1-\eta_1^2)I \end{bmatrix}, \Phi_{2,k} = \begin{bmatrix} 0 & \mathbf{0} & \mathbf{0} \\ \mathbf{0} & -I & \mathbf{0} \\ \mathbf{0} & \mathbf{0} & \eta_2^2 I \end{bmatrix}.$$

Together with the cost function and the constraints, Problem 3.1 can be equivalently transformed into the following problem.

Problem 3.2

$$\min_{y_k^a} \ \mathbb{E}\left\{\sum_{k=1}^{\bar{N}} \begin{bmatrix} 1 \\ e_k^a \\ y_k^a \end{bmatrix}^T R_k \begin{bmatrix} 1 \\ e_k^a \\ y_k^a \end{bmatrix}\right\}$$
$$s.t. \ \ e_{k+1}^a = (A_k - L_k C_k)e_k^a - L_k y_k^a,$$
$$\mathbb{E}\left\{\begin{bmatrix} 1 \\ e_k^a \\ y_k^a \end{bmatrix}^T \Phi_{i,k} \begin{bmatrix} 1 \\ e_k^a \\ y_k^a \end{bmatrix}\right\} \le 0, \ i = 1, 2,$$
$$y_k^a \sim \mathbb{P}_k^a(e_k^a),$$
$$e_0^a \sim \mathbb{P}_0^{ea}.$$

Remark 3.6 In Problem 3.2, it is known that R_k and $\Phi_{i,k}$, $i = 1, 2$ are symmetric and indefinite matrices, so Problem 3.2 is a type of non-convex problem containing a non-convex cost function and non-convex inequality constraints. When the subscript i in $\Phi_{i,k}$ is larger than 2, then Problem 3.2 is more general and difficult to solve than Problem 3.1.

Theorem 3.1 *Let the moment matrices* Ψ_k *be defined as*

$$\Psi_k = \begin{bmatrix} \psi_k^{11} & \psi_k^{12} & \psi_k^{13} \\ \psi_k^{21} & \Psi_k^{22} & \Psi_k^{23} \\ \psi_k^{31} & \Psi_k^{32} & \Psi_k^{33} \end{bmatrix} \triangleq \mathbb{E}\left\{\begin{bmatrix} 1 \\ e_k^a \\ y_k^a \end{bmatrix}\begin{bmatrix} 1 \\ e_k^a \\ y_k^a \end{bmatrix}^T\right\}. \tag{3.11}$$

There exists a stochastic attack strategy y_k^a generating these moment matrices Ψ_k, if and only if the conditions

$$\psi_k^{11} = 1,\ \Psi_k \succeq 0,\ \tilde{F}_k(\Psi_k, \Psi_{k+1}) = \boldsymbol{0}, \tag{3.12}$$

are met. Consequently, the non-convex dynamic program, Problem 3.2 can be equivalently transformed into the following convex optimization problem,

Problem 3.3

$$\begin{aligned} &\min_{(\Psi_k)} \sum_{k=1}^{\tilde{N}} tr(R_k \Psi_k) \\ &s.t.\ (3.12),\ tr(\Phi_{i,k}\Psi_k) \leq 0,\ i = 1, 2, \end{aligned}$$

where $\Psi_k = \Psi_k^T \in \mathbb{R}^{(1+n_x+n_u)\times(1+n_x+n_u)}$, $\tilde{F}_k(\Psi_k, \Psi_{k+1}) = \Gamma_k^T \Omega_k \Gamma_k$, and

$$\Omega_k = \begin{bmatrix} -\Psi_k & & \\ & \psi_{k+1}^{11} & \psi_{k+1}^{12} \\ & \psi_{k+1}^{21} & \psi_{k+1}^{22} \end{bmatrix},\ \Gamma_k = \begin{bmatrix} 1 & 0 \\ 0 & (A_k - L_k C_k)^T \\ 0 & -L_k^T \\ 1 & 0 \\ 0 & I \end{bmatrix}.$$

Proof From the definition of the moment matrices (3.11), one can extract that $\psi_k^{11} = 1$. And, for any non-zero vector $\xi \in \mathbb{R}^{(1+n_x+n_u)\times 1}$, one has

$$\begin{aligned} \mathbb{E}\left\{\xi^T \begin{bmatrix} 1 \\ e_k^a \\ y_k^a \end{bmatrix}\begin{bmatrix} 1 \\ e_k^a \\ y_k^a \end{bmatrix}^T \xi\right\} &= \mathbb{E}\left\{\left(\xi^T \begin{bmatrix} 1 \\ e_k^a \\ y_k^a \end{bmatrix}\right)\left(\begin{bmatrix} 1 \\ e_k^a \\ y_k^a \end{bmatrix}^T \xi\right)\right\}, \\ &= \mathbb{E}\left\{\left(\xi^T \begin{bmatrix} 1 \\ e_k^a \\ y_k^a \end{bmatrix}\right)^2\right\} \geq 0, \end{aligned}$$

therefore, combined with the definition of positive semi-definite matrix, $\Psi_k \succeq 0$.

Subsequently, consider the equation

$$\begin{aligned} &\begin{bmatrix} 1 & \boldsymbol{0} \\ \boldsymbol{0} & (A_k - L_k C_k)^T \\ \boldsymbol{0} & -L_k^T \end{bmatrix}^T (-\Psi_k) \begin{bmatrix} 1 & \boldsymbol{0} \\ \boldsymbol{0} & (A_k - L_k C_k)^T \\ \boldsymbol{0} & -L_k^T \end{bmatrix} \\ &= -\mathbb{E}\begin{bmatrix} 1 & \boldsymbol{0} \\ \boldsymbol{0} & (A_k - L_k C_k)^T \\ \boldsymbol{0} & -L_k^T \end{bmatrix}^T \begin{bmatrix} 1 \\ e_k^a \\ y_k^a \end{bmatrix}\begin{bmatrix} 1 \\ e_k^a \\ y_k^a \end{bmatrix}^T \begin{bmatrix} 1 & \boldsymbol{0} \\ \boldsymbol{0} & (A_k - L_k C_k)^T \\ \boldsymbol{0} & -L_k^T \end{bmatrix}, \\ &= -\mathbb{E}\begin{bmatrix} 1 \\ e_{k+1}^a \end{bmatrix}\begin{bmatrix} 1 \\ e_{k+1}^a \end{bmatrix}^T = -\begin{bmatrix} \psi_{k+1}^{11} & \psi_{k+1}^{12} \\ \psi_{k+1}^{21} & \Psi_{k+1}^{22} \end{bmatrix}, \end{aligned}$$

then $\tilde{F}_k(\Psi_k, \Psi_{k+1}) = \mathbf{0}$ can be obtained. Thus, for any attack strategy y_k^a, the condition (3.12) is satisfied.

Moreover, for the cost function of Problem 3.2, one has

$$\begin{aligned}
\mathbb{E}\left\{\sum_{k=1}^{\bar{N}}\begin{bmatrix}1\\e_k^a\\y_k^a\end{bmatrix}^T R_k \begin{bmatrix}1\\e_k^a\\y_k^a\end{bmatrix}\right\} &= tr\left(\mathbb{E}\left\{\sum_{k=1}^{\bar{N}}\begin{bmatrix}1\\e_k^a\\y_k^a\end{bmatrix}^T R_k \begin{bmatrix}1\\e_k^a\\y_k^a\end{bmatrix}\right\}\right), \\
&= \mathbb{E}\left\{\sum_{k=1}^{\bar{N}} tr\left(\begin{bmatrix}1\\e_k^a\\y_k^a\end{bmatrix}^T R_k \begin{bmatrix}1\\e_k^a\\y_k^a\end{bmatrix}\right)\right\}, \\
&= \mathbb{E}\left\{\sum_{k=1}^{\bar{N}} tr\left(R_k \begin{bmatrix}1\\e_k^a\\y_k^a\end{bmatrix}\begin{bmatrix}1\\e_k^a\\y_k^a\end{bmatrix}^T\right)\right\}, \\
&= \sum_{k=1}^{\bar{N}} tr\left(R_k \mathbb{E}\left\{\begin{bmatrix}1\\e_k^a\\y_k^a\end{bmatrix}\begin{bmatrix}1\\e_k^a\\y_k^a\end{bmatrix}^T\right\}\right) = \sum_{k=1}^{\bar{N}} tr(R_k \Psi_k).
\end{aligned}$$

Similar to the above, for the inequality conditions of Problem 3.2, one has

$$\mathbb{E}\left\{\begin{bmatrix}1\\e_k^a\\y_k^a\end{bmatrix}^T \Phi_{i,k} \begin{bmatrix}1\\e_k^a\\y_k^a\end{bmatrix}\right\} = tr(\Phi_{i,k}\Psi_k),\ i = 1, 2.$$

Since the equality constraint of Problem 3.2 can be satisfied by the condition $\tilde{F}_k(\Psi_k, \Psi_{k+1}) = \mathbf{0}$, Problem 3.2 can be transformed into Problem 3.3. Then, the following Theorem 3.2 proves the existence of a stochastic injection attack input y_k^{a*} in Problem 3.2, which produces a solution Ψ_k^* corresponding to Problem 3.3, that is omitted here. Thus Problems 3.2 and 3.3 are equivalent. □

Remark 3.7 Note that the constraint (3.12) in Theorem 3.1 is not a strong constraint for the following reasons. According to the definition of the moment matrices, it is natural to know that $\psi_k^{11} = 1$. At the same time, combined with the proof process of Theorem 3.1 and the definition of the positive semi-definite matrix, it can be known that the properties of the moment matrices themselves satisfy $\Psi_k \succeq 0$. The constraint $\tilde{F}_k(\Psi_k, \Psi_{k+1}) = \mathbf{0}$ corresponds to the equation $e_{k+1}^a = (A_k - L_k C_k)e_k^a - L_k y_k^a$, that is, it is a constraint implied by Problem 3.2 and no additional restriction. Furthermore, it is noteworthy that, in contrast to Problem 3.2, which is optimized directly over a stochastic attack input y_k^a, Problem 3.3 is optimized by utilizing the moment matrices generated through the stochastic attack input y_k^a.

3.4 Optimal Stochastic Injection Attack Strategy

In this section, the optimal stochastic injection attack strategy and algorithm are given.

Theorem 3.2 *Under Assumptions 3.1-3.3, if the condition (3.12) holds, the optimal stochastic injection attack strategy y_k^{a*} for Problem 3.1 is described as*

$$y_k^{a*} = \begin{bmatrix} \psi_k^{31*} & \Psi_k^{32*} \end{bmatrix} \begin{bmatrix} \psi_k^{11*} & \psi_k^{12*} \\ \psi_k^{21*} & \Psi_k^{22*} \end{bmatrix}^{-1} \begin{bmatrix} 1 \\ e_k^{a*} \end{bmatrix} + \varpi_k, \tag{3.13}$$

where ψ_k^{ij*} and Ψ_k^{ij*}, $i, j = 1, 2, 3$ are elements of the optimal moment matrices Ψ_k^*, which are obtained from a solution to Problem 3.3. e_k^{a*} is the optimal state residual value. The random variables $\varpi_k \sim \mathbb{P}_k^{\varpi}$ have zero mean and variance as

$$\Psi_k^{\varpi} = \Psi_k^{33*} - \begin{bmatrix} \psi_k^{31*} & \Psi_k^{32*} \end{bmatrix} \begin{bmatrix} \psi_k^{11*} & \psi_k^{12*} \\ \psi_k^{21*} & \Psi_k^{22*} \end{bmatrix}^{-1} \begin{bmatrix} \psi_k^{13*} \\ \Psi_k^{23*} \end{bmatrix}, \tag{3.14}$$

and ϖ_k is independent from e_j^a for $j \leq k$.

Remark 3.8 It can be observed that the optimal stochastic injection attack input y_a^* is divided into two parts, the specific meanings of the first part is the main part of the attack input for degrading the state estimation performance of the system, and the another is the optimal random noise part, which is employed to make the designed strategy stochastically and difficult to be predicted.

Proof As the moment matrices Ψ_k are generated by y_k^a, combined with the definition of the moment matrices (3.11), one can have

$$\mathbb{E}\left\{\begin{bmatrix} 1 \\ e_k^{a*} \end{bmatrix}\begin{bmatrix} 1 \\ e_k^{a*} \end{bmatrix}^T\right\} = \begin{bmatrix} \psi_k^{11*} & \psi_k^{12*} \\ \psi_k^{21*} & \Psi_k^{22*} \end{bmatrix}, \tag{3.15}$$

and

$$\begin{aligned} \Psi_k^* = \mathbb{E}\left\{\begin{bmatrix} 1 \\ e_k^{a*} \\ y_k^{a*} \end{bmatrix}\begin{bmatrix} 1 \\ e_k^{a*} \\ y_k^{a*} \end{bmatrix}^T\right\} &= \begin{bmatrix} \psi_k^{11*} & \psi_k^{12*} & \psi_k^{13*} \\ \psi_k^{21*} & \Psi_k^{22*} & \Psi_k^{23*} \\ \psi_k^{31*} & \Psi_k^{32*} & \Psi_k^{33*} \end{bmatrix}, \\ &= \begin{bmatrix} 1 & \mathbb{E}\{(e_k^{a*})^T\} & \mathbb{E}\{(y_k^{a*})^T\} \\ \mathbb{E}\{e_k^{a*}\} & \mathbb{E}\{(e_k^{a*})(e_k^{a*})^T\} & \mathbb{E}\{(e_k^{a*})(y_k^{a*})^T\} \\ \mathbb{E}\{y_k^{a*}\} & \mathbb{E}\{(y_k^{a*})(e_k^{a*})^T\} & \mathbb{E}\{(y_k^{a*})(y_k^{a*})^T\} \end{bmatrix}. \end{aligned} \tag{3.16}$$

Therefore, it is only necessary to prove that Eq. (3.16) is satisfied when the optimal stochastic injection attack strategy is designed as (3.13). Then, considering (3.15) and the symmetry property of the moment matrices Ψ_k, the proof process can be simplified to show that the following equations holds.

Firstly, as ϖ_k are zero-mean variables, when adopting (3.13) as the optimal stochastic injection attack strategy, then $\mathbb{E}\{y_k^{a*}\}$ can be described as

$$\begin{aligned}\mathbb{E}\{y_k^{a*}\} &= \mathbb{E}\left\{\begin{bmatrix}\psi_k^{31*} & \Psi_k^{32*}\end{bmatrix}\begin{bmatrix}\psi_k^{11*} & \psi_k^{12*}\\ \psi_k^{21*} & \Psi_k^{22*}\end{bmatrix}^{-1}\begin{bmatrix}1\\ e_k^{a*}\end{bmatrix}+\varpi_k\right\},\\ &= \begin{bmatrix}\psi_k^{31*} & \Psi_k^{32*}\end{bmatrix}\begin{bmatrix}\psi_k^{11*} & \psi_k^{12*}\\ \psi_k^{21*} & \Psi_k^{22*}\end{bmatrix}^{-1}\begin{bmatrix}1\\ \mathbb{E}\{e_k^{a*}\}\end{bmatrix},\\ &= \begin{bmatrix}\psi_k^{31*} & \Psi_k^{32*}\end{bmatrix}\begin{bmatrix}\psi_k^{11*} & \psi_k^{12*}\\ \psi_k^{21*} & \Psi_k^{22*}\end{bmatrix}^{-1}\begin{bmatrix}\psi_k^{11*}\\ \psi_k^{21*}\end{bmatrix},\\ &= \begin{bmatrix}\psi_k^{31*} & \Psi_k^{32*}\end{bmatrix}\begin{bmatrix}1\\ \mathbf{0}\end{bmatrix},\\ &= \psi_k^{31*}.\end{aligned} \tag{3.17}$$

Subsequently, since ϖ_k and e_k^a are stochastically independent of each other, $\mathbb{E}\{y_k^{a*}(e_k^{a*})^T\}$ can be formulated as

$$\begin{aligned}\mathbb{E}\{y_k^{a*}(e_k^{a*})^T\} &= \mathbb{E}\left\{\left(\begin{bmatrix}\psi_k^{31*} & \Psi_k^{32*}\end{bmatrix}\begin{bmatrix}\psi_k^{11*} & \psi_k^{12*}\\ \psi_k^{21*} & \Psi_k^{22*}\end{bmatrix}^{-1}\begin{bmatrix}1\\ e_k^{a*}\end{bmatrix}+\varpi_k\right)(e_k^{a*})^T\right\},\\ &= \begin{bmatrix}\psi_k^{31*} & \Psi_k^{32*}\end{bmatrix}\begin{bmatrix}\psi_k^{11*} & \psi_k^{12*}\\ \psi_k^{21*} & \Psi_k^{22*}\end{bmatrix}^{-1}\begin{bmatrix}\mathbb{E}\{(e_k^{a*})^T\}\\ \mathbb{E}\{e_k^{a*}(e_k^{a*})^T\}\end{bmatrix},\\ &= \begin{bmatrix}\psi_k^{31*} & \Psi_k^{32*}\end{bmatrix}\begin{bmatrix}\psi_k^{11*} & \psi_k^{12*}\\ \psi_k^{21*} & \Psi_k^{22*}\end{bmatrix}^{-1}\begin{bmatrix}\psi_k^{12*}\\ \Psi_k^{22*}\end{bmatrix},\\ &= \begin{bmatrix}\psi_k^{31*} & \Psi_k^{32*}\end{bmatrix}\begin{bmatrix}0\\ I\end{bmatrix},\\ &= \Psi_k^{32*}.\end{aligned} \tag{3.18}$$

Moreover, together with the variance (3.14) of ϖ_k, $\mathbb{E}\{y_k^{a*}(y_k^{a*})^T\}$ can be depicted as

$$
\begin{aligned}
\mathbb{E}\{y_k^{a*}(y_k^{a*})^T\} &= \mathbb{E}\left\{\left(\begin{bmatrix}\psi_k^{31*} & \Psi_k^{32*}\end{bmatrix}\begin{bmatrix}\psi_k^{11*} & \psi_k^{12*}\\ \psi_k^{21*} & \Psi_k^{22*}\end{bmatrix}^{-1}\begin{bmatrix}1\\ e_k^{a*}\end{bmatrix}+\varpi_k\right)\left(\begin{bmatrix}\psi_k^{31*} & \Psi_k^{32*}\end{bmatrix}\right.\right.\\
&\quad\left.\left.\times\begin{bmatrix}\psi_k^{11*} & \psi_k^{12*}\\ \psi_k^{21*} & \Psi_k^{22*}\end{bmatrix}^{-1}\begin{bmatrix}1\\ e_k^{a*}\end{bmatrix}+\varpi_k\right)^T\right\},\\
&= \begin{bmatrix}\psi_k^{31*} & \Psi_k^{32*}\end{bmatrix}\begin{bmatrix}\psi_k^{11*} & \psi_k^{12*}\\ \psi_k^{21*} & \Psi_k^{22*}\end{bmatrix}^{-1}\mathbb{E}\left\{\begin{bmatrix}1\\ e_k^{a*}\end{bmatrix}\begin{bmatrix}1\\ e_k^{a*}\end{bmatrix}^T\right\}\begin{bmatrix}\psi_k^{11*} & \psi_k^{12*}\\ \psi_k^{21*} & \Psi_k^{22*}\end{bmatrix}^{-1}\\
&\quad\times\begin{bmatrix}\psi_k^{13*}\\ \Psi_k^{23*}\end{bmatrix}+\begin{bmatrix}\psi_k^{31*} & \Psi_k^{32*}\end{bmatrix}\begin{bmatrix}\psi_k^{11*} & \psi_k^{12*}\\ \psi_k^{21*} & \Psi_k^{22*}\end{bmatrix}^{-1}\begin{bmatrix}\mathbb{E}\{(\varpi_k)^T\}\\ \mathbb{E}\{e_k^{a*}(\varpi_k)^T\}\end{bmatrix}\\
&\quad+\begin{bmatrix}\mathbb{E}\{\varpi_k\} & \mathbb{E}\{\varpi_k(e_k^{a*})^T\}\end{bmatrix}\begin{bmatrix}\psi_k^{11*} & \psi_k^{12*}\\ \psi_k^{21*} & \Psi_k^{22*}\end{bmatrix}^{-1}\begin{bmatrix}\psi_k^{13*}\\ \Psi_k^{23*}\end{bmatrix}+\mathbb{E}\{\varpi_k(\varpi_k)^T\},\\
&= \begin{bmatrix}\psi_k^{31*} & \Psi_k^{32*}\end{bmatrix}\begin{bmatrix}\psi_k^{11*} & \psi_k^{12*}\\ \psi_k^{21*} & \Psi_k^{22*}\end{bmatrix}^{-1}\begin{bmatrix}\psi_k^{13*}\\ \Psi_k^{23*}\end{bmatrix}+\Psi_k^{\varpi}=\Psi_k^{33*}.
\end{aligned}
\tag{3.19}
$$

It can be seen from (3.17), (3.18), and (3.19) that the optimal moment matrices Ψ_k^* are generated by the optimal stochastic attack strategy designed as (3.13). This is the end of the proof. □

The optimal stochastic injection attack strategy is given in Algorithm 1.

Algorithm 1: Optimal Stochastic Injection Attack Strategy.

Initialize: Obtain the system parameter matrices A_k, B_k, C_k, L_k, set the termination time $\bar{N}$, give suitable scalars $\eta_1 > 0$, $\eta_2 > 0$, the initial value of the attack input y_0^a, and the matrices $R_{i,k}$ which meet the conditions of $R_{i,k} \succeq 0$, $i = 1, 2, 3$, $k = 1, ..., \bar{N}$.

Step 1. Measure the initial value $e_0^a \sim \mathbb{P}_0^{ea}$, and calculate the initial moment matrix Ψ_0, the matrices R_k, Γ_k, $\Phi_{1,k}$, and $\Phi_{2,k}$;

Step 2. Apply the semi-definite programming to solve Problem 3.3 and obtain the optimal moment matrix Ψ_k^*;

while $k \leq \bar{N}$ **do**

  Step 3. Use the optimal moment matrix Ψ_k^*, e_k^{a*}, (3.13) and (3.14) to obtain the optimal stochastic injection attack strategy y_k^{a*};

  Step 4. Perform injection attack $\tilde{y}_k \leftarrow y_k + y_k^{a*}$;

  Step 5. Update the virtual residual system (3.6) and store e_{k+1}^{a*}.

Remark 3.9 Two points should be emphasized. Firstly, the proposed injection attack strategy y_k^{a*} is stochastic, the stochastic nature makes this strategy more unpredictable than previous deterministic approaches. Secondly, since the initial value of the attack is set by the attacker, it can be selected as random distribution, and the corresponding initial value e_0^a is also stochastic.

Remark 3.10 The algorithm of the designed attack strategy is mainly divided into two parts. One part is the solution for the optimal moment matrices, that is, semi-definite programming is used to solve Problem 3.3. The other part is to solve the optimal stochastic injection attack input based on the obtained moment matrix. Therefore, the computational complexity of the designed attack strategy is mainly caused by the semi-definite programming and the matrix inversion in Eq. (3.13). According to the reference [14], the computational complexity of semi-positive programming using an interior point method is $O\left(n^{3.5}\log(\frac{1}{\epsilon})\right)$, where n is the dimension of the matrix and ϵ is the objective accuracy. According to the literature [15], the computational complexity of the matrix inversion using the Gauss-Jordan elimination method is $O(n^3)$, where n is the dimension of the matrix. Therefore, the computational complexity of the method proposed in this chapter is $O\left((n_x + n_y + 1)^{3.5}\log(\frac{1}{\epsilon}) + (n_x + 1)^3\right)$.

3.5 Simulation and Experiment

In this section, the numerical simulation and physical experiment are implemented to verify the effectiveness of the proposed method.

3.5.1 *Numerical Simulation*

The autonomous networked mass-spring-damper system as shown in Fig. 3.1 is studied in this subsection. This mechanical system has a wide range of applications in car suspensions and buffers. A sensor is used to observe the position of the mass block in the system, and it is transmitted to the remote observer through a wireless network. Table 3.1 provides the symbols, physical meanings, values, and units of the autonomous networked mass-spring-damper system.

By discretizing the system corresponding to the parameters from Table 3.1, the discrete-time autonomous networked mass-spring-damper system can be expressed as

$$\begin{aligned} x_{k+1} &= Ax_k + Bu_k + \omega_k, \\ y_k &= Cx_k + v_k, \end{aligned}$$

where $A = [0.9951\ 0.0974; -0.0974\ 0.9464]$, $B = [0.0049;\ 0.0974]$, $C = [1\ 0]$. The state variable $x_k = [x_{1,k}; x_{2,k}]$ stands for the position and velocity of the mass block, $x(0) = [1; 0]$, the control input u_k is the input force, and the position of the mass block is selected as the system output y_k, which is obtained through the smart wireless position sensor and transmitted to the remote observer through the wireless network. The sampling time is $T_s = 0.1s$, and an LQG control strategy is used in the local controller of the discrete-time autonomous system with the control gain $K = [-19.6639\ \ -4.7733]$. The Luenberger-based observer is applied for the

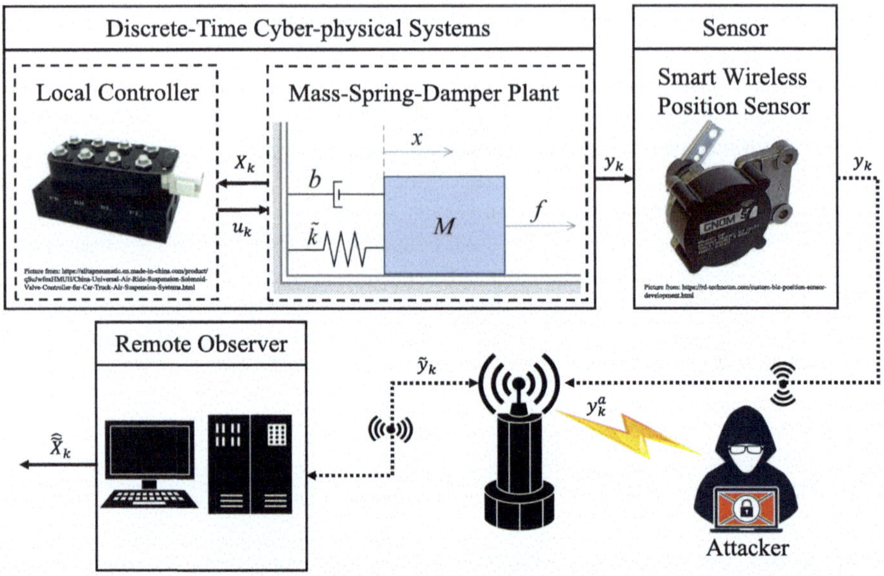

Fig. 3.1 Autonomous networked mass-spring-damper system under attack

Table 3.1 Autonomous networked mass-spring-damper system specifications

Symbol	Description	Value/Unit
M	Mass	1 kg
$\tilde{k}$	Spring constant	1 N/m
b	Damping constant	0.5 Ns/m
f	Input force	-N
x	The displacement of the mass	-m
ω	The process noise	$\tilde{N}(0,0.1)$
υ	The measurement noise	$\tilde{N}(0,0.01)$

purpose of providing estimates of unknown data about the velocity of the mass block, where the observer gain is $L = [0.1500; -0.0750]$ and the eigenvalues of the matrix $(A - LC)$ are 0.8761 and 0.9154, which are within the unit circle.

Subsequently, the virtual residual system can be described as

$$\begin{aligned} e_{k+1}^a &= \begin{bmatrix} 0.8596 & 0.0905 \\ -0.0226 & 0.9501 \end{bmatrix} e_k^a - \begin{bmatrix} 0.1500 \\ -0.0750 \end{bmatrix} y_k^a, \\ z_k^a &= \begin{bmatrix} 1 & 0 \end{bmatrix} e_k^a + y_k^a, \end{aligned}$$

and the specific description of the parameters adopted in the attack strategy is as follows. The attack duration is $\bar{N} = 100$, the parameters $\eta_1 = 0.01$, $\eta_2 = 1$, $R_1 =$

0.01, $R_2 = 100$, $R_3 = 0.01I_2$ are chosen. The initial attack input value $y_0^a = 10$, and the initial value $e_0^a = [-1.5; 0.75]$. The optimization problem, Problem 3.3, is solved using the semi-definite programming toolkit YALMIP 3 in MATLAB [16]. Meanwhile, the random attack input $y_k^{ar} \sim \tilde{N}(0, 0.75)$ and extreme attack input $y_k^{ae} = 0.75$ are adopted to compare with the optimal attack input y_k^{a*} of Theorem 3.2. Note that 0.75 is the mean of the absolute values of the optimal attack input y_k^{a*}. To reduce the influence of randomness on the experimental results, the simulation has been repeated 10,000 times and averaged. The results are shown in Figs. 3.2-3.7.

From Fig. 3.2, it can be observed that the optimal stochastic injection attack strategy developed in this chapter induces a significantly larger deviation in the state residual compared to a random attack. In contrast to the extreme attack, the sum of the norm for the state residual in Fig. 3.3 demonstrates that the proposed attack yields superior effects throughout the attack duration. Regarding the stealthiness of the attack, Fig. 3.4 shows that the proposed attack strategy maintains the output residuals within ± 0.001 m, rendering it challenging to detect with the residual detector. Conversely, both random and extreme attacks are easily detected and trigger alarms. And the attack inputs are presented in Fig. 3.5. Additionally, Fig. 3.6 indicates that the proposed attack strategy satisfies condition (3.10a). Figure 3.7 illustrates that the proposed attack strategy consistently ensures that the state residual deviation exceeds threshold $\eta_2^2 \mathbb{E}\{\| y_k^a \|^2\}$. Finally, the optimal objective value, determined through semi-definite programming, is -8.91.

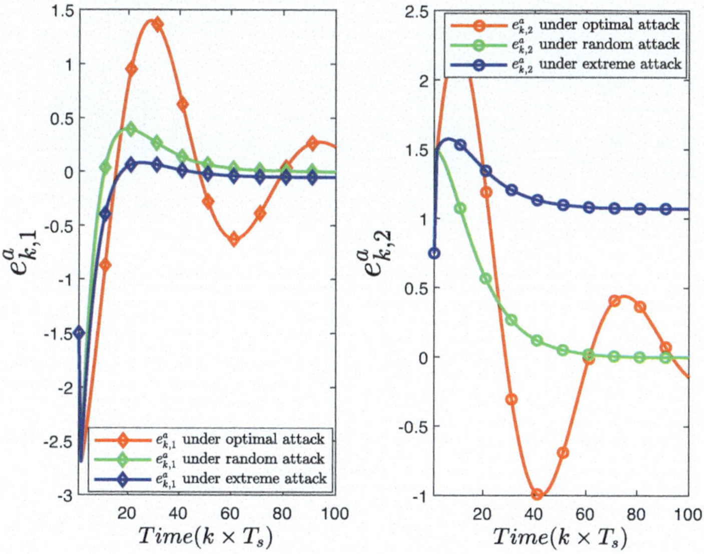

Fig. 3.2 The amplitudes of the state residual e_k^a under the optimal, random, and extreme attacks

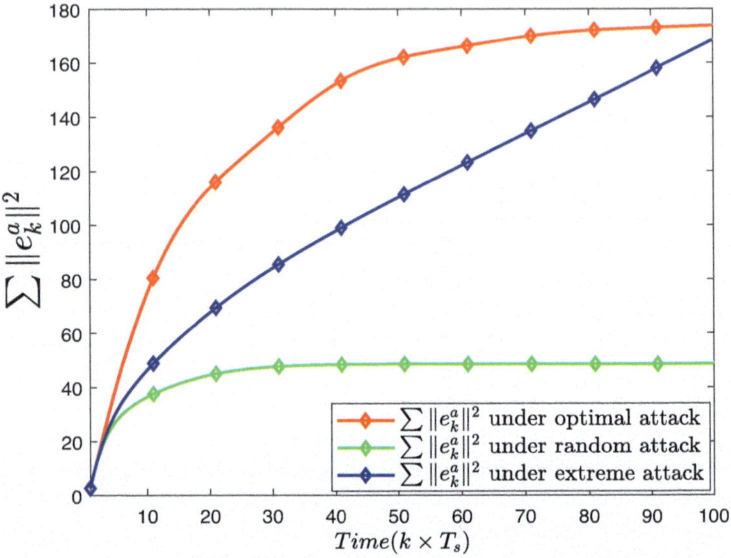

Fig. 3.3 The sum of the norm for the state residual e_k^a under the optimal, random, and extreme attacks

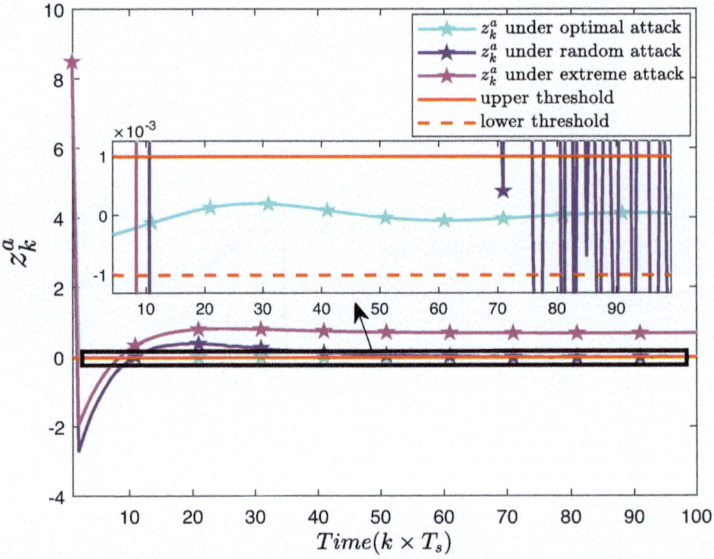

Fig. 3.4 The amplitudes of the output residual z_k^a under the optimal, random, and extreme attacks

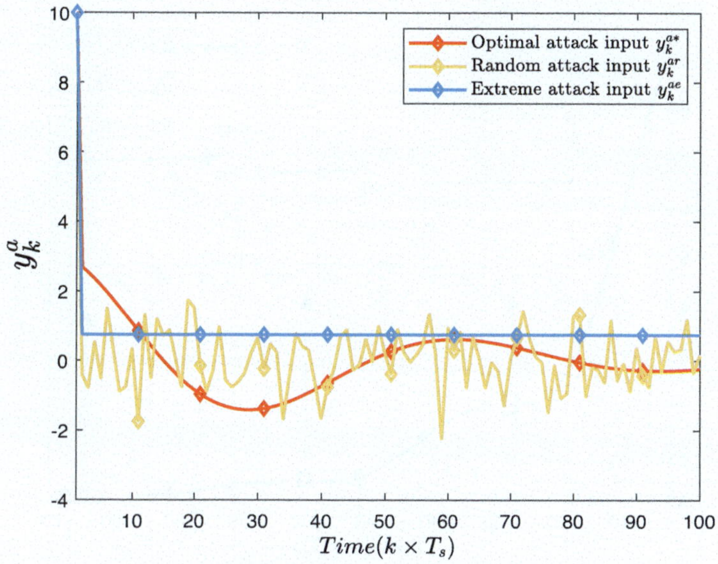

Fig. 3.5 The attack input y_k^{a*} under the optimal, random, and extreme attacks

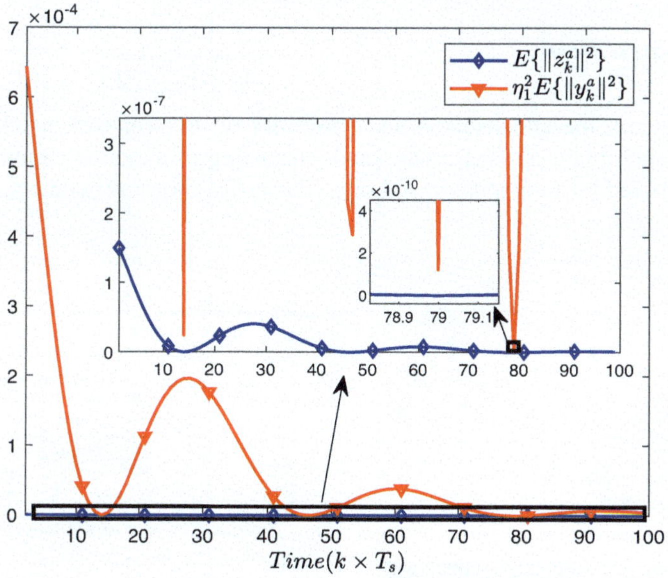

Fig. 3.6 The curves of the relationship between $\mathbb{E}\{\|z_k^a\|^2\}$ and $\eta_1^2\mathbb{E}\{\|y_k^a\|^2\}$

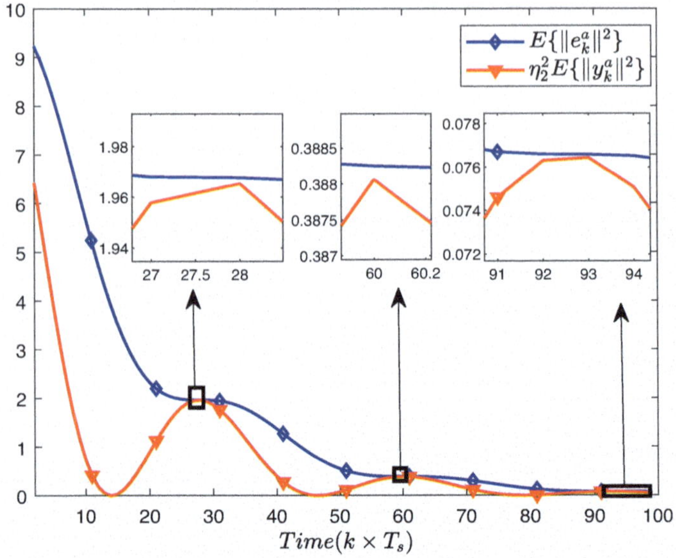

Fig. 3.7 The curves of the relationship between $\mathbb{E}\{\|e_k^a\|^2\}$ and $\eta_2^2\mathbb{E}\{\|y_k^a\|^2\}$

3.5.2 Experiment

To further verify the effectiveness and feasibility of the proposed attack strategy, an experimental platform was constructed, consisting of a motion capture system, two Linux-based host computers, and three leader-follower quadcopters, as shown in Fig. 3.8.

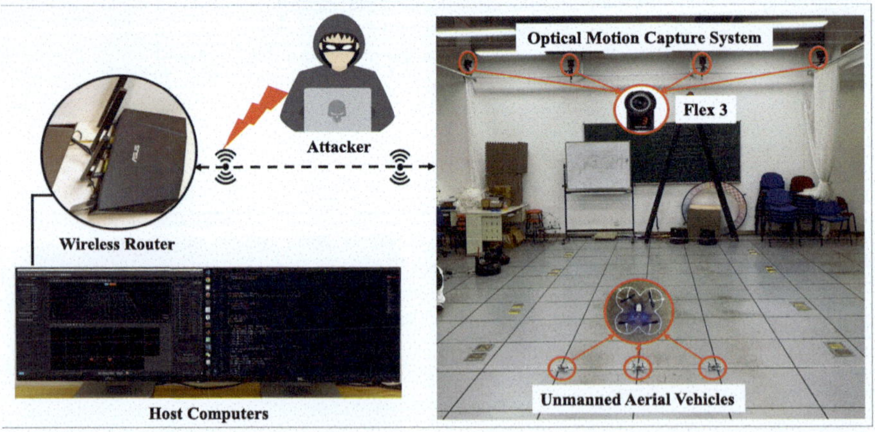

Fig. 3.8 Schematic diagram of experimental platform for the leader-follower model involving three quadrotors

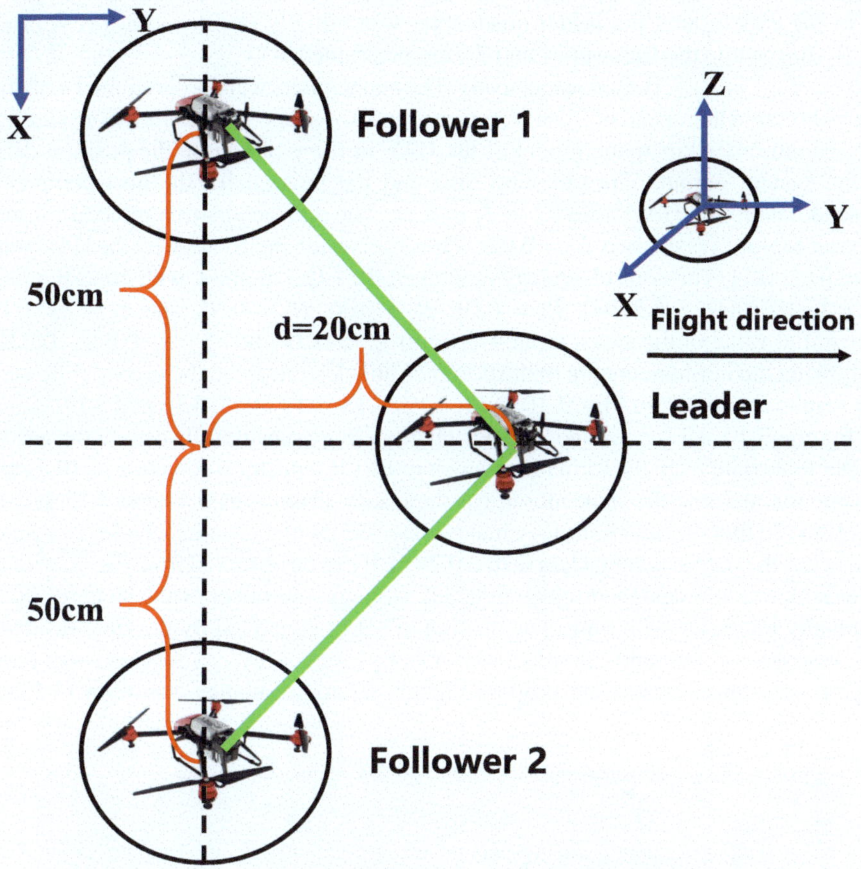

Fig. 3.9 Schematic diagram of three leader-follow mode UAVs formation flight

The quadcopters are arranged in a V-shaped formation, as shown in Fig. 3.9. Position data for the quadcopters is captured by the motion capture system, supported by OptiTrack, and then transmitted to the host computer via a wireless network. The host computer then relays the leader's position information to the two followers over the wireless network. The leader's trajectory is computed by the host computer and transmitted to the quadcopter, while the trajectories of the two follower quadcopters are calculated locally based on the received leader position data and local estimation data. In this experiment, only the position data in the y-direction is subjected to attack. This selective approach prevents potential collisions and allows for precise observation of trajectory deviations following the attack. The attacker intercepts data through the wireless network and employs a cover attack technique [2] to tamper with the leader's position data transmitted from the host to the followers, thus implementing the proposed attack.

In the experiment, the leader quadcopter was initially positioned at $(x, y, z) = (0, 0, 0)m$, while the followers 1 and 2 were positioned at $(x, y, z) = (-0.5, 0, 0)m$ and $(x, y, z) = (0.5, 0, 0)m$ respectively. The quadcopters were programmed to fly in the y-direction at a speed of $0.1m/s$, maintaining a V-shaped formation. Considering the second-order kinematic model of the UAV in the y-direction, the position data is the system output. The sampling time and the interval for the host computer to send data to the UAV were set to $T_s = 0.1s$. The data acquisition interval for the motion capture system was $T_c = 0.02s$. The observation matrix used by the UAV was $L = [0.2; 0.1]$. The control strategy employed the LQG method, with control gains $K = [-19.7069 \ \ -5.2394]$, $R_1 = 0.01$, $R_2 = 100$, $R_3 = 0.01$, and $\eta_1 = \eta_2 = 1$. The initial value of the attack input was set to $y_0^a = 0.2$, with $e_0^a = [-0.04; -0.02]$. Following the implementation of the designed attack strategy on the second follower, the results are shown in Figs. 3.10, 3.11, 3.12, 3.13, 3.14, 3.15, 3.16 and 3.17.

Figures 3.10, 3.11, 3.12, and 3.13 reveal the strategy successfully caused the quadcopter to deviate from the formation, which results in a position deviation of 10.7 cm, despite maintaining the same altitude as the leader. From Figs. 3.14 and 3.15, it can be observed that the attack strategy makes the residual of the state increase over time and keeps the output residual less than ±0.24 m. It can be observed from Fig. 3.14 that after adopting the designed injection attack strategy, the offset of the state residual gradually increases with time. The maximum offset generated by the state residual $e_{k,1}^a$ reaches −1.1 m, and the maximum offset generated by $e_{k,2}^a$ reaches −0.1 m/s. Figure 3.16 demonstrates the proposed attack strategy satisfies condition (3.10a).

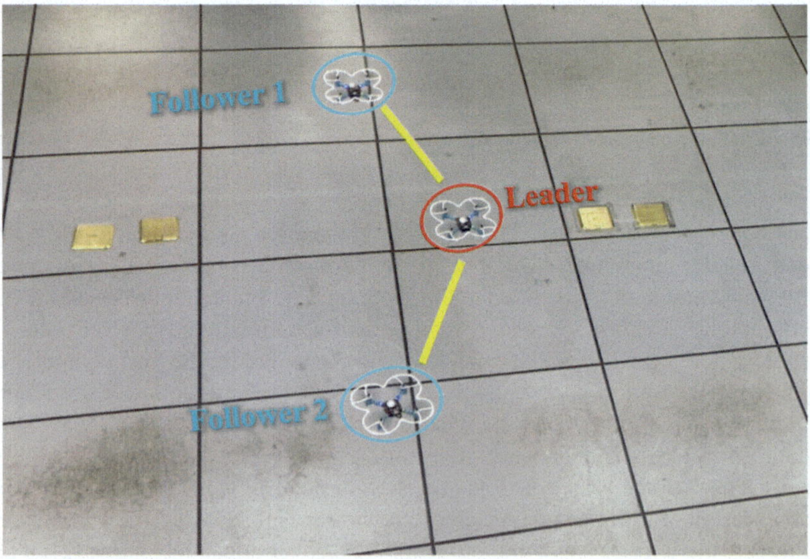

Fig. 3.10 The snapshot of the healthy flight processes

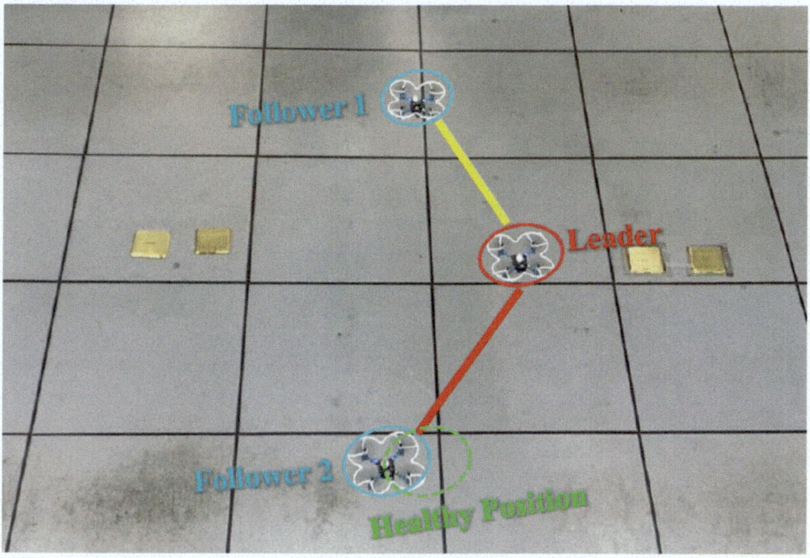

Fig. 3.11 The snapshot of the attacked flight processes

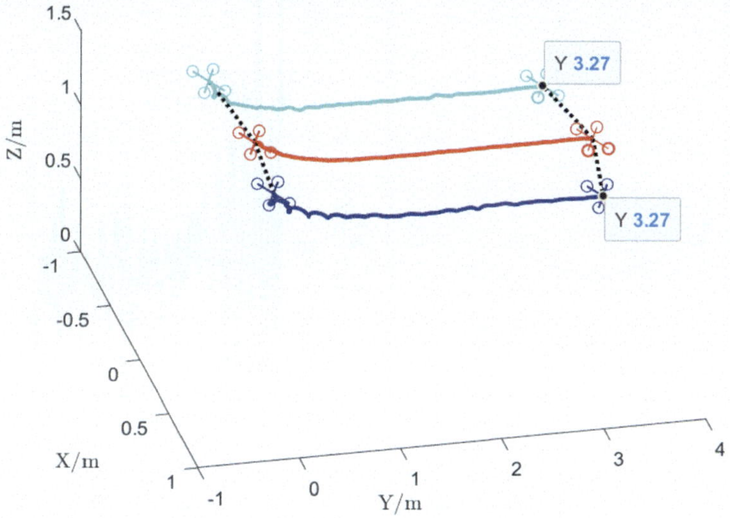

Fig. 3.12 The trajectories of the healthy flight processes

Furthermore, Fig. 3.17 shows that the proposed attack strategy always ensures that the state residual deviation exceeds the threshold $\eta_2^2\mathbb{E}\{\| y_k^a \|^2\}$. And the optimal objective value is -31.41. The details are provided in the experimental video included in the supplementary materials.

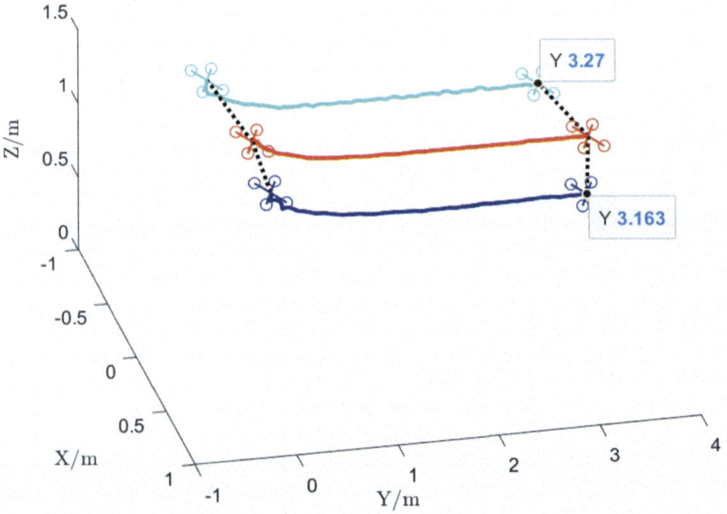

Fig. 3.13 The trajectories of the attacked flight processes

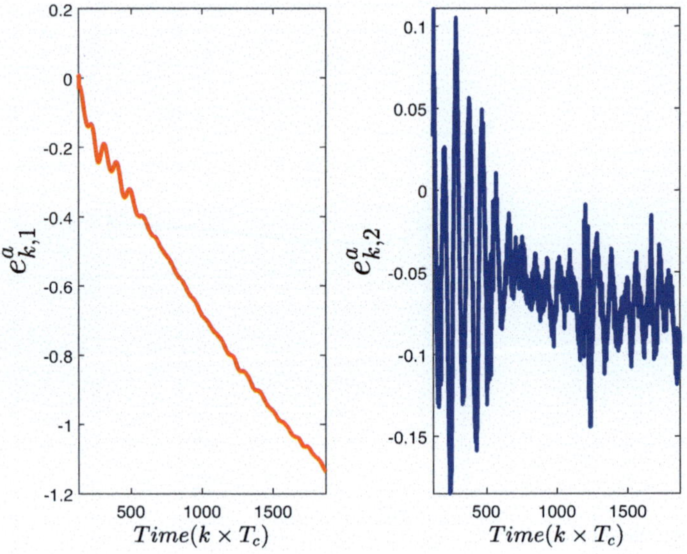

Fig. 3.14 The amplitudes of the state residual e_k^a employing the Theorem 3.2

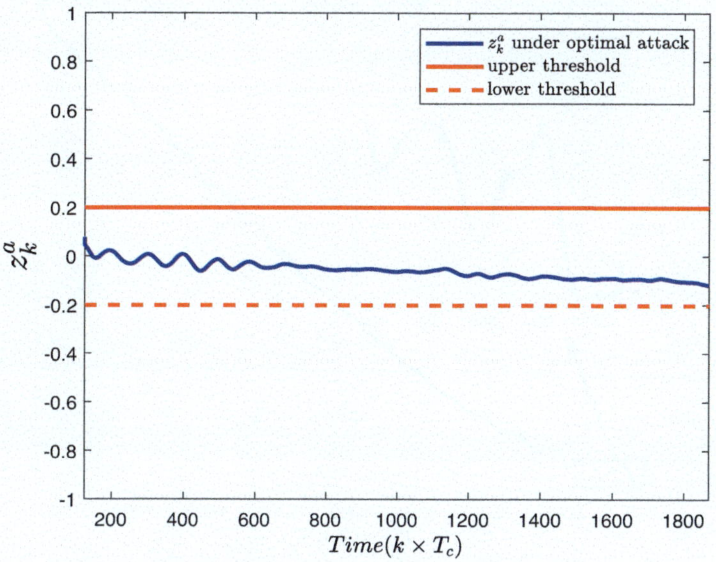

Fig. 3.15 The amplitudes of the output residual z_k^a employing the Theorem 3.2

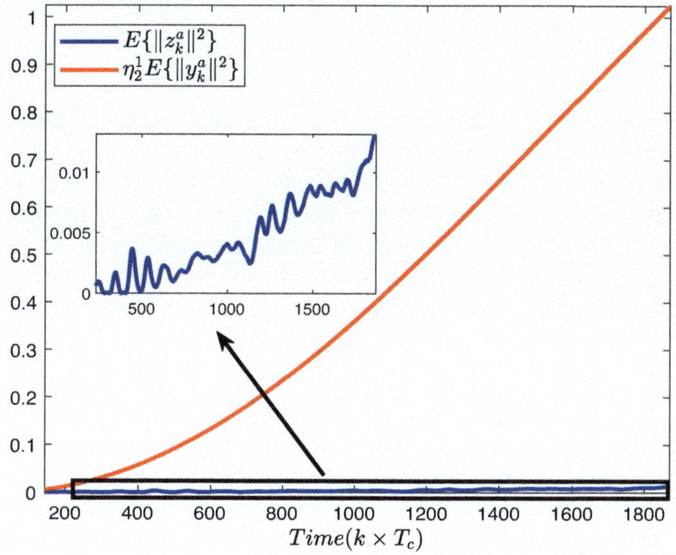

Fig. 3.16 The curves of the relationship between $\mathbb{E}\{\|z_k^a\|^2\}$ and $\eta_1^2\mathbb{E}\{\|y_k^a\|^2\}$

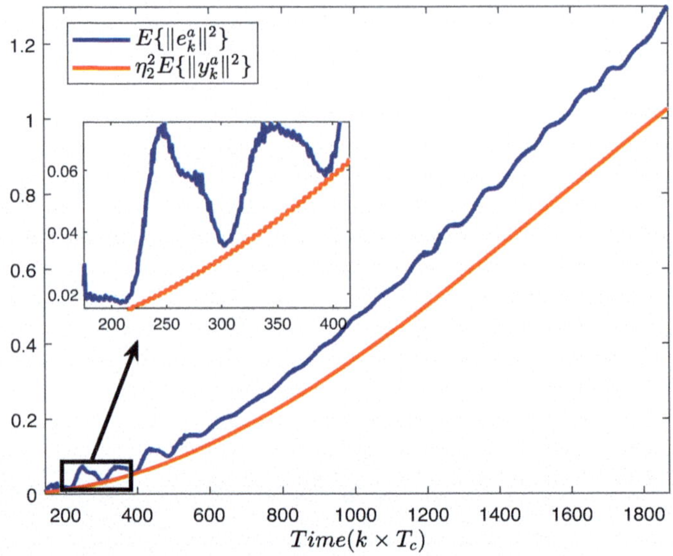

Fig. 3.17 The curves of the relationship between $\mathbb{E}\{\|e_k^a\|^2\}$ and $\eta_2^2\mathbb{E}\{\|y_k^a\|^2\}$

3.6 Conclusion

This chapter focuses on the problem of the designing stochastic injection attack strategy for linear discrete time-varying system. To illustrate the variations in estimation errors pre- and post-attack, a virtual residual system is established. To take into account both the stealthiness and the destructiveness in the design process of the attack, a non-convex objective function and constraint condition are chosen. On this basis, a new stochastic injection attack strategy is devised by converting the original non-convex attack optimization problem into a convex one using moment matrices. Subsequently, the effectiveness of the designed strategy is verified using a networked spring-damped system and the quadcopter formation experiment. Future work will explore switching forms of the injection attack and communication network delays.

References

1. Chen, Y., Kar, S., Moura, J.M.: Cyber-physical attacks with control objectives. IEEE Trans. Autom. Control **63**(5), 1418–1425 (2018)
2. Mikhaylenko, D., Zhang, P.: Stealthy local covert attacks on cyber-physical systems. IEEE Trans. Autom. Control **67**(12), 6778–6785 (2022)
3. Lu, A.Y., Yang, G.H.: False data injection attacks against state estimation without knowledge of estimators. IEEE Trans. Autom. Control **67**(9), 4529–4540 (2022)
4. Alisic, R., Kim, J., Sandberg, H.: Model-free undetectable attacks on linear systems using LWE-based encryption. IEEE Control Syst. Lett. **7**, 1249–1254 (2023)

5. Zhang, T.Y., Ye, D.: False data injection attacks with complete stealthiness in cyber–physical systems: A self-generated approach. Automatica **120**, 109117 (2020)
6. Wu, G., Wang, G., Sun, J., Chen, J.: Optimal partial feedback attacks in cyber-physical power systems. IEEE Trans. Autom. Control **65**(9), 3919–3926 (2020)
7. Gao, S., Zhang, H., Wang, Z., Huang, C.: A class of optimal switching mixed data injection attack in cyber-physical systems. IEEE Robot. Autom. Lett. **6**(2), 1598–1605 (2021)
8. An, D., Zhang, F., Yang, Q., Zhang, C.: Data integrity attack in dynamic state estimation of smart grid: Attack model and countermeasures. IEEE Trans. Autom. Sci. Eng. **19**(3), 1631–1644 (2023)
9. Zhang, K., Keliris, C., Parisini, T., Polycarpou, M.M.: Stealthy integrity attacks for a class of nonlinear cyber-physical systems. IEEE Trans. Autom. Control **67**(12), 6723–6730 (2022)
10. Wu, G., Wang, G., Sun, J., Xiong, L.: Optimal switching attacks and countermeasures in cyber-physical systems. IEEE Trans. Syst. Man Cybern. -Syst. **51**(8), 4825–4835 (2021)
11. Hu, L., Wang, Z., Han, Q.L., Liu, X.: State estimation under false data injection attacks: Security analysis and system protection. Automatica **87**, 176–183 (2018)
12. Huang, X.Z., Tian, Y.P.: Asynchronous distributed localization in networks with communication delays and packet losses. Automatica **96**, 134–140 (2018)
13. Wang, X.L., Yang, G.H.: Piecewise attack strategy design for T-S fuzzy cyber–physical systems. IEEE Trans. Syst. Man Cybern. -Syst. **52**(10), 6477–6486 (2022)
14. Luo, Z.Q., Ma, W.K., So, A.M.C., Ye, Y., Zhang, S.: Semidefinite relaxation of quadratic optimization problems. IEEE Signal Process. Mag. **27**(3), 20–34 (2010)
15. Ma, J., Li, Y.: Gauss-Jordan elimination method for computing all types of generalized inverses related to the 1-inverse. J. Comput. Appl. Math. **321**, 26–43 (2017)
16. Lofberg, J.: YALMIP: A toolbox for modeling and optimization in MATLAB. In: Proceedings of the 2004 IEEE International Conference on Robotics and Automation, pp. 284–289 (2004)

Chapter 4
Data-Driven FDI Attack Strategy Against Linear CPSs via Iterative Learning

Abstract The system security of linear cyber-physical systems (CPSs) with repetitive movement characteristics under injection attacks is investigated in this chapter. The optimal data-driven injection attack strategies through iterative learning are proposed without utilizing the knowledge of the system model. A switching attack strategy with high stealthiness, flexibility, and low energy consumption is established. Finding the optimal attack strategy that maximizes the quadratic cost function of the system is the main objective of this chapter. Different from the existing attack strategies, the proposed one applies the attack input data from the previous iteration at the same sampling time to update the current attack input, thereby improving the attack effect. Furthermore, in order to overcome the difficulty that the attacker is unaware of the attacked system model, a system identification method based on input and output (I/O) data is used to construct a virtual model. An improved projection estimation approach is developed to estimate the parameters of the virtual model with the assistance of the matrix inverse lemma. Finally, a networked direct current (DC) motor simulation is applied to verify the effectiveness of the presented schemes.

4.1 Introduction

The design of the data-driven FDI attack strategies is the natural development of the existing classical model-based FDI attack strategies. Due to the upgrading of security protection mechanism in CPSs, the repair of network bugs, and the development of artificial intelligence technology, the ability of the attackers to obtain information from the attacked systems is seriously weakened. Therefore, the attacked systems are unknown to the attackers on some occasions. It is natural that the attacker has the ability to access the I/O data of the attacked system from the open wireless transmission channel, and then this wireless channel is utilized to inject the designed false data into the system. Among the existing works on the design of FDI attack strategies, the attacker designs the attack strategies based on the premise that the attacked system model is roughly or completely known, such as [1–4]. This assumption does not always hold in practical applications. Specifically, during actual chemical production processes, various data are not available due to extreme operating conditions,

S. Gao et al., *Cyberattack Strategies Against Cyber-Physical Systems: From a Control Perspective*, https://doi.org/10.1007/978-981-95-9645-4_4

including high temperatures and pressures. For instance, the concentration of carbon monoxide in blast furnace iron-making cannot be measured owing to the high furnace temperatures and the contact state of the seed crystals with the liquid surface in single crystal furnaces is difficult to be detected. Moreover, the previously mentioned injection attack strategies are dependent on the system matrices, as is shown in [5, 6]. Due to the uncertainty of the system, the effect of these attack strategies, which relies on the accurate model of the system, will be biased, or even fail to work. For this reason, it is urgent to relax the ability of the attackers to access the information from the attacked systems.

In view of the model-free systems, data-driven identification is frequently adopted to overcome the difficulty of unknown system parameters [7, 8]. In the past decade, data-driven approaches have played a vital role and attracted extensive attention for solving unmodeled dynamics problems. The nonlinear non-affine systems are dealt with by a series of adaptive control schemes based on I/O data, which are presented in [9–11]. Unfortunately, not until very recently, did the design of injection attack strategy under model free have attracted much attention. This problem is challenging for the design of injection attack strategies against model-free CPSs, which can make the strategies more widely applicable. Furthermore, many research interests are devoted to CPSs which exhibits the characteristics of repetitive task execution [12–14]. These can be considered as time-dimensional dynamics and repeated over these dynamics. Therefore, iterative learning is expected to be applied to the design of injection attack strategies for some new attempts, as previous attack information can be employed to improve the effect of the attack. This chapter mainly discusses the design of FDI attack strategy based on iterative learning without using the system model. The ability of the attacker accessing information of the attacked systems is restricted; specifically, it can only obtain the system input and output data. In addition, the attenuation effect of the information transmission process is considered. Meanwhile, there are multiple channels from the controller to the actuator. When these channels are selectively destroyed by the attacker, it will increase the flexibility and stealthiness of the attack. For example, the optimal location switching and non-switching attack problems based on feedback way are investigated in [1]. Then, on the basis of that, the optimal switching attack under mixed data is presented in [4]. Although these switching attack strategies have greater flexibility, higher defense difficulty, and lower energy consumption than non-switching ones, the calculation of switching sequences depends on the model parameters of the attacked system. Consequently, this motivates us to investigate the design of switching FDI attack strategy based on iterative learning without using the system model.

The new data-driven FDI attack based on iterative learning is constructed for CPSs with characteristics of repetitive movements, and it has the advantage of relaxing the ability of attacker to flit the data or parameters from the attacked CPSs. The main contributions of this chapter are summarized as follows:

(1) The I/O data-driven system identification is adopted to overcome the difficulties of linear time-varying unknown cyber-physical systems. On the foundation of the system identification, a virtual linear attacked model based on I/O data is

established to relax the ability of the attacker to gain access to the stolen data from the attacked CPSs.

(2) The accuracy of the virtual linear attacked model is guaranteed by utilizing a presented improved projection estimation method. The identification parameter matrix is estimated by solving the optimization problem for the estimation, and the corresponding estimation scheme is given. Furthermore, a step factor is introduced into the application to make the estimation method more general.

(3) Compared with the attack strategies in [1, 2], and [5] which rely on the ability of attackers to steal system information, the attack strategies designed in this chapter remove this constraint. Meanwhile, a quadratic cost is defined in which the attacker makes the system output deviate from the desired value with low energy consumption. Furthermore, the optimal FDI attack and switching FDI attack strategies are derived by solving the corresponding optimal attack problem.

The remainder of this chapter is organized as follows. In Sect. 4.2, the problem formulation for a class of linear discrete time-varying cyber-physical systems is presented. The attack problems transformation via data-driven model are developed in Sect. 4.3. The FDI attack and switching FDI attack strategies are developed in Sect. 4.4. Then, in Sect. 4.5, the efficiency of the proposed schemes is illustrated by a networked DC motor example. Finally, Sect. 4.6 concludes this chapter.

4.2 Problem Formulation

4.2.1 System Model

A class of linear discrete time-varying cyber-physical systems is considered by

$$\begin{cases} x(k+1) = A(k)x(k) + B(k)u(k), \\ y(k) = C(k)x(k), \end{cases} \tag{4.1}$$

where $x(k) \in \mathbb{R}^n$, $u(k) \in \mathbb{R}^l$, and $y(k) \in \mathbb{R}^m$ are the process state, the control input, and the measurement output vector at time k. $A(k) \in \mathbb{R}^{n\times n}$, $B(k) \in \mathbb{R}^{n\times l}$, and $C(k) \in \mathbb{R}^{m\times n}$ are the unknown matrices.

For the system (4.1), the following assumptions are needed.

Assumption 4.1 The system (4.1) is asymptotically stable, and the initial state of the system, i.e., $x(0) = x_0$ is a constant vector.

Assumption 4.2 The norm of unknown coefficient matrices $A(k)$, $B(k)$, and $C(k)$ is bounded, i.e., $\|A(k)\| \le \bar{A}$, $\|B(k)\| \le \bar{B}$, and $\|C(k)\| \le \bar{C}$, where $\bar{A}$, $\bar{B}$, and $\bar{C}$ are constants.

For the CPSs with repetitive characteristics under unknown system matrices, through iterative learning theory, the system (4.1) can be re-expressed as

$$\begin{cases} x_i(k+1) = A(k)x_i(k) + B(k)u_i(k), \\ y_i(k) = C(k)x_i(k), \end{cases} \tag{4.2}$$

where $x_i(k) \in \mathbb{R}^n$, $u_i(k) \in \mathbb{R}^l$, and $y_i(k) \in \mathbb{R}^m$ are the process state, the control input, and the measurement output vector in the ith iteration at time k.

Remark 4.1 Assumption 4.1 for (4.2) can be naturally re-described as that the initial value of the system state under iteration is a constant vector, i.e., $x_i(0) = x_0$. Furthermore, it should be noted that for systems with repeated operations, the coefficient matrices of the system are maintained in each iteration during the same period. Therefore, the coefficient matrices in the system (4.2) remain consistent with the system (4.1).

Healthy cyber-physical systems typically have geographically dispersed controllers and actuators. To facilitate control, the controller transmits control inputs through open wireless channels, known as controller-to-actuator (C-A) channels, to the actuators. These control inputs are also delivered to a detector equipped on the plant. Moreover, the plant, with its repeated operation characteristics, sends output signals from the sensor to the controller via an open wireless sensor-to-controller (S-C) channel, along with the detector. Using the received control input and system output data, the detector assesses whether the system is operating abnormally. This is a critical function for healthy cyber-physical systems as it enables prompt detection and response to unusual system behavior. The block diagram of the healthy system is shown in Fig. 4.1.

For the detector equipped on the plant, this chapter applies the bad data detection (BDD) scheme to detect whether the output of the system is abnormal. The same technique is used in [15, 16]. The Luenberger-based observer is employed in the detector, which can be specified as

$$\begin{cases} x_i^d(k+1) = A(k)x_i^d(k) + B(k)u_i(k) + L(k)(y_i(k) - y_i^d(k)), \\ y_i^d(k) = C(k)x_i^d(k), \end{cases} \tag{4.3}$$

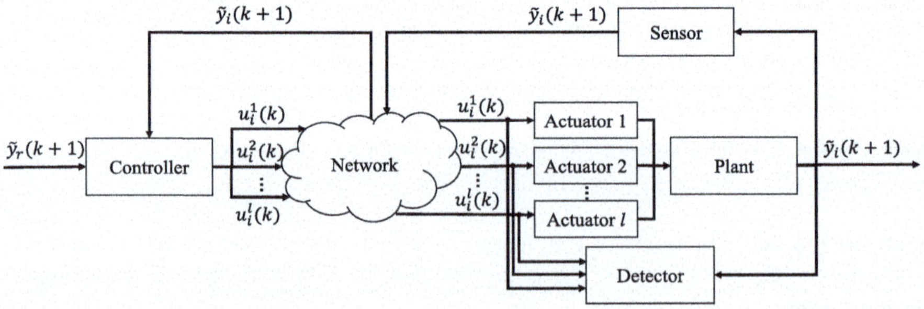

Fig. 4.1 The schematic diagram of the healthy system

where $(\cdot)^d$ represents the observation and $x_i^d(0) = x_i(0)$. $L(k) \in \mathbb{R}^{n \times m}$ is chosen such that $(A(k) - L(k)C(k)))$ is Schur. The residual between the system output measurement and the observation is $r_i(k+1) = y_i(k+1) - y_i^d(k+1)$. The 2-norm of this residual is monitored by the detector with a predetermined threshold Υ, and the detection mechanism is as follows:

$$\begin{cases} \text{when } \|r_i(k+1)\| \leq \Upsilon, \text{ system is normal,} \\ \text{when } \|r_i(k+1)\| > \Upsilon, \text{ system is abnormal.} \end{cases} \tag{4.4}$$

4.2.2 *Attacked Models*

In this subsection, the attacked models under non-switching FDI attack and switching FDI attack are given. Before introducing the attacked models, the assumption about the attacker needs to be presented first.

Assumption 4.3 For the injection attacks, the attackers have the capacity for injecting the designed false data inputs $u_{a_i}(k)$ and the control input signals into the actuators synchronously.

Remark 4.2 Assumption 4.3 elucidates the capability of malevolent entities to infiltrate the system. In actuality, considering the pronounced openness and vulnerability of wireless transmission channels, the presumption concerning the attacker's capacity to breach the system is well-founded. Similar assumptions have also been employed in prior studies such as [1, 2, 4]. Furthermore, a cover attack technology [17] can be leveraged to synchronously inject attack input and control input into the targeted system.

4.2.2.1 Attacked Model Under FDI Attack

The attacker carries out the FDI attack on all wireless transmission channels connecting the controller and the actuators. The attacker gains unauthorized access to the transmitted control and output signals, employing the iterative learning approach to design the attack strategy. Furthermore, the signals received by the actuators undergo tampering as the attacker applies the specifically designed attack signals. The schematic diagram of the linear CPSs under the FDI attack is presented in Fig. 4.2.

The false data injection attack is described as

$$\tilde{u}_i(k) = u_i(k) + D_a(k)u_{a_i}(k), \tag{4.5}$$

where $\tilde{u}_i(k) \in \mathbb{R}^l$ is the attacked l-dimensional control input in the ith iteration at time k, $u_{a_i}(k) \in \mathbb{R}^l$ is the attack input in the ith iteration at time k, and $u_{a_i}(0) = 0$. $D_a(k) \in \mathbb{R}^{l \times l}$ is the matrix representing the attack direction at time k.

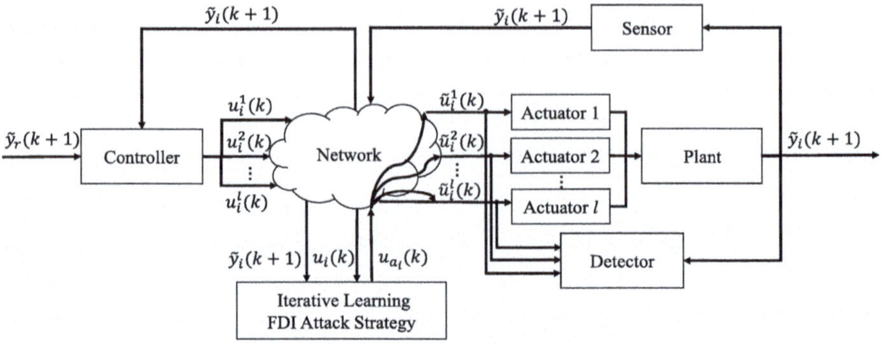

Fig. 4.2 The attacked cyber-physical system under the FDI attack

Remark 4.3 Since the dimension of the attack input $u_{a_i}(k)$ is generally less than or equal to the control input $u_i(k)$, when the dimension of the attack input is less than the control input, the dimension expansion can be used, that is, $u_{a_i}(k) \in \mathbb{R}^p$ is expanded to $u_{a_i}(k) \in \mathbb{R}^l$ by adding zero elements, where $p < l$.

Subsequently, combining (4.2) and (4.5), the attacked system can be expressed as

$$\begin{cases} \tilde{x}_i(k+1) = A(k)\tilde{x}_i(k) + B(k)u_i(k) + B_a(k)u_{a_i}(k) \\ \tilde{y}_i(k) = C(k)\tilde{x}_i(k), \end{cases} \tag{4.6}$$

where $\tilde{x}_i(k)$ and $\tilde{y}_i(k)$ are the attacked state and output in the ith iteration at time k, and $B_a(k) = B(k)D_a(k)$.

4.2.2.2 Attacked Model Under Switching FDI Attack

This part focuses on the implementation of a switching FDI attack against a specific channel within the set of the C-A wireless channels. The selection of the target channel is determined through the switching sequence designed via real-time online computation. Unlike the previous attack, where the FDI attack was conducted on all C-A wireless channels simultaneously, the switching FDI attack strategy increases stealthiness, flexibility, and applicability. This strategy is particularly advantageous when only a subset of the C-A wireless channels is accessible, allowing for targeted exploitation and enhancing the overall effectiveness of the attack. The schematic diagram of the linear CPSs under the switching FDI attack is depicted in Fig. 4.3.

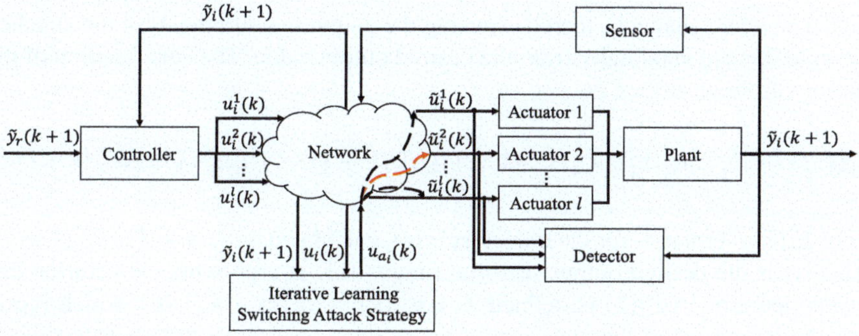

Fig. 4.3 The attacked cyber-physical system under the switching FDI attack

The switching FDI attack is described as

$$\tilde{u}_i(k) = u_i(k) + \sum_{h=1}^{l} \theta_h(k) D_{a_h}(k) u_{a_i}(k), \tag{4.7}$$

where $\tilde{u}_i(k)$, $u_i(k)$, and $u_{a_i}(k)$ are defined as in (4.5). The matrix $D_{a_h}(k) \in \mathbb{R}^{l \times l}$ is the attack direction matrix at time k. The switching input $\Theta(k) = [\theta_1(k), \theta_2(k), ..., \theta_l(k)]^T \in \tilde{\Theta}(k)$, $\theta_h(k) \in \{0, 1\}$, and the set is $\tilde{\Theta}(k) = \{(1; 0; ...; 0), (0; 1; ...; 0), ..., (0; 0; ...; 1)\}$.

Similarly, by combining (4.2) and (4.7), the expression for the switching attacked system can be represented as follows:

$$\begin{cases} \tilde{x}_i(k+1) = A(k)\tilde{x}_i(k) + B(k)u_i(k) + \sum_{h=1}^{l} \theta_h(k) B(k) D_{a_h}(k) u_{a_i}(k), \\ \tilde{y}_i(k) = C(k)\tilde{x}_i(k). \end{cases} \tag{4.8}$$

where $\tilde{x}_i(k)$ and $\tilde{y}_i(k)$ are defined in the system (4.6).

4.2.3 *Optimal FDI Attack Problem*

This chapter focuses on the design of the attack strategy that deteriorates the control performance of linear discrete time-varying cyber-physical systems from the perspective of the attacker. The attacker does not have knowledge of the state or coefficient matrices $A(k)$, $B(k)$, and $C(k)$ of the system (4.2). Nevertheless, the attacker has the ability to steal the control inputs from the C-A wireless channels as well the system output measurement from the S-C wireless channel. Specifically, the main investigated problem is how the attacker can design an attack strategy with minimal energy consumption to tamper with the control input transmitted through the C-A wireless channels, thereby causing the output measurements of the attacked system to deviate maximally from the desired output value. The cost function of the attacker can be described as

$$J_1(\tilde{E}_i(k+1), \Delta u_{a_i}(k)) = \frac{1}{2}(\tilde{E}_i^T(k+1)Q\tilde{E}_i(k+1) - \eta^2 \Delta u_{a_i}^T(k) R \Delta u_{a_i}(k)), \quad (4.9)$$

where $\tilde{E}_i(k+1)$ stands for the trajectory error, calculated by $\tilde{y}_r(k+1) - \tilde{y}_i(k+1)$, $\tilde{y}_r(k+1)$ is the desired output vector at time $k+1$, Δ represents the iteration difference operator, $\Delta u_{a_i}(k)$ is defined as $\Delta u_{a_i}(k) \triangleq u_{a_i}(k) - u_{a_{i-1}}(k)$, which represents the increment of the attack input under two consecutive iterations at time k, $Q \in \mathbb{R}^{m\times m}$ and $R \in \mathbb{R}^{l\times l}$ satisfy $Q \succeq 0$ and $R \succ 0$, and the prescribed constant disturbance attenuation level is denoted by $\eta > 0$.

Remark 4.4 In the cost function (4.9), the increment of the attack input under the iteration is used as the optimized variable, and $\Delta u_{a_i}^T(k) R \Delta u_{a_i}(k)$ is the energy term consumed by the attacker. Since $\Delta u_{a_i}(k)$ is optimized from the first iteration, the optimal value of the attack input at the ith iteration can be calculated by $u_{a_i}^*(k) = \sum_{j=1}^{i} \Delta u_{a_j}^*(k)$. Thus, it makes sense to obtain the optimal attack input by optimizing $\Delta u_{a_i}(k)$. The similar incremental approach is also applied in [18, 19].

According to the attacked model under FDI attack (4.6) and the cost function of the attacker (4.9), the optimal FDI attack problem in linear CPSs is formulated as follows.

Problem 4.1

$$\max_{\Delta u_{a_i}(k)} \quad J_1(\tilde{E}_i(k+1), \Delta u_{a_i}(k))$$

$$\text{s.t.} \quad \begin{cases} \tilde{x}_i(k+1) = A(k)\tilde{x}_i(k) + B(k)u_i(k) + B_a(k)u_{a_i}(k) \\ \tilde{y}_i(k) = C(k)\tilde{x}_i(k), \\ Q \succeq 0, R \succ 0. \end{cases}$$

Similarly, on the basis of the attacked model under switching FDI attack (4.8) and the cost function of the attacker (4.9), the optimal switching FDI attack problem in linear CPSs can be formulated as

Problem 4.2

$$\max_{\Delta u_{a_i}(k)} \quad J_1(\tilde{E}_i(k+1), \Delta u_{a_i}(k))$$

$$\text{s.t.} \quad \begin{cases} \tilde{x}_i(k+1) = A(k)\tilde{x}_i(k) + B(k)u_i(k) + \sum_{h=1}^{l} \theta_h(k)B(k) \\ \qquad \times D_{a_h}(k)u_{a_i}(k), \\ \tilde{y}_i(k) = C(k)\tilde{x}_i(k), \\ Q \succeq 0, R \succ 0. \end{cases}$$

Remark 4.5 Due to the constrained capacity of the attackers to pilfer system information, their awareness of the knowledge of the system (4.2) remains limited. Put differently, the attacker lacks knowledge of the system state $\tilde{x}_i(k)$ and the system matrices $A(k)$, $B(k)$, and $C(k)$. Consequently, the resolution of Problems 4.1 and 4.2 becomes infeasible. To overcome this obstacle, a virtual linear attacked model via the input/output data is constructed in Subsect. 4.3 and uses this data-driven model to transform the above unsolvable problems into the equivalent forms.

4.3 Attack Problems Transformation via Data-Driven Model

This section primarily discusses the input/output data-driven virtual linear attacked models established to address the unsolvable Problems 4.1 and 4.2 caused by unknown system states $\tilde{x}_i(k)$ and system parameter matrices $A(k)$, $B(k)$, and $C(k)$. Subsequently, the transformation of the original attack problem based on this data-driven model is presented.

4.3.1 Virtual Linear Attacked Models Based on I/O Data

As the establishment processes of data-driven virtual linear attacked models under non-switching and switching FDI attacks are similar, the following example demonstrates the development of a data-driven virtual linear attacked model based on I/O data under non-switching FDI attacks. In this chapter, the attacker can steal the I/O data from the attacked system through the C-A and S-C channels and then use this data to reconstruct the attacked model under non-switching FDI attack (4.6). This can be described in detail in the following Lemma 4.1.

Lemma 4.1 *A virtual linear attacked model to describe the I/O behavior of the attacked model under non-switching FDI attack (4.6) between two consecutive iterations is built as*

$$\triangle\tilde{y}_i(k+1) = \Gamma(k)\triangle U_i(k) + \Gamma(k)\bar{D}_a(k)\triangle U_{a_i}(k), \tag{4.10}$$

and the norm of matrix $\Gamma(k)$ *is bounded, where*

$$\triangle U_i(k) = U_i(k) - U_{i-1}(k),\ \triangle U_{a_i}(k) = U_{a_i}(k) - U_{a_{i-1}}(k),$$

$$\gamma(t) = \prod_{j=t+1}^{k} C(k+1)A(j)B(t),\ \triangle \tilde{y}_i(k+1) = \tilde{y}_i(k+1) - \tilde{y}_{i-1}(k+1),$$

$$\Gamma(k) = [\gamma(k), \gamma(k-1), ..., \gamma(0)],\ \bar{D}_a(k) = \text{diag}(D_a(k), D_a(k-1), ..., D_a(0)),$$
$$U_i(k) = [u_i(k), u_i(k-1), ..., u_i(0)]^T,\ U_{a_i}(k) = [u_{a_i}(k), u_{a_i}(k-1), ..., u_{a_i}(0)]^T,$$

and $\triangle \tilde{y}_i(k+1) \in \mathbb{R}^m$, $\triangle U_i(k) \in \mathbb{R}^{(l*(k+1))\times 1}$, $\triangle U_{a_i}(k) \in \mathbb{R}^{(l*(k+1))\times 1}$, $\Gamma(k) \in \mathbb{R}^{m\times(l*(k+1))}$, $U_i(k) \in \mathbb{R}^{(l*(k+1))\times 1}$, $U_{a_i}(k) \in \mathbb{R}^{(l*(k+1))\times 1}$, $\bar{D}_a(k) \in \mathbb{R}^{(l*(k+1))\times(l*(k+1))}$.

Proof According to the attacked model under non-switching FDI attack (4.6), the attacked output $\tilde{y}_i(k+1)$ can be written as

$$\begin{aligned}
\tilde{y}_i(k+1) &= C(k+1)\tilde{x}_i(k+1) \\
&= C(k+1)A(k)\tilde{x}_i(k) + C(k+1)B(k)u_i(k) + C(k+1)B(k)D_a(k)u_{a_i}(k) \\
&= \prod_{d=0}^{k} C(k+1)A(k-d)\tilde{x}_i(0) + C(k+1)\sum_{d=0}^{k}\prod_{j=0}^{d-1} A(k-j)B(k-d)u_i(k-d) \\
&\quad + \sum_{d=0}^{k}\prod_{j=0}^{d-1} C(k+1)A(k-j)B(k-d)D_a(k-d)u_{a_i}(k-d).
\end{aligned}$$

Subsequently, one has

$$\begin{aligned}
\triangle \tilde{y}_i(k+1) &= \tilde{y}_i(k+1) - \tilde{y}_{i-1}(k+1) \\
&= C(k+1)\sum_{d=0}^{k}\prod_{j=0}^{d-1} A(k-j)B(k-d)\triangle u_i(k-d) + \sum_{d=0}^{k}\prod_{j=0}^{d-1} C(k+1) \\
&\quad \times A(k-j)B(k-d)D_a(k-d)\triangle u_{a_i}(k-d) \\
&= \Gamma(k)\triangle U_i(k) + \Gamma(k)\bar{D}_a(k)\triangle U_{a_i}(k),
\end{aligned}$$

therefore, the virtual linear attacked model (4.10) is derived. Then, by applying the Cauchy inequality, the norm of matrix $\Gamma(k)$ can be described as

$$\begin{aligned}
\|\Gamma(k)\| &= \sum_{d=0}^{k}\prod_{j=0}^{d-1} \|C(k+1)A(k-j)B(k-d)\| \\
&\leq \sum_{d=0}^{k}\prod_{j=0}^{k-1} \|C(k+1)\|\|A(k-j)\|\|B(k-d)\|.
\end{aligned}$$

Consequently, under Assumption 4.2, the norm of matrix $\Gamma(k)$ is bounded for $k < \infty$, and this proof is completed. □

Remark 4.6 For the attacker, the attacked system represents a model-free system, with only knowledge of it being a linear discrete time-varying system and I/O data. Consequently, Lemma 4.1, which establishes a virtual linear attacked model based on input and output data, can be derived naturally. Nevertheless, the matrix $\Gamma(k)$ in the system (4.10) remains undisclosed. Therefore, a data-driven system identification approach is required to liberate the virtual linear attacked model (4.10) from reliance on the system parameter matrices.

Since the attacker lacks knowledge of the parameter matrices $A(k)$, $B(k)$, and $C(k)$ in system (4.2), the matrix $\Gamma(k)$ in the virtual linear attacked model (4.10) also remains unknown to the attacker. To address this limitation, a data-driven system identification method, namely the improved projection estimation method, is employed. This method utilizes I/O data to estimate the matrix $\Gamma(k)$. Regarding the matrix $\Gamma(k)$ in the system (4.10), its estimation $\hat{\Gamma}_i(k)$ adheres to the principle of estimation error, as depicted by the following expression,

$$\begin{aligned} J_2(\hat{\Gamma}_i(k)) =& \|\Delta \tilde{y}_{i-1}(k+1) - \hat{\Gamma}_i(k)\Delta U_{i-1}(k) - \hat{\Gamma}_i(k)\bar{D}_a(k)\Delta U_{a_{i-1}}(k)\|^2 \\ &+ \lambda\|\hat{\Gamma}_i(k) - \hat{\Gamma}_{i-1}(k)\|^2, \end{aligned} \tag{4.11}$$

where $\hat{\Gamma}_i(k)$ is the estimation of $\Gamma(k)$ in the ith iteration at time k and λ is a positive weight coefficient.

The cost function $J_2(\hat{\Gamma}_i(k))$ primarily quantifies the discrepancy between $\hat{\Gamma}_i(k)$ and $\Gamma_i(k)$, encompassing the variation of the estimation between two consecutive iterations. Consequently, the minimization of $J_2(\hat{\Gamma}_i(k))$ aims to enhance the accuracy of $\hat{\Gamma}_i(k)$. The optimal estimation of the matrix $\Gamma(k)$ can be formally described as Theorem 4.1.

Theorem 4.1 *Minimizing the estimation error of $\hat{\Gamma}_i(k)$ can be illustrated by the following optimal problem,*

$$\begin{aligned} &\min_{\hat{\Gamma}_i(k)} \quad J_2(\hat{\Gamma}_i(k)), \\ &s.t. \quad \Delta \tilde{y}_i(k+1) = \hat{\Gamma}_i(k)\Delta U_i(k) + \hat{\Gamma}_i(k)\bar{D}_a(k)\Delta U_{a_i}(k), \end{aligned}$$

then the optimal estimation $\hat{\Gamma}_i^*(k)$ can be obtained as

$$\begin{aligned} \hat{\Gamma}_i^*(k) =& \hat{\Gamma}_{i-1}^*(k) + (\Delta \tilde{y}_{i-1}(k+1) - \hat{\Gamma}_{i-1}^*(k)\Delta U_{i-1}(k) - \hat{\Gamma}_{i-1}^*(k)\bar{D}_a(k) \\ &\times \Delta U_{a_{i-1}}(k))(\Delta U_{i-1}(k) + \bar{D}_a(k)\Delta U_{a_{i-1}}(k))^T[(\lambda + \|\Delta U_{i-1}(k) \\ &+ \bar{D}_a(k)\Delta U_{a_{i-1}}(k)\|^2)I_1 - (\Delta U_{i-1}(k) + \bar{D}_a(k)\Delta U_{a_{i-1}}(k)) \\ &\times (\Delta U_{i-1}(k) + \bar{D}_a(k)\Delta U_{a_{i-1}}(k))^T]/(\lambda^2 + \lambda\|(\Delta U_{i-1}(k) \\ &+ \bar{D}_a(k)\Delta U_{a_{i-1}}(k))\|^2), \end{aligned} \tag{4.12}$$

and the virtual linear attacked model (4.10) can be re-written as

$$\Delta\hat{\tilde{y}}_i(k+1) = \hat{\Gamma}_i^*(k)\Delta U_i(k) + \hat{\Gamma}_i^*(k)\bar{D}_a(k)\Delta U_{a_i}(k), \tag{4.13}$$

where I_1 is a $(l*(k+1)) \times (l*(k+1))$-dimensional identity matrix.

Proof The cost function (4.11) can be expressed as

$$\begin{aligned} J_2(\hat{\Gamma}_i(k)) =& \|\Delta\tilde{y}_{i-1}(k+1) - \hat{\Gamma}_{i-1}(k)\Delta U_{i-1}(k) - \hat{\Gamma}_{i-1}(k)\bar{D}_a(k)\Delta U_{a_{i-1}}(k) + \hat{\Gamma}_{i-1}(k) \\ &\times \Delta U_{i-1}(k) + \hat{\Gamma}_{i-1}(k)\bar{D}_a(k)\Delta U_{a_{i-1}}(k) - \hat{\Gamma}_i(k)\Delta U_{i-1}(k) - \hat{\Gamma}_i(k)\bar{D}_a(k) \\ &\times \Delta U_{a_{i-1}}(k)\|^2 + \lambda\|\hat{\Gamma}_i(k) - \hat{\Gamma}_{i-1}(k)\|^2 \\ =& \|\Delta\tilde{y}_{i-1}(k+1) - \hat{\Gamma}_{i-1}(k)\Delta U_{i-1}(k) - \hat{\Gamma}_{i-1}(k)\bar{D}_a(k)\Delta U_{a_{i-1}}(k) - (\hat{\Gamma}_i(k) \\ &- \hat{\Gamma}_{i-1}(k))(\Delta U_{i-1}(k) - \bar{D}_a(k)\Delta U_{a_{i-1}}(k))\|^2 + \lambda\|\hat{\Gamma}_i(k) - \hat{\Gamma}_{i-1}(k)\|^2. \end{aligned}$$

Then, the first-order partial derivative of the value function $J_2(\hat{\Gamma}_i(k))$ with respect to $\hat{\Gamma}_i(k)$ is

$$\begin{aligned} \frac{\partial J_2(\hat{\Gamma}_i(k))}{\hat{\Gamma}_i(k)} =& -2[\Delta\tilde{y}_{i-1}(k+1) - \hat{\Gamma}_{i-1}(k)\Delta U_{i-1}(k) - \hat{\Gamma}_{i-1}(k)\bar{D}_a(k)\Delta U_{a_{i-1}}(k) \\ &- (\hat{\Gamma}_i(k) - \hat{\Gamma}_{i-1}(k))(\Delta U_{i-1}(k) - \bar{D}_a(k)\Delta U_{a_{i-1}}(k))](\Delta U_{i-1}(k) \\ &- \bar{D}_a(k)\Delta U_{a_{i-1}}(k))^T + 2\lambda(\hat{\Gamma}_i(k) - \hat{\Gamma}_{i-1}(k)). \end{aligned}$$

Through the optimal theory[20], $\frac{\partial J_2(\hat{\Gamma}_i(k))}{\hat{\Gamma}_i(k)} = \mathbf{0}$ is applied. Thus, one can obtain

$$\begin{aligned} \hat{\Gamma}_i^*(k) =& \hat{\Gamma}_{i-1}^*(k) + (\Delta\tilde{y}_{i-1}(k+1) - \hat{\Gamma}_{i-1}^*(k)\Delta U_{i-1}(k) - \hat{\Gamma}_{i-1}^*(k)\bar{D}_a(k) \\ &\times \Delta U_{a_{i-1}}(k))(\Delta U_{i-1}(k) + \bar{D}_a(k)\Delta U_{a_{i-1}}(k))^T[\lambda I_1 + (\Delta U_{i-1}(k) \\ &+ \bar{D}_a(k)\Delta U_{a_{i-1}}(k))(\Delta U_{i-1}(k) + \bar{D}_a(k)\Delta U_{a_{i-1}}(k))^T]^{-1}. \end{aligned} \tag{4.14}$$

The matrix inversion lemma [21] is applied to the second term on the right side of (4.14) to obtain (4.12). This proof is completed. □

Remark 4.7 In the estimation process of $\hat{\Gamma}_i(k)$, there exists an estimation error between $\hat{\Gamma}_i(k)$ and $\Gamma(k)$. As the cost function J_2 is iteratively optimized, the estimation error gradually diminishes to zero. When the estimation error reaches zero, we have $\hat{\Gamma}_i(k) = \Gamma(k)$. Consequently, $\Delta\tilde{y}_i(k+1)$ and $\Delta\hat{\tilde{y}}_i(k+1)$ become equivalent.

To achieve the optimal estimation of $\hat{\Gamma}_i^*(k)$ as expressed in equation (4.12), the improved projection estimation approach based on the matrix inverse lemma [21] is proposed in Theorem 4.1. In this approach, the determination of the weight coefficient λ leads to fixed estimated iteration steps. However, to enhance flexibility, the step factor μ, which allows a more generalized and flexible estimation and satisfies $0 < \mu < 2$, is introduced. Mathematically, the introduction of the step factor can be represented as follows:

$$\begin{aligned}\hat{\Gamma}_i^*(k) =&\hat{\Gamma}_{i-1}^*(k) + \mu(\Delta\tilde{y}_{i-1}(k+1) - \hat{\Gamma}_{i-1}^*(k)\Delta U_{i-1}(k) - \hat{\Gamma}_{i-1}^*(k)\bar{D}_a(k)\\ &\times \Delta U_{a_{i-1}}(k))(\Delta U_{i-1}(k) + \bar{D}_a(k)\Delta U_{a_{i-1}}(k))^T[(\lambda + \|\Delta U_{i-1}(k)\\ &+ \bar{D}_a(k)\Delta U_{a_{i-1}}(k)\|^2)I_1 - (\Delta U_{i-1}(k) + \bar{D}_a(k)\Delta U_{a_{i-1}}(k))\\ &\times (\Delta U_{i-1}(k) + \bar{D}_a(k)\Delta U_{a_{i-1}}(k))^T]/(\lambda^2 + \lambda\|(\Delta U_{i-1}(k)\\ &+ \bar{D}_a(k)\Delta U_{a_{i-1}}(k))\|^2).\end{aligned} \tag{4.15}$$

By appropriately adjusting the step factor μ, it becomes possible to expedite the reduction of estimation error and enhance accuracy. The step factor is employed in the computation of $\hat{\Gamma}_i(k)$ in Section IV. This approach has been utilized in prior works such as [10, 11].

Similar to the above I/O data-driven virtual linear attack model under non-switching FDI attacks, the I/O data-driven virtual linear attack model under switching FDI attacks is constructed as follows:

$$\Delta\tilde{y}_i(k+1) =\Phi(k)\Delta U_i(k) + \sum_{h=1}^{l} \theta_h(k)\Phi(k)\bar{D}_{a_h}(k)\Delta U_{a_i}(k), \tag{4.16}$$

where $\Phi(k)$ is the unknown system parameter matrix and $\bar{D}_{a_h}(k) = \text{diag}(D_{a_h}(k), D_{a_h}(k-1), \ldots, D_{a_h}(0))$ with $\bar{D}_{a_h}(k) \in \mathbb{R}^{(l*(k+1))\times(l*(k+1))}$. Subsequently, the data-driven system description for identifying system (4.16) based on the projection estimation method is described as

$$\Delta\hat{\tilde{y}}_i(k+1) =\hat{\Phi}_i(k)\Delta U_i(k) + \sum_{h=1}^{l} \theta_h(k)\hat{\Phi}_i(k)\bar{D}_{a_h}(k)\Delta U_{a_i}(k), \tag{4.17}$$

where $\hat{\Phi}_i(k)$ is the estimation of $\Phi(k)$ in the ith iteration at time k, the optimal estimation $\hat{\Phi}_i^*(k)$ can be calculated by

$$\begin{aligned}\hat{\Phi}_i^*(k) = \hat{\Phi}_{i-1}^*(k) + \mu\Bigg(&\Delta\tilde{y}_{i-1}(k+1) - \hat{\Phi}_{i-1}^*(k)\Delta U_{i-1}(k) - \sum_{h=1}^{l}\theta_h(k)\hat{\Phi}_{i-1}^*(k)\\ &\times \bar{D}_{a_h}(k)\Delta U_{a_{i-1}}(k)\Bigg)\Bigg(\Delta U_{i-1}(k) + \sum_{h=1}^{l}\theta_h(k)\bar{D}_{a_h}(k)\Delta U_{a_{i-1}}(k)\Bigg)^T\\ &\times\Bigg[\Bigg(\lambda + \|\Delta U_{i-1}(k) + \sum_{h=1}^{l}\theta_h(k)\bar{D}_{a_h}(k)\Delta U_{a_{i-1}}(k)\|^2\Bigg)I_1\\ &-\Bigg(\Delta U_{i-1}(k) + \sum_{h=1}^{l}\theta_h(k)\bar{D}_{a_h}(k)\Delta U_{a_{i-1}}(k)\Bigg)\Bigg(\Delta U_{i-1}(k)\\ &+\sum_{h=1}^{l}\theta_h(k)\bar{D}_{a_h}(k)\Delta U_{a_{i-1}}(k)\Bigg)^T\Bigg]\Bigg/\Bigg(\lambda^2 + \lambda\|(\Delta U_{i-1}(k)\end{aligned}$$

$$+\sum_{h=1}^{l}\theta_h(k)\bar{D}_{a_h}(k)\triangle U_{a_{i-1}}(k))\|^2\Big), \tag{4.18}$$

and I_1 is defined in Theorem 4.1.

4.3.2 *Optimal Data-Driven FDI Attack Problems*

As the estimation error becomes zero after a sufficient number of iterations, the trajectory error in the objective function J_1 can be replaced with $\tilde{E}_i^p(k+1)$. Specifically, $\tilde{E}_i^p(k+1)$ is defined as $\tilde{E}_i^p(k+1) \triangleq \tilde{y}_r(k+1) - \tilde{y}_{i-1}(k+1) - \triangle\hat{\tilde{y}}_i(k+1)$. By utilizing $\tilde{E}_i^p(k+1)$ along with Theorem 4.1, the initially unsolvable Problem 4.1 resulting from the unknown matrix $\Gamma(k)$ can be transformed into the following solvable problem.

Problem 4.3

$$\max_{\triangle u_{a_i}(k)} \quad J_1(\tilde{E}_i^p(k+1), \triangle u_{a_i}(k))$$

$$\text{s.t.} \quad \begin{cases} \triangle\hat{\tilde{y}}_i(k+1) = \hat{\Gamma}_i(k)\triangle U_i(k) + \hat{\Gamma}_i(k)\bar{D}_a(k)\triangle U_{a_i}(k), \\ Q \succeq 0,\ R \succ 0. \end{cases}$$

Similarly, by utilizing the data-driven system (4.17) and the definition of $\tilde{E}_i^p(k+1)$, the previously unsolvable Problem 4.2 can be transformed into the following solvable problem.

Problem 4.4

$$\max_{\triangle u_{a_i}(k)} \quad J_1(\tilde{E}_i^p(k+1), \triangle u_{a_i}(k))$$

$$\text{s.t.} \quad \begin{cases} \triangle\hat{\tilde{y}}_i(k+1) = \hat{\Phi}_i(k)\triangle U_i(k) + \sum_{h=1}^{l}\theta_h(k)\hat{\Phi}_i(k)\bar{D}_{a_h}(k)\triangle U_{a_i}(k), \\ Q \succeq 0,\ R \succ 0. \end{cases}$$

Remark 4.8 The relationship between Problem 4.1 and Problem 4.3 can be elucidated as follows. Initially, Problem 4.3 serves as an approximate optimization problem for Problem 4.1, owing to the presence of iteration error. However, as the iteration progresses, the iteration error tends to zero. Consequently, when the iteration error reaches zero, $\tilde{E}_i^p(k+1)$ and $\tilde{E}_i(k+1)$ become equivalent, thereby establishing the equivalence between Problem 4.3 and Problem 4.1. The relationship between Problem 4.2 and Problem 4.4 is analogous to that of Problem 4.1 and Problem 4.3, omitted here.

4.4 Data-Driven FDI Attack Strategies via Iterative Learning

In this section, the optimal data-driven FDI attack strategies and their algorithmic implementations based on I/O data in non-switching and switching forms are presented.

4.4.1 Data-Driven FDI Attack Strategy

The optimal data-driven FDI attack strategy and its algorithm based on I/O data in the non-switching form are obtained by solving the data-driven optimization Problem 4.3.

Theorem 4.2 *Suppose that Assumptions 4.1-4.3 hold, if*

$$(\bar{D}_a^T(k)\hat{\Gamma}_i^T(k)Q\hat{\Gamma}_i(k)\bar{D}_a(k) - \eta^2 P^T(k)RP(k)) \prec 0, \tag{4.19}$$

then the optimal FDI attack input $u_{a_i}^*(k)$ for Problem 4.3 can be designed as

$$u_{a_i}^*(k) = u_{a_{i-1}}(k) + P(k)\Delta U_{a_i}^*(k) - \Lambda_i(k), \tag{4.20}$$

which is stealthy with respect to the detection mechanism (4.4), where the optimal attack iteration difference $\Delta U_{a_i}^*(k)$ is obtained as

$$\begin{aligned}\Delta U_{a_i}^*(k) =&(\bar{D}_a^T(k)\hat{\Gamma}_i^{*T}(k)Q\hat{\Gamma}_i^*(k)\bar{D}_a(k) - \eta^2 P^T(k)RP(k))^{-1}[\bar{D}_a^T(k)\\ &\times \hat{\Gamma}_i^{*T}(k)Q(\tilde{E}_{i-1}(k+1) - \hat{\Gamma}_i^*(k)\Delta U_i(k)) - \eta^2 P^T(k)R\Lambda_i(k)],\end{aligned} \tag{4.21}$$

with $P(k) = [I_2, I_2, ..., I_2] \in \mathbb{R}^{l\times(l*(k+1))}$, $\Lambda_i(k) = \sum_{j=1}^{k} \Delta u_{a_i}(k-j) \in \mathbb{R}^{l\times 1}$, and I_2 is a $(l \times l)$-dimensional identity matrix.

Proof Jointing the definitions of $\tilde{E}_i^p(k+1)$ and (4.13), $\tilde{E}_i^p(k+1)$ can be re-expressed as

$$\tilde{E}_i^p(k+1) = \tilde{y}_r(k+1) - \tilde{y}_{i-1}(k+1) - \hat{\Gamma}_i(k)\Delta U_i(k) - \hat{\Gamma}_i(k)\bar{D}_a(k)\Delta U_{a_i}(k),$$

where $U_{a_i}(k) = [u_{a_i}(k), u_{a_i}(k-1), ..., u_{a_i}(0)]^T$, $\Delta U_{a_i}(k) = [\Delta u_{a_i}(k), \Delta u_{a_i}(k-1), ..., \Delta u_{a_i}(0)]^T$. Then, according to the least squares, one has

$$\Delta u_{a_i}(k) = P(k)\Delta U_{a_i}(k) - \Lambda_i(k), \tag{4.22}$$

where $P(k)$ and $\Lambda_i(k)$ are defined in Theorem 4.2.

Consequently, the cost function (4.9) can be rewritten as

$$\begin{aligned} J_1 =& \frac{1}{2}\{(\tilde{y}_r(k+1) - \tilde{y}_{i-1}(k+1) - \hat{\Gamma}_i^*(k)\Delta U_i(k) - \hat{\Gamma}_i^*(k)\bar{D}_a(k)\Delta U_{a_i}(k))^T Q \\ &\times (\tilde{y}_r(k+1) - \tilde{y}_{i-1}(k+1) - \hat{\Gamma}_i^*(k)\Delta U_i(k) - \hat{\Gamma}_i^*(k)\bar{D}_a(k)\Delta U_{a_i}(k)) \\ &- \eta^2(P(k)\Delta U_{a_i}(k) - \Lambda_i(k))^T R(P(k)\Delta U_{a_i}(k) - \Lambda_i(k))\}. \end{aligned} \tag{4.23}$$

Reorganizing (4.23), one has

$$\begin{aligned} &-\frac{1}{2}(\tilde{y}_r(k+1) - \tilde{y}_{i-1}(k+1) - \hat{\Gamma}_i^*(k)\Delta U_i(k))^T Q\hat{\Gamma}_i^*(k)\bar{D}_a(k)\Delta U_{a_i}(k) - \frac{1}{2}\Delta U_{a_i}^T(k) \\ &\times \bar{D}_a^T(k)\hat{\Gamma}_i^{*T}(k)Q(\tilde{y}_r(k+1) - \tilde{y}_{i-1}(k+1) - \hat{\Gamma}_i^*(k)\Delta U_i(k)) + \frac{1}{2}\Delta U_{a_i}^T(k)\bar{D}_a^T(k) \\ &\times \hat{\Gamma}_i^{*T}(k)Q\hat{\Gamma}_i^*(k)\bar{D}_a(k)\Delta U_{a_i}(k) - \frac{1}{2}\eta^2\Delta U_{a_i}^T(k)P^T(k)RP(k)\Delta U_{a_i}(k)) + \frac{1}{2}\eta^2\Delta U_{a_i}^T(k) \\ &\times P^T(k)R\Lambda_i(k) + \frac{1}{2}\eta^2\Lambda_i^T(k)RP(k)\Delta U_{a_i}(k) - \frac{1}{2}\eta^2\Lambda_i^T(k)R\Lambda_i(k) + \frac{1}{2}(\tilde{y}_r(k+1) \\ &- \tilde{y}_{i-1}(k+1) - \hat{\Gamma}_i^*(k)\Delta U_i(k))^T Q(\tilde{y}_r(k+1) - \tilde{y}_{i-1}(k+1) - \hat{\Gamma}_i^*(k)\Delta U_i(k)). \end{aligned}$$

Through the optimal theory[20], $\frac{\partial J_1}{\partial \Delta U_{a_i}(k)} = \mathbf{0}$ is applied,

$$\begin{aligned} &(\bar{D}_a^T(k)\hat{\Gamma}_i^T(k)Q\hat{\Gamma}_i(k)\bar{D}_a(k) - \eta^2 P^T(k)RP(k))\Delta U_{a_i}(k) - \bar{D}_a^T(k)\hat{\Gamma}_i^T(k) \\ &\times Q(\tilde{y}_r(k+1) - \tilde{y}_{i-1}(k+1) - \hat{\Gamma}_i(k)\Delta U_i(k)) + \eta^2 P^T(k)R\Lambda_i(k) = \mathbf{0}, \end{aligned} \tag{4.24}$$

then, the optimal attack iteration difference $\Delta U_{a_i}^*(k)$ in (4.21) can be obtained. Together with (4.21) and (4.22), the optimal false data injection attack input $u_{a_i}^*(k)$ in (4.20) can be derived.

In view of Maximum-Minimum Theorem[22], the second-order partial derivative condition

$$\frac{\partial^2 J_1}{\partial \Delta U_{a_i}^2(k)} = (\bar{D}_a^T(k)\hat{\Gamma}_i^T(k)Q\hat{\Gamma}_i(k)\bar{D}_a(k) - \eta^2 P^T(k)RP(k)) \prec 0, \tag{4.25}$$

is adopted to ensure optimality.

The following is the proof of the stealthiness of the attack. Under the attack defined as (4.5), the corresponding dynamic equation (4.3) and output measurement residual of detector is updated as

$$\begin{cases} \tilde{x}_i^d(k+1) = A(k)\tilde{x}_i^d(k) + B(k)u_i(k) + B_a(k)u_{a_i}(k) \\ \qquad\qquad +L(k)(\tilde{y}_i(k) - \tilde{y}_i^d(k)), \\ \tilde{y}_i^d(k) = C(k)\tilde{x}_i^d(k), \\ \tilde{r}_i(k+1) = \tilde{y}_i(k+1) - \tilde{y}_i^d(k+1). \end{cases} \tag{4.26}$$

Combining (4.6) and (4.26), the output measurement residual $\tilde{r}_i(k+1)$ can be be derived as follows:

$$\begin{aligned}
\tilde{r}_i(k+1) &= \tilde{y}_i(k+1) - \tilde{y}_i^d(k+1) \\
&= C(k+1)(\tilde{x}_i(k+1) - \tilde{x}_i^d(k+1)) \\
&= C(k+1)[A(k)\tilde{x}_i(k) + B(k)u_i(k) + B_a(k)u_{a_i}(k) - A(k)\tilde{x}_i^d(k) \\
&\quad - B(k)u_i(k) - B_a(k)u_{a_i}(k) - L(k)(\tilde{y}_i(k) - \tilde{y}_i^d(k))] \\
&= C(k+1)(A(k) - L(k)C(k))(\tilde{x}_i(k) - \tilde{x}_i^d(k)) \\
&= C(k+1)\prod_{s=0}^{k}(A(s) - L(s)C(s))(\tilde{x}_i(0) - \tilde{x}_i^d(0)).
\end{aligned}$$

According to Assumption 4.1, the initial states of both the attacked system and the detector are

$$\tilde{x}_i(0) = x_i(0) = \tilde{x}_i^d(0) = x_i^d(0) = x_0,$$

thus, the residual $\tilde{r}_i(k+1)$ is

$$\begin{aligned}
\tilde{r}_i(k+1) &= C(k+1)\prod_{s=0}^{k}(A(s) - L(s)C(s))(x_i(0) - x_i^d(0)), \\
&= \mathbf{0},
\end{aligned}$$

that is, $\|\tilde{r}_i(k+1)\| = 0 < \Upsilon$, and it can be observed that the output measurement residual $\tilde{r}_i(k+1)$ is independent of the attack input $u_{a_i}(k)$. Therefore, the attack (4.20) is covert to the detection mechanism (4.4). This proof is completed. □

The application of Theorem 4.2 can be translated into the data-driven FDI attack algorithm via iterative learning, i.e., Algorithm 2.

Remark 4.9 For the optimal FDI attack strategy designed by Theorem 4.2, it is crucial to ensure the satisfaction of condition $(\bar{D}_a^T(k)\hat{\Gamma}_i^T(k)Q\hat{\Gamma}_i(k)\bar{D}_a(k) - \eta^2 P^T(k)RP(k)) \prec 0$ for all $i = 1, 2, ..., Na$ and $k = 0, 1, ..., N$ in practical applications, where N_a and N are the numbers of iterations and the attack duration, respectively. This condition is necessary to guarantee the uniqueness of the optimal attack strategy. However, during the execution of the attack, it is not always the case that the condition is satisfied. Therefore, the attacker needs to adjust the attack matrix $D_a(k)$ to achieve the satisfaction of the condition. The specific process outlining this adjustment is elaborated in Algorithm 2.

Algorithm 2: The Data-Driven FDI Attack via Iterative Learning.

Initialization: Set the attack duration N and the number of iterations N_a, appropriate attack matrix $D_a(k)$, suitable coefficients λ and η, satisfying $\lambda > 0$, and $0 \leq \eta \leq 1$, and the appropriate matrices Q and R which meet the conditions of $Q \geq 0$ and $R > 0$. Preset negative eigenvalue vectors $\beta_k \in \mathbb{R}^{(l*(k+1))\times 1}$, $k = 0, 1, ..., N$;

for $i = 1$ *to* N_a **do**

  while $k \leq N$ **do**

    Step 1 (Data collection): Collect data for $\tilde{y}_r(k+1)$, $\tilde{y}_{i-1}(k+1)$, and $u_i(k)$;

    Step 2 (Matrix update): Update the matrices $\hat{\Gamma}_{i-1}(k)$, $\Delta U_{i-1}(k)$, $\Delta U_{a_{i-1}}(k)$, $\Lambda_i(k)$, $\bar{D}_a(k)$ and $P(k)$; Compute the estimation $\hat{\Gamma}_i(k)$ via (4.15), and then calculate the condition $(\bar{D}_a^T(k)\hat{\Gamma}_i^T(k) \times Q\hat{\Gamma}_i(k)\bar{D}_a(k) - \eta^2 P^T(k)RP(k))$;

    Step 3 (Condition judgment): Calculate the eigenvalues of the condition $(\bar{D}_a^T(k)\hat{\Gamma}_i^T(k) \times Q\hat{\Gamma}_i(k)\bar{D}_a(k) - \eta^2 P^T(k)RP(k))$, and assign them to the vector $Eig(k)$;

    if *all*$(Eig(k) < 0)$ `// all elements of` $Eig(k)$ `are less than zero`

    then

      1). Update eigenvalue vector $\beta_k = Eig(k)$, and fill the columns $(l+1)$ to $(l*(k+2))$ of β_{k+1} with β_k;

      2). Goto **Step 4**;

    else

      `// the adjustment of the attack matrix`

      1). Set $D_a(k) = \text{diag}(\alpha_1, \alpha_2, ..., \alpha_l)$, and compute the eigenvalue vector $Eig(k)$;

      2). Solve the equation $Eig(k) - \beta_k = 0$ via the newton-raphson method [23], then the solution $\alpha_1^*, \alpha_2^*, ..., \alpha_l^*$ can be obtained.

      3). Reset $D_a(k) = \text{diag}(\alpha_1^*, \alpha_2^*, ..., \alpha_l^*)$ and repeat **Step 2**;

    Step 4 (Strategy generation): Compute the optimal attack iteration difference $\Delta U_{a_i}(k)$ using (4.21), then the optimal FDI attack input $u_{a_i}^*(k)$ can be generated by calculating (4.20);

    Step 5 (Attack implementation): $\tilde{u}_i(k) \longleftarrow u_i(k) + D_a(k)u_{a_i}(k)$.

4.4.2 *Data-Driven Switching FDI Attack Strategy*

The optimal data-driven switching FDI attack strategy and its algorithm based on I/O data are obtained by solving the data-driven optimization Problem 4.4.

Theorem 4.3 *Suppose that Assumptions 4.1-4.3 hold, if $\Xi_i(k) \prec 0$, then the optimal switching FDI attack input $u_{a_i}^*(k)$ for Problem 4.4 can be designed as*

$$u_{a_i}^*(k) = u_{a_{i-1}}(k) + \tilde{P}(k)\Delta U_{a_i}^*(k) - \tilde{\Lambda}_i(k), \tag{4.27}$$

which is stealthy with respect to the detection mechanism (4.4), and the optimal switching sequence is

$$h^* = \mathop{\arg\max}_{h\in\{1,2,...,l\}} \{\Psi_i^h(k)\}, \tag{4.28}$$

where the optimal attack iteration difference $\triangle U^*_{a_i}(k)$ is obtained as

$$\triangle U^*_{a_i}(k) = \Xi_i^{-1}(k)\Bigg[\sum_{h=1}^{l}\theta_h(k)\bar{D}^T_{a_h}(k)\hat{\Phi}_i^{*T}(k)Q(\tilde{E}_{i-1}(k+1) - \hat{\Phi}_i^*(k)\triangle U_i(k)) - \eta^2\tilde{P}^T(k)R\tilde{\Lambda}_i(k)\Bigg], \tag{4.29}$$

the matrix $\Xi_i(k)$ is calculated through

$$\Xi_i(k) = \left(\sum_{h=1}^{l}\theta_h(k)\hat{\Phi}_i^*(k)\bar{D}_{a_h}(k)\right)^T Q\left(\sum_{h=1}^{l}\theta_h(k)\hat{\Phi}_i^*(k)\bar{D}_{a_h}(k)\right) - \eta^2\tilde{P}^T(k)R\tilde{P}(k),$$

and the matrix $\Psi_i^h(k)$ is denoted as

$$\begin{aligned}\Psi_i^h(k) =& [(\tilde{E}_{i-1}(k+1) - \hat{\Phi}_i^*(k)\triangle U_i(k))^T Q\hat{\Phi}_i^*(k)\bar{D}_{a_h}(k) - \eta^2\tilde{\Lambda}_i^T(k)R\tilde{P}(k)]\\ &\times(\eta^2\tilde{P}^T(k)R\tilde{P}(k) - \bar{D}_{a_h}(k)\hat{\Phi}_i^{*T}(k)Q\hat{\Phi}_i^*(k)\bar{D}_{a_h}(k))^{-1}[\bar{D}^T_{a_h}(k)\\ &\times\hat{\Phi}_i^{*T}(k)Q(\tilde{E}_{i-1}(k+1) - \hat{\Phi}_i^*(k)\triangle U_i(k)) - \eta^2\tilde{P}^T(k)R\tilde{\Lambda}_i(k)],\end{aligned} \tag{4.30}$$

with $\tilde{P}(k) = [I_2, I_2, ..., I_2] \in \mathbb{R}^{l\times(l*(k+1))}$, $\tilde{\Lambda}_i(k) = \sum_{j=1}^{k}\triangle u_{a_i}(k-j) \in \mathbb{R}^{l\times 1}$, and I_2 is a $(l \times l)$-dimensional identity matrix.

Proof Since the solution of the estimation of $\Phi(k)$, the stealthiness of the attack, the optimal attack iteration difference $\triangle U^*_{a_i}(k)$, and the optimal switching attack input $u^*_{a_i}(k)$ share similarities with the proof presented in Theorem 4.2, the proof is omitted here. The following presents the proof that the acquisition of the optimal switching condition.

Combining the definitions of $\tilde{E}_i^p(k+1)$ and the system (4.17), one has

$$\tilde{E}_i^p(k+1) = \tilde{y}_r(k+1) - \tilde{y}_{i-1}(k+1) - \hat{\Phi}_i^*(k)\triangle U_i(k) - \sum_{h=1}^{l}\theta_h(k)\hat{\Phi}_i^*(k) \times \bar{D}_{a_h}(k)\triangle U_{a_i}(k).$$

Subsequently, the cost function (4.9) can be rewritten as

$$
\begin{aligned}
J_1 =& \frac{1}{2}\Bigg[\Bigg(\tilde{y}_r(k+1) - \tilde{y}_{i-1}(k+1) - \hat{\Phi}_i^*(k)\Delta U_i(k) - \sum_{h=1}^{l}\theta_h(k)\hat{\Phi}_i^*(k)\bar{D}_{a_h}(k) \\
& \times \Delta U_{a_i}(k)\Bigg)^T Q\Bigg(\tilde{y}_r(k+1) - \tilde{y}_{i-1}(k+1) - \hat{\Phi}_i^*(k)\Delta U_i(k) - \sum_{h=1}^{l}\theta_h(k) \\
& \cdot \hat{\Phi}_i^*(k)\bar{D}_{a_h}(k)\Delta U_{a_i}(k)\Bigg) - \eta^2(\tilde{P}(k)\Delta U_{a_i}(k) - \tilde{\Lambda}_i(k))^T R(\tilde{P}(k) \\
& \times \Delta U_{a_i}(k) - \tilde{\Lambda}_i(k))\Bigg].
\end{aligned} \tag{4.31}
$$

Reorganizing (4.31), one has

$$
\begin{aligned}
& -\frac{1}{2}(\tilde{E}_{i-1}(k+1) - \hat{\Phi}_i^*(k)\Delta U_i(k))^T Q \sum_{h=1}^{l}\theta_h(k)\hat{\Phi}_i^*(k)\bar{D}_{a_h}(k)\Delta U_{a_i}(k) - \frac{1}{2}\sum_{h=1}^{l}\theta_h(k) \\
& \cdot \Delta U_{a_i}^T(k)\bar{D}_{a_h}^T(k)\hat{\Phi}_i^{*T}(k)Q(\tilde{E}_{i-1}(k+1) - \hat{\Phi}_i^*(k)\Delta U_i(k)) + \frac{1}{2}\Delta U_{a_i}^T(k)\Bigg(\sum_{h_1=1}^{l}\sum_{h_2=1}^{l} \\
& \theta_{h_1}(k)\theta_{h_2}(k)\bar{D}_{a_{h_1}}^T(k)\hat{\Phi}_i^{*T}(k)Q\hat{\Phi}_i^*(k)\bar{D}_{a_{h_2}}(k) - \eta^2\tilde{P}^T(k)R\tilde{P}(k)\Bigg)\Delta U_{a_i}(k) \\
& + \frac{1}{2}\eta^2\Delta U_{a_i}^T(k)\tilde{P}^T(k)R\tilde{\Lambda}_i(k) + \frac{1}{2}\eta^2\tilde{\Lambda}_i^T(k)R\tilde{P}(k)\Delta U_{a_i}(k) - \frac{1}{2}\eta^2\tilde{\Lambda}_i^T(k)R\tilde{\Lambda}_i(k) \\
& + \frac{1}{2}(\tilde{E}_{i-1}(k+1) - \hat{\Phi}_i^*(k)\Delta U_i(k))^T Q(\tilde{E}_{i-1}(k+1) - \hat{\Phi}_i^*(k)\Delta U_i(k)).
\end{aligned}
$$

Substituting (4.29) into the above equation, after simplification, one has

$$
\begin{aligned}
J_1 =& \frac{1}{2}(\tilde{E}_{i-1}(k+1) - \hat{\Phi}_i^*(k)\Delta U_i(k))^T Q(\tilde{E}_{i-1}(k+1) - \hat{\Phi}_i^*(k)\Delta U_i(k)) \\
& - \frac{1}{2}\eta^2\tilde{\Lambda}_i^T(k)R\tilde{\Lambda}_i(k) - \frac{1}{2}\Bigg[(\tilde{E}_{i-1}(k+1) - \hat{\Phi}_i^*(k)\Delta U_i(k))^T Q\hat{\Phi}_i^*(k) \\
& \times \sum_{h=1}^{l}\theta_h(k)\bar{D}_{a_h}(k) - \eta^2\tilde{\Lambda}_i^T(k)R\tilde{P}(k)\Bigg]\Bigg(\sum_{h_1=1}^{l}\sum_{h_2=1}^{l}\theta_{h_1}(k)\theta_{h_2}(k)\bar{D}_{a_{h_1}}^T(k) \\
& \times \hat{\Phi}_i^{*T}(k)Q\hat{\Phi}_i^*(k)\bar{D}_{a_{h_2}}(k) - \eta^2\tilde{P}^T(k)R\tilde{P}(k)\Bigg)^{-1}\Bigg[\sum_{h=1}^{l}\theta_h(k)\bar{D}_{a_h}^T(k) \\
& \times \hat{\Phi}_i^{*T}(k)Q(\tilde{E}_{i-1}(k+1) - \hat{\Phi}_i^*(k)\Delta U_i(k)) - \eta^2\tilde{P}^T(k)R\tilde{\Lambda}_i(k)\Bigg].
\end{aligned} \tag{4.32}
$$

Thus, maximizing J_1 with respect to $\theta_h(k)$ can be simplified to maximize

$$
\begin{aligned}
\tilde{J}_1 = &- \Bigg[(\tilde{E}_{i-1}(k+1) - \hat{\Phi}_i^*(k)\Delta U_i(k))^T Q \hat{\Phi}_i^*(k) \sum_{h=1}^{l} \theta_h(k) \bar{D}_{a_h}(k) \\
&- \eta^2 \tilde{\Lambda}_i^T(k) R \tilde{P}(k) \Bigg] \Bigg(\sum_{h_1=1}^{l} \sum_{h_2=1}^{l} \theta_{h_1}(k)\theta_{h_2}(k) \bar{D}_{a_{h_1}}^T(k) \hat{\Phi}_i^{*T}(k) Q \hat{\Phi}_i^*(k) \\
&\times \bar{D}_{a_{h_2}}(k) - \eta^2 \tilde{P}^T(k) R \tilde{P}(k) \Bigg)^{-1} \Bigg[\sum_{h=1}^{l} \theta_h(k) \bar{D}_{a_h}^T(k) \hat{\Phi}_i^{*T}(k) Q \\
&\times (\tilde{E}_{i-1}(k+1) - \hat{\Phi}_i^*(k)\Delta U_i(k)) - \eta^2 \tilde{P}^T(k) R \tilde{\Lambda}_i(k) \Bigg].
\end{aligned} \tag{4.33}
$$

Further, (4.33) can be formulated in compact form as follows:

$$
\begin{aligned}
\tilde{J}_1 = &- [(\tilde{E}_{i-1}(k+1) - \hat{\Phi}_i^*(k)\Delta U_i(k))^T Q \hat{\Phi}_i^*(k) \Theta^T(k) \bar{\bar{D}}_a(k) - \eta^2 \tilde{\Lambda}_i^T(k) \\
&\times R \tilde{P}(k)](\bar{\bar{D}}_a^T(k) \Theta(k) \hat{\Phi}_i^{*T}(k) Q \hat{\Phi}_i^*(k) \Theta^T(k) \bar{\bar{D}}_a(k) - \eta^2 \tilde{P}^T(k) R \\
&\times \tilde{P}(k))^{-1} [\bar{\bar{D}}_a^T(k) \Theta(k) \hat{\Phi}_i^{*T}(k) Q (\tilde{E}_{i-1}(k+1) - \hat{\Phi}_i^*(k)\Delta U_i(k)) \\
&- \eta^2 \tilde{P}^T(k) R \tilde{\Lambda}_i(k)],
\end{aligned} \tag{4.34}
$$

where $\Theta(k)$ is defined in (4.7) and $\bar{\bar{D}}_a(k) = [\bar{D}_{a_1}(k); \bar{D}_{a_2}(k); ...; \bar{D}_{a_l}(k)]$.

The embedded technology [1] is adopted in (4.33) and combining with the condition of $\Xi_i(k) \prec 0$, one can have the Hessian matrix [22],

$$
\frac{\partial^2 \tilde{J}_1}{\partial \Theta^2(k)} \succeq 0.
$$

Consequently, the optimal switching condition can be obtained as (4.28). This proof is completed. □

The application of Theorem 4.3 can be converted to a data-driven switching FDI attack algorithm via iterative learning, that is, Algorithm 3.

Remark 4.10 Distinct from the injection attack devised for all C-A wireless channels in Theorem 4.2, Theorem 4.3 aims at executing an injection attack on a specifically selected C-A wireless channel. Theorem 4.3 possesses a broader applicability compared to Theorem 4.2, particularly in scenarios where simultaneous access to certain C-A channels is not feasible. Furthermore, as the switching sequence (4.28) is computed in real time, it becomes arduous for defenders to anticipate the next C-A wireless channel that will be targeted, thereby enhancing the attack's stealthiness and defense complexity. The relationship between Algorithm 3 and Algorithm 2 is that Algorithm 3 is an extension of Algorithm 2. Different from Algorithm 2, another main focus of Algorithm 3 is the computation of the switching sequence, that is, the selection of the attacking communication channel from the controller to the actuator.

Algorithm 3: The Data-Driven Switching FDI Attack via Iterative Learning.

Initialization: Set the attack duration N and the number of iterations N_a, appropriate attack direction matrix $D_{a_h}(k)$, $h = 1, 2, ..., l$, suitable coefficients λ and η, satisfying $\lambda > 0$, and $0 \leq \eta \leq 1$, and the appropriate matrices Q and R which meet the conditions of $Q \geq 0$ and $R > 0$. Preset negative eigenvalue vectors $\tilde{\beta}_k^h \in \mathbb{R}^{(l*(k+1))\times 1}$, $k = 0, 1, ..., N$, $h = 1, 2, ..., l$;
for $i = 1$ *to* N_a **do**
  while $k \leq N$ **do**
    Step 1 (Data collection): Collect data for $\tilde{y}_r(k+1)$, $\tilde{y}_{i-1}(k+1)$, and $u_i(k)$;
    for $h = 1$ *to* l **do**
      Step 2 (Matrix update): Update the matrices $\hat{\Phi}_{i-1}(k)$, $\Delta U_{i-1}(k)$, $\Delta U_{a_{i-1}}(k)$, $\tilde{\Lambda}_i(k)$, $\bar{D}_{a_h}(k)$ and $\tilde{P}(k)$; Compute the estimation $\hat{\Phi}_i(k)$ via (4.18), and then calculate the matrix $\Xi_i(k)$;
      Step 3 (Condition judgment): Calculate the eigenvalues of the matrix $\Xi_i(k)$, and assign them to the vector $\tilde{E}ig(k)$;
      if *all*($\tilde{E}ig(k) < 0$) `// all elements of` $\tilde{E}ig(k)$ `are less than zero`
      then
        Update eigenvalue vector $\tilde{\beta}_k^h = Eig(k)$, and fill the columns $(l+1)$ to $(l*(k+2))$ of $\tilde{\beta}_{k+1}^h$ with $\tilde{\beta}_k^h$;
      else
        `// the adjustment of the attack direction matrix`
        1). Set $D_{a_h}(k) = \text{diag}(\tilde{\alpha}_1^h, \tilde{\alpha}_2^h, ..., \tilde{\alpha}_l^h)$, and compute the eigenvalue vector $\tilde{E}ig(k)$;
        2). Solve the equation $\tilde{E}ig(k) - \beta_k^h = 0$ via the newton-raphson method [23], then the solution $\alpha_1^{h*}, \alpha_2^{h*}, ..., \alpha_l^{h*}$ can be obtained.
        3). Reset $D_{a_h}(k) = \text{diag}(\alpha_1^{h*}, \alpha_2^{h*}, ..., \alpha_l^{h*})$ and repeat **Step 2**;
    Step 4 (Optimal switching sequence calculation): for $h = 1$ *to* l **do**
      Compute the matrix $\Psi_i^h(k)$ via (4.30);
      if $\Psi_i^h(k) > \Psi_i^{h-1}(k)$ **then**
        $h^* = h$;
      else
        $h^* = h - 1$;
      Set $\theta_{h*}(k) = 1$, while keeping all other settings at zero.
    Step 5 (Strategy generation): Compute the optimal attack iteration difference $\Delta U_{a_i}(k)$ using (4.29), then the optimal switching FDI attack input $u_{a_i}^*(k)$ can be generated by calculating (4.27);
    Step 6 (Attack implementation): $\tilde{u}_i(k) \longleftarrow u_i(k) + \sum_{h=1}^{l} \theta_h(k) D_{a_h}(k) u_{a_i}(k)$.

4.5 Numerical Simulation

In this section, a simulation of a networked DC motor is employed to verify the effectiveness of the proposed algorithms. The schematic diagram of the attacked networked DC motor is presented in Fig. 4.4, while Fig. 4.5 demonstrates the circuit diagram of the motor. Table 4.1 provides a comprehensive overview of the symbols, their physical interpretations, associated values, and units pertaining to the specifications of the motor.

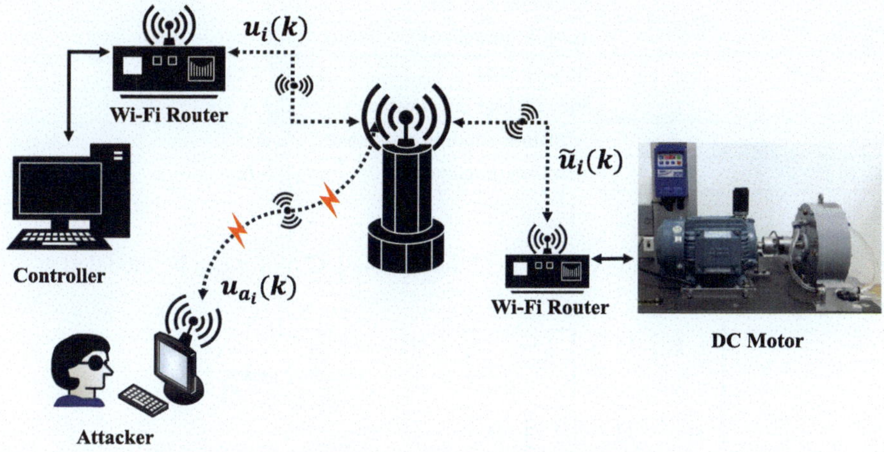

Fig. 4.4 The schematic diagram of networked DC motor under attack

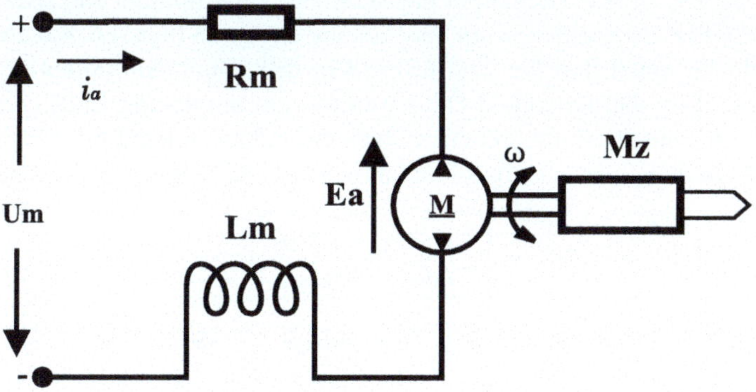

Fig. 4.5 The circuit diagram of the DC motor

Table 4.1 Networked DC Motor Specifications

Symbol	Description	Value/Unit
U_m	Armature voltage	V
E_a	Back-emf voltage	V
M_z	Load torque	$N \cdot m$
M_d	Motor drive torque	$N \cdot m$
i_a	Armature current	A
ω	Motor angular velocity	rad/s
ω_r	Desired motor angular velocity	rad/s
C_e	Back-emf constant	0.1
C_m	Torque constant	0.1
J	Rotor inertia	0.02 kg · m^2
f	Viscous braking constant	0.2 $N \cdot m \cdot s$
R_m	Armature winding resistance	2 Ω
L_m	Armature inductance	0.5 H

Through the application of electrical theory, the DC motor can be modeled as

$$\begin{cases} x(k+1) = \begin{bmatrix} -\frac{R_m}{L_m} & -\frac{C_e}{L_m} \\ \frac{C_m}{J} & -\frac{f}{J} \end{bmatrix} x(k) + \begin{bmatrix} \frac{1}{L_m} & 0 \\ 0 & -\frac{1}{J} \end{bmatrix} u(k), \\ y(k+1) = \begin{bmatrix} 0 & 1 \end{bmatrix} x(k+1), \end{cases}$$

where $x = [i_a; \omega]$ are the states of the DC motor, and $u = [U_m; M_z]$ are the control inputs for the DC motor. The output measurement, control input and error curve of the networked DC motor that is not attacked are shown in Figs. 4.6, 4.6 and 4.8.

It should be noted that the data utilized in this simulation comprises the I/O data generated by the networked DC motor system employing LQR control. The networked DC motor exhibits repetitive behavior, characterized by a cycle time of 40s. The desired angular velocity of the networked DC motor is defined as

$$\omega_r(k) = \begin{cases} 1\ rad/s, k \geq 20s, \\ 2\ rad/s, 20s < k \leq 40s. \end{cases}$$

The attacked weighting matrices are chosen as $Q = \text{diag}(1, 1)$ and $R = \text{diag}(0.06, 0.06)$, and the signal interference attenuation factor is $\eta = 0.9$. The non-switching attack direction matrix is defined as $D_a = [-1, 0.001; 0.001, 4]$, while the switching attack direction matrices are represented by $D_{a_1} = [-1, 0.001; 0, 0]$ and $D_{a_2} = [0, 0; 0.001, 4]$. The factors λ and μ are assigned values of 0.1 and 1.5, respectively. The sampling time is denoted as $\tau = 1s$. Subsequently, the designed attack strategies are implemented during a 40s simulation under 100 iterations. The simulation results are depicted in Figs. 4.8, 4.9, 4.10, 4.11, 4.12, 4.13, 4.14, 4.15, 4.16, 4.17, 4.18, 4.19, 4.20, 4.21, 4.22, 4.23, 4.24, 4.25, 4.26, and 4.27. It should be

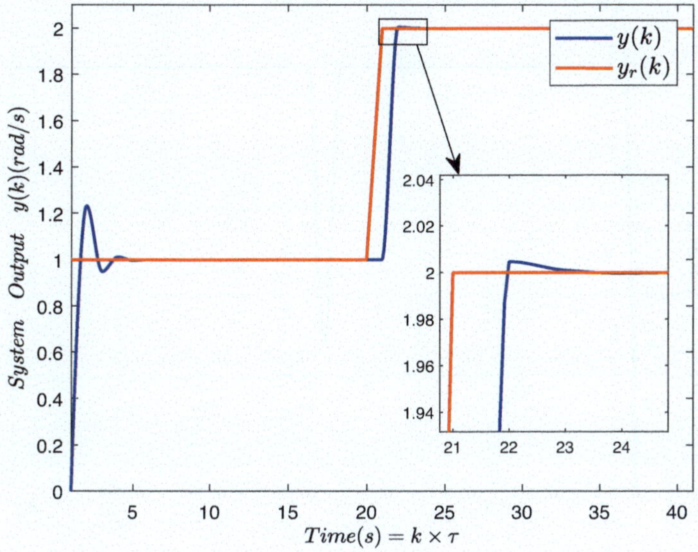

Fig. 4.6 The networked DC motor output $y(k)$ without being attacked and the desired output $y_r(k)$

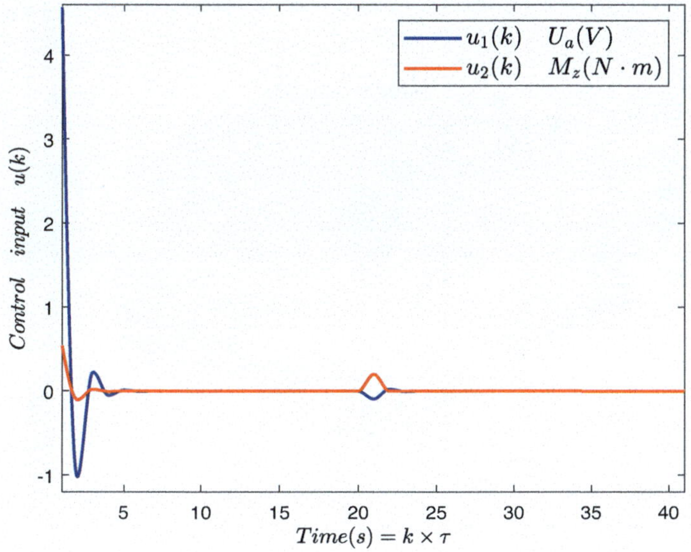

Fig. 4.7 The networked DC motor control input $u(k)$ without being attacked

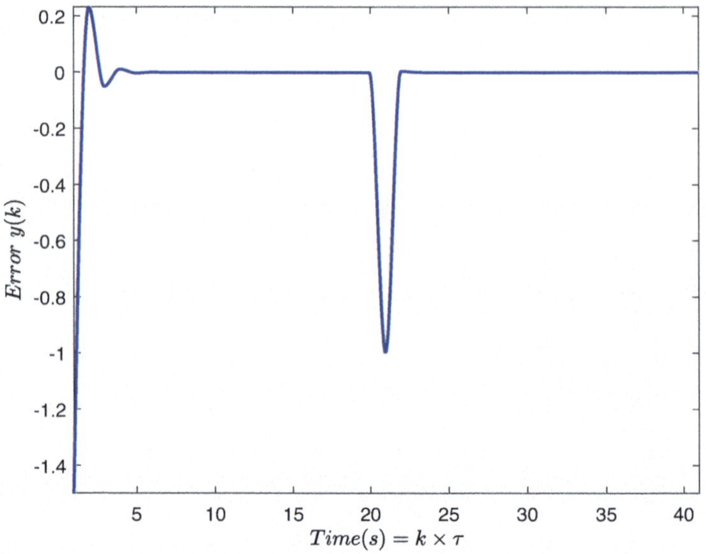

Fig. 4.8 The error between output $y(k)$ of networked DC motor system and the desired output $y_r(k)$

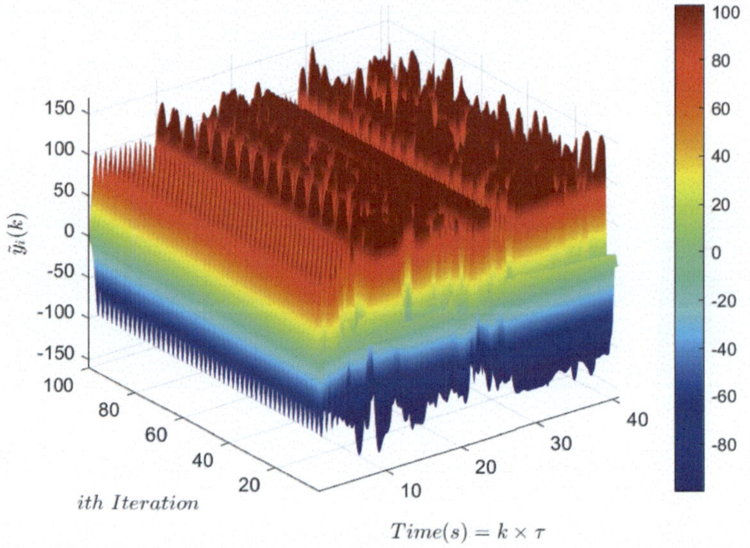

Fig. 4.9 The system output $\tilde{y}_i(k)$ under non-switching attack for the ith iteration in [0,40s]

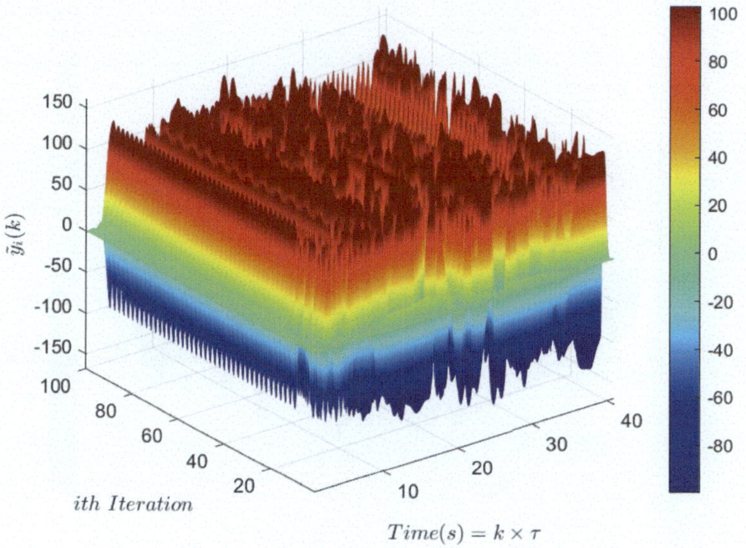

Fig. 4.10 **b** The system output $\tilde{y}_i(k)$ under switching attack for the ith iteration in [0,40s]

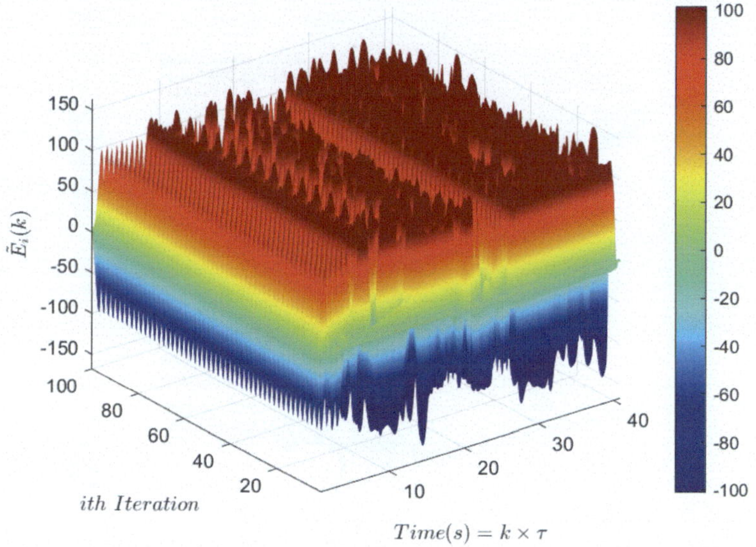

Fig. 4.11 The system output error $\tilde{E}_i(k)$ under non-switching attack for the ith iteration in [0,40s]

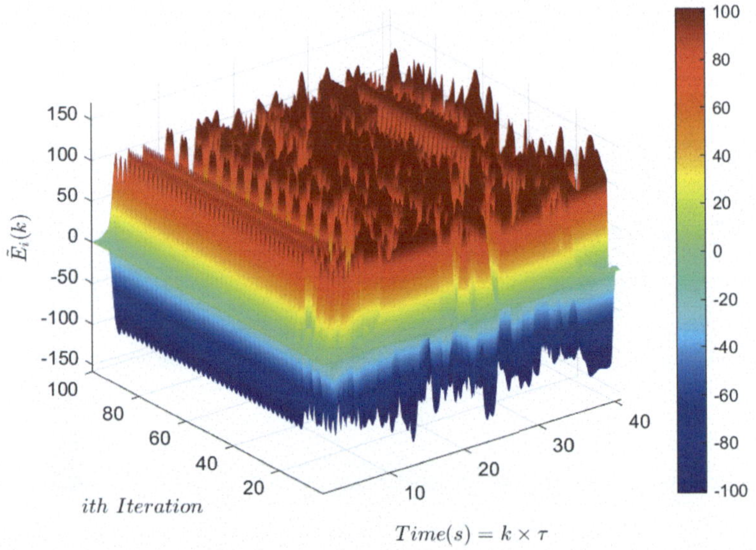

Fig. 4.12 The system output error $\tilde{E}_i(k)$ under switching attack for the ith iteration in [0,40s]

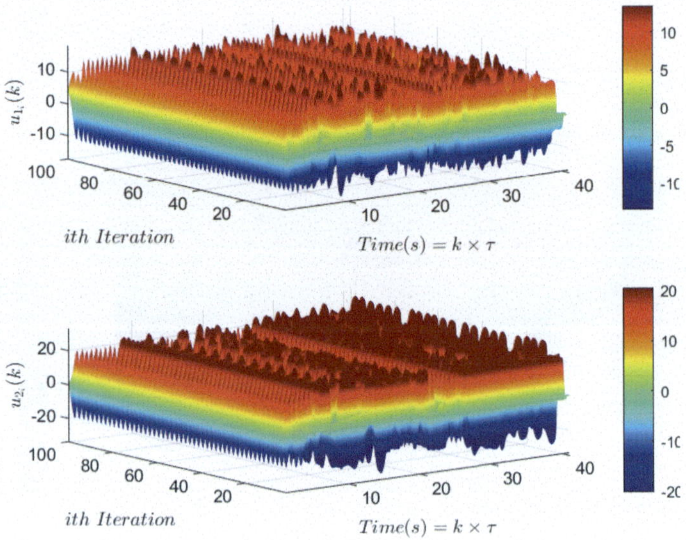

Fig. 4.13 The attacked control input $U_i(k) = [u_{1_i}(k); u_{2_i}(k)]$ under non-switching attack for the ith iteration in [0, 40s]

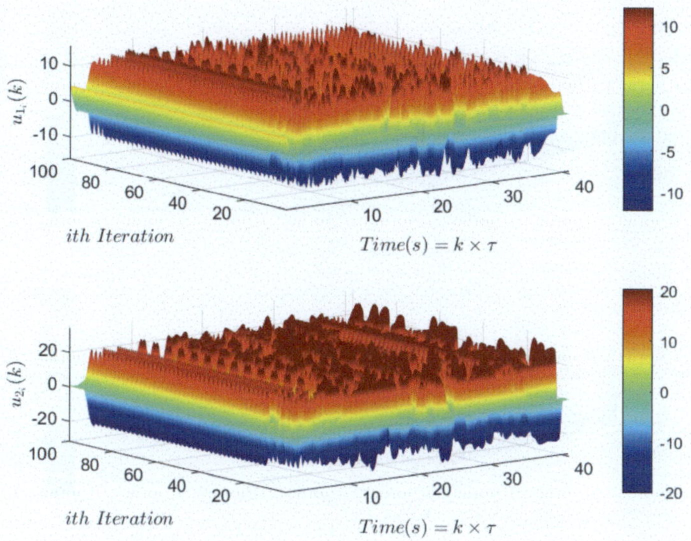

Fig. 4.14 The attacked control input $U_i(k) = [u_{1_i}(k);\ u_{2_i}(k)]$ under switching attack for the ith iteration in [0, 40s]

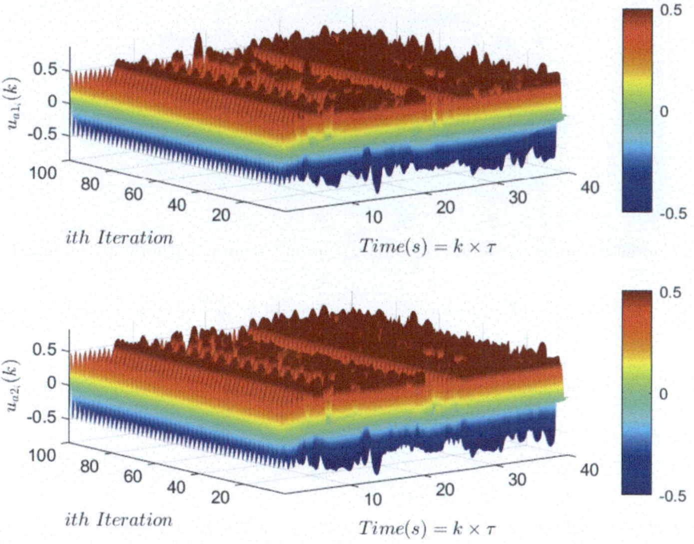

Fig. 4.15 The attacked input $U_{a_i}(k) = [u_{a1_i}(k);\ u_{a2_i}(k)]$ under non-switching attack for the ith iteration in [0, 40s]

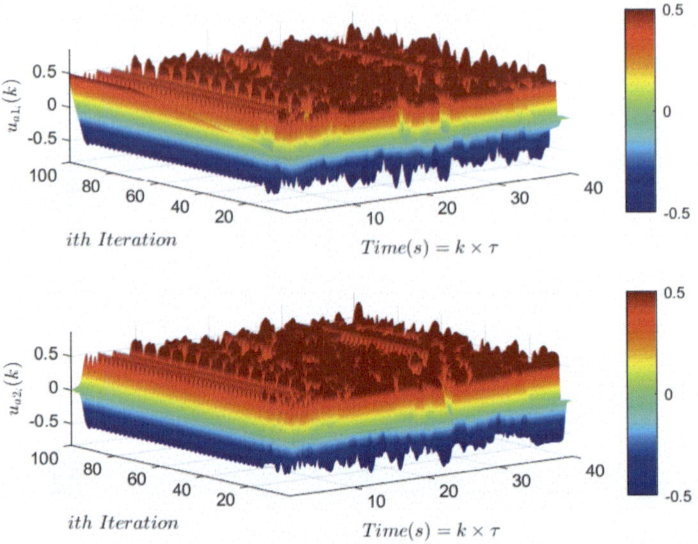

Fig. 4.16 The attacked input $U_{a_i}(k) = [u_{a1_i}(k); u_{a2_i}(k)]$ under switching attack for the ith iteration in [0, 40s]

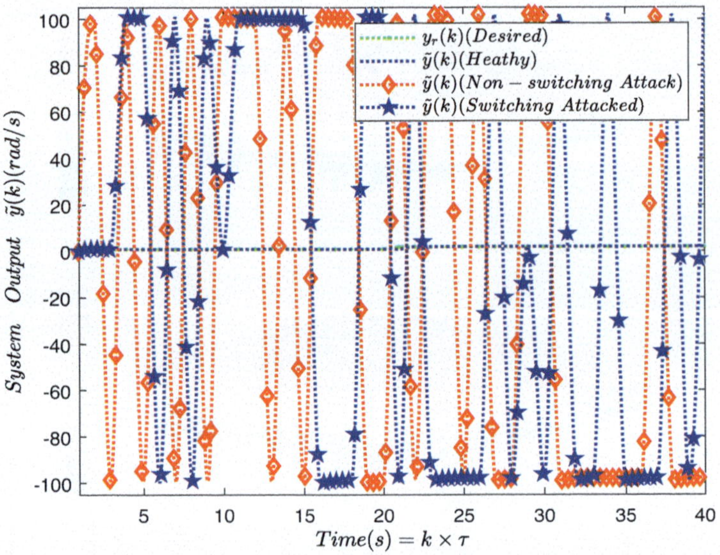

Fig. 4.17 The system output $y(k)$ without being attacked, the desired output $y_r(k)$, the system output $\tilde{y}(k)$ under the non-switching and switching attacks for the 100-th iteration in [0, 40]s

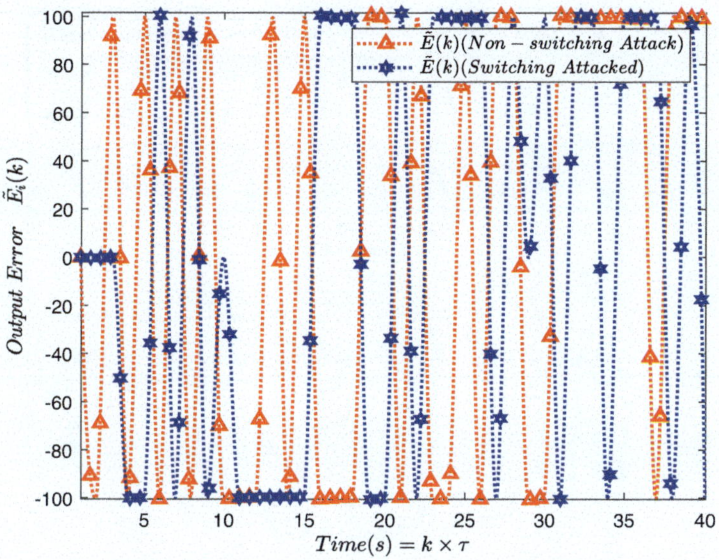

Fig. 4.18 The system output error $\tilde{E}(k)$ under the non-switching and switching attacks for the 100th iteration in [0, 40]s

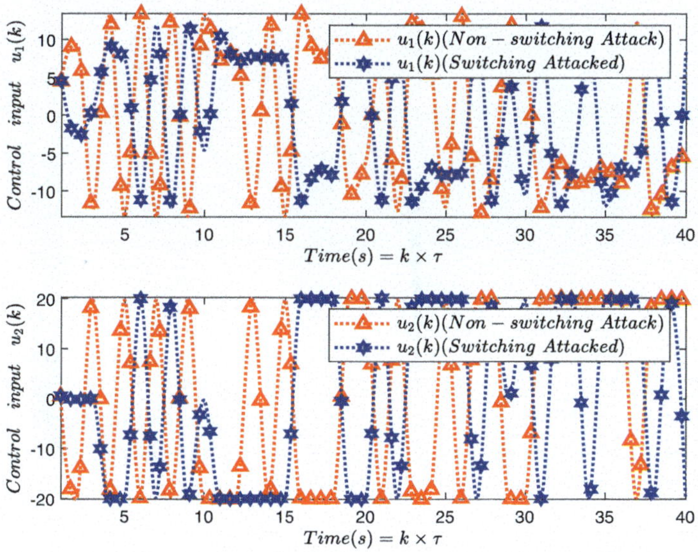

Fig. 4.19 The attacked control input $U(k) = [u_1(k); u_2(k)]$ for the 100th iteration in [0, 40]s

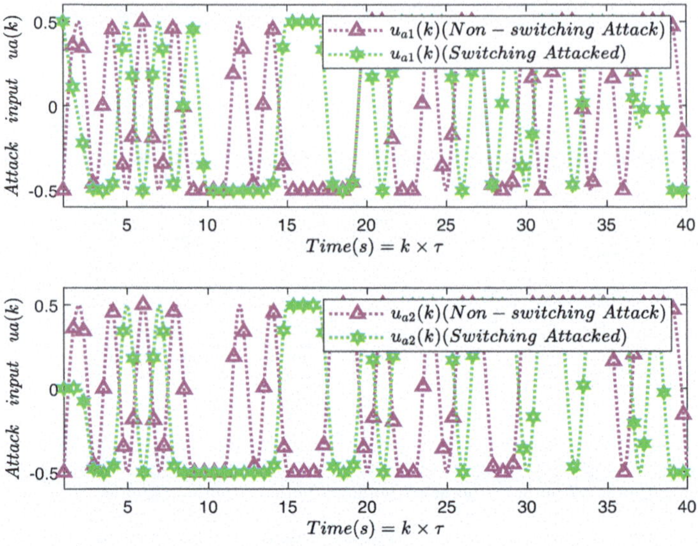

Fig. 4.20 The attacked input $U_a(k) = [u_{a1}(k); u_{a2}(k)]$ for the 100th iteration in [0, 40]s

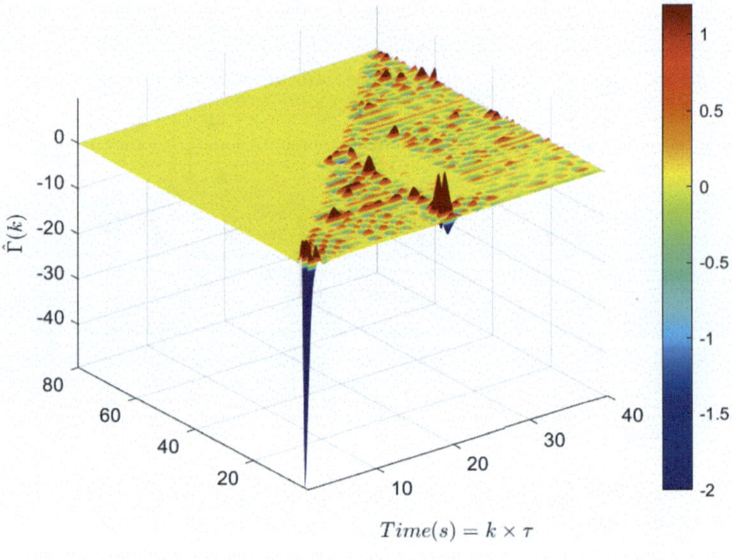

Fig. 4.21 The estimate of $\hat{\Gamma}(k)$ for the 100th iteration in [0, 40]s

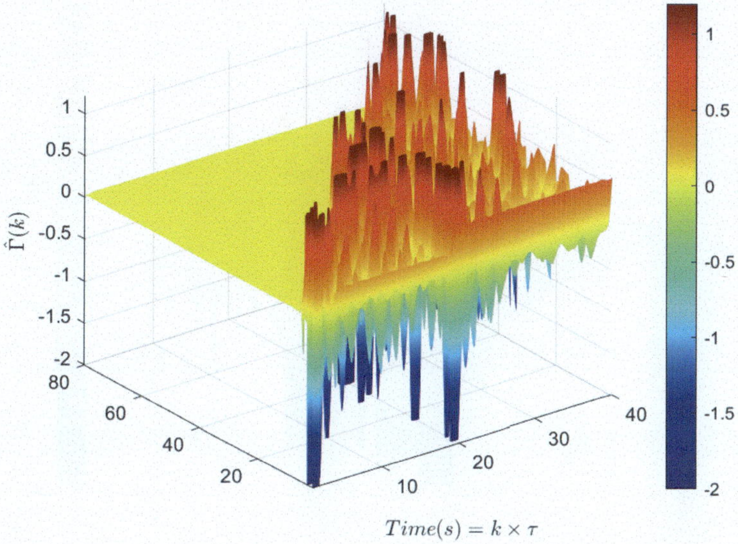

Fig. 4.22 The local amplification of estimate of $\hat{\Gamma}(k)$ for the 100th iteration in [0, 40]s

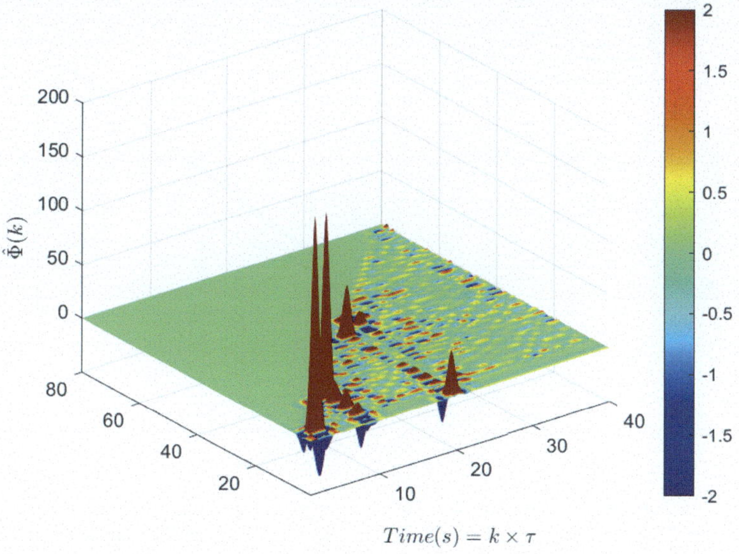

Fig. 4.23 The estimate of $\hat{\Phi}(k)$ for the 100th iteration in [0, 40]s

emphasized that attackers have limited access to system information and are only allowed to obtain system output and control input. To facilitate clarity and comprehension, the FDI attack formulated in Theorem 4.2 and the switching FDI attack developed in Theorem 4.3 are collectively denoted as the non-switching attack and the switching attack, respectively, below.

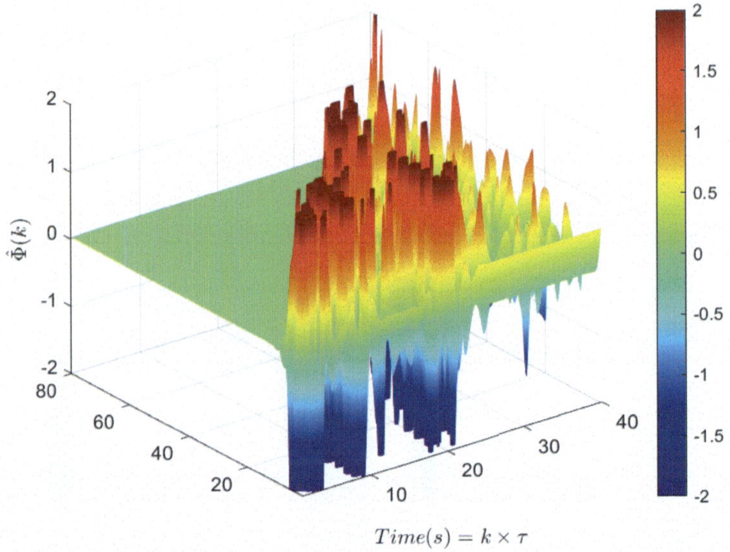

Fig. 4.24 The local amplification of estimate of $\hat{\Phi}(k)$ for the 100th iteration in [0, 40]s

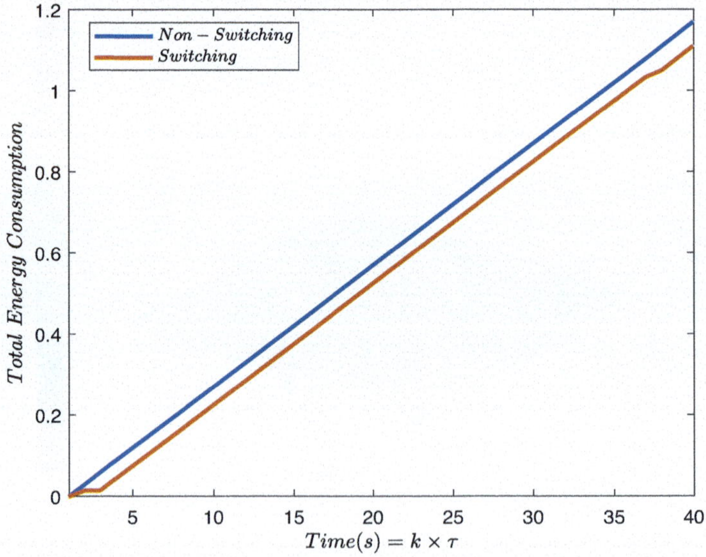

Fig. 4.25 Total energy consumption of the non-switching and switching attacks for the 100th iteration in [0, 40]s

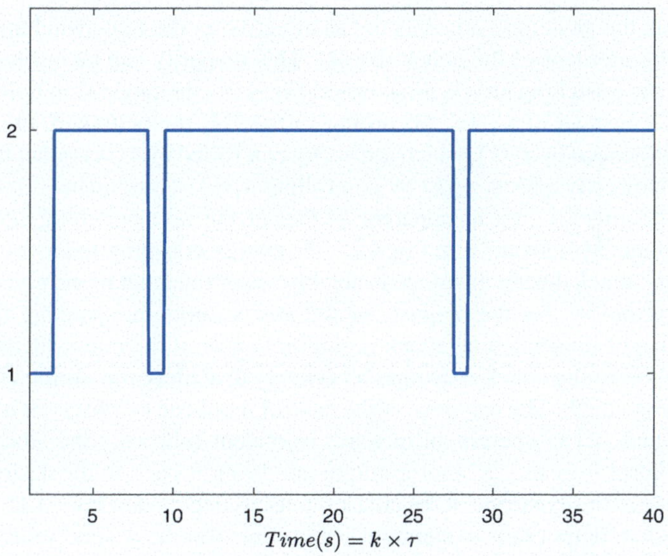

Fig. 4.26 The optimal switching sequence h^* for the 100th iteration in [0, 40]s

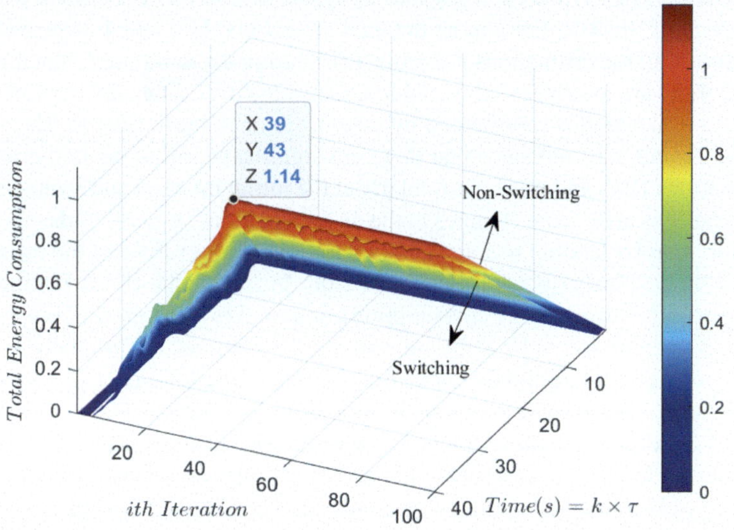

Fig. 4.27 Total energy consumption of the non-switching and switching attacks for the ith iteration in [0, 40]s

A comparison between Figs. 4.6 and 4.17, as well as Figs. 4.8 and 4.18, reveals that both the non-switching FDI attack and the switching attack exhibit similar overall effects. Specifically, the output of the DC motor deviates from the desired value, resulting in non-zero output errors. Furthermore, from Figs. 4.17 and 4.18, it can be observed that the switching attack is not as effective as the non-switching attack at the beginning moment of the attack (0∼3s). This disparity can be attributed to the fact that, in the switching attack, as shown in Fig. 4.26, the target is initially directed toward $u_1(k)$ instead of $u_2(k)$. According to the DC motor model, the output is primarily influenced by $x_2(k)$, which is directly controlled by $u_2(k)$ rather than $u_1(k)$. And $u_1(k)$ indirectly affects $x_2(k)$ by controlling $x_1(k)$. Consequently, the effect of the switching attack is not as apparent as that of the non-switching attack at the beginning stage. Moreover, from Fig. 4.25, the energy consumption associated with the switching attack during this stage is notably lower than that of the non-switching attack. Nevertheless, for the majority of the attack duration, spanning from 4s to 40s, the effect of the switching attack is similar to that of the non-switching attack. Overall, the switching attack consumes less energy to achieve the similar effect as the non-switching attack. The response of the control input can be observed in Figs. 4.7, 4.13, 4.14, and ,4.19, wherein an increase is evident following the attack, aiming to regain control over the DC motor output and bring it back to the desired output. However, due to the existence of the attack inputs, as depicted in Figs. 4.15, 4.16, and 4.20, the system output fails to maintain stable control at the desired value, resulting in significant oscillations.

Figures 4.21 and 4.23 present the estimations $\hat{\Gamma}(k)$ and $\hat{\Phi}(k)$ for the virtual linear attacked models (4.13) and (4.17), respectively, during the 100-th iteration within the time interval of [0, 40]s. For a more detailed view, Figs. 4.22 and 4.24 provide local amplifications of the estimations $\hat{\Gamma}(k)$ and $\hat{\Phi}(k)$ during the same stage. Notably, since $\hat{\Gamma}(k)$ and $\hat{\Phi}(k)$ are matrices with variable dimensions over time, any portion beyond the current dimension is represented by a zero vector in these figures. The analysis of Figs. 4.21 and 4.23 indicates that there are significant variations in the values of the estimations $\hat{\Gamma}(k)$ and $\hat{\Phi}(k)$, particularly at the initial moment and around the 20s. These variations arise from changes in the desired angular velocity of the networked DC motor, thereby leading to substantial alterations in the values of the estimations $\hat{\Gamma}(k)$ and $\hat{\Phi}(k)$. Furthermore, it is evident from Fig. 4.27 that the switching attack strategy exhibits lower energy consumption compared to the non-switching strategy throughout the duration of [0, 40]s and across 100 iterations. Additionally, a notable observation is that the optimal attack strategy, where the total energy consumption will not change, is achieved at around 43 iterations.

Table 4.2 presents a comparison between the proposed attack strategies and existing strategies [1, 2, 5]. In contrast to the non-switching injection attack strategy for linear CPSs [5], the attack strategy presented in Theorem 4.2 has an advantage in terms of the information acquisition of attackers. In other words, unlike the strategy [5] which requires the process state, control input, system output, and parameter matrices A, B, and C of the attacked system, the proposed strategy only requires the control input and system output of the attacked system transmitted through the wireless channels. Moreover, in comparison to the non-switching injection attack strategy [5], the switching attack strategy proposed in Theorem 4.3 exhibits superior adaptability, especially when all channels cannot be accessed at the same time. Furthermore, contrasted with the existing switching injection attack strategies [1, 2], the proposed switching attack strategy is based on the input and output data and gets rid

Table 4.2 The comparison between different injection attack strategies

Strategies	Switch	Effective under model free
Strategy in [1]	Yes	No
Strategy in [2]	Yes	No
Strategy in [5]	No	No
Theorem 4.2 **(ours)**	No	Yes
Theorem 4.3 **(ours)**	Yes	Yes

of the dependence on the system model. Meanwhile, the optimal switching sequence can be calculated online without the assistance of the system parameter matrices.

4.6 Conclusion

The problems of data injection attacks based on iterative learning for linear CPSs have been solved in this chapter. A class of linear discrete time-varying CPSs is considered as a physical plant, in which the system matrices $A(t)$, $B(t)$, $C(t)$ and the state $x_i(k)$ are unknown, and the attacker can only obtain the control input and system output of the attacked system. Therefore, under the framework of iterative learning theory, a virtual linear attacked system model based on input and output data is established to solve the problem of limited access to system information for the attacker. The optimal switching attack strategy and the switching sequence are obtained by applying the improved projection estimation approach and optimality principle. Finally, the effectiveness of the proposed methods is validated through the networked DC motor example. In future research work, the focus will be on the design of attack strategies for CPS in scenarios involving packet loss, delay, and asynchronous clocks without relying on the knowledge of the attacked system model.

References

1. Wu, G., Sun, J., Chen, J.: Optimal data injection attacks in cyber-physical systems. IEEE Trans. Cybern. **48**(12), 3302–3312 (2018)
2. Wu, G., Wang, G., Sun, J., Xiong, L.: Optimal switching attacks and countermeasures in cyber-physical systems. IEEE Trans. Syst. Man Cybern. -Syst. **51**(8), 4825–4835 (2021)
3. Li, Y., Shi, D., Chen, T.: False data injection attacks on networked control systems: A Stackelberg game analysis. IEEE Trans. Autom. Control **63**(10), 3503–3509 (2018)
4. Gao, S., Zhang, H., Wang, Z., Huang, C.: A class of optimal switching mixed data injection attack in cyber-physical systems. IEEE Robot. Autom. Lett. **6**(2), 1598–1605 (2021)
5. Zhang, X.G., Yang, G.H.: Optimal sensor attacks in cyber-physical systems with round-robin protocol. Inf. Sci. **548**, 85–100 (2021)
6. Li, Y.G., Yang, G.H.: Optimal stealthy innovation-based attacks with historical data in cyber-physical systems. IEEE Trans. Syst. Man Cybern. -Syst. **51**(6), 3401–3411 (2019)

7. Tanaskovic, M., Fagiano, L., Novara, C., Morari, M.: Data-driven control of nonlinear systems: An on-line direct approach. Automatica **75**, 1–10 (2017)
8. Hou, Z., Chi, R., Gao, H.: An overview of dynamic-linearization-based data-driven control and applications. IEEE Trans. Ind. Electron. **64**(5), 4076–4090 (2017)
9. Hou, Z., Xiong, S.: On model-free adaptive control and its stability analysis. IEEE Trans. Autom. Control **64**(11), 4555–4569 (2019)
10. Hou, Z., Jin, S.: Model free adaptive control: theory and applications. CRC Press, Boca Raton, FL, USA (2019)
11. Chi, R., Lv, Y., Hou, Z.: Compensation-based data-driven ILC with input and output package dropouts. Int. J. Robust Nonlinear Control **30**(3), 950–965 (2020)
12. Kumar, S., Savur, C., Sahin, F.: Survey of human–robot collaboration in industrial settings: Awareness, intelligence, and compliance. IEEE Trans. Syst. Man Cybern. -Syst. **51**(1), 280–297 (2021)
13. Wang, Z., Shen, J., Fu, X., Chen, B.: Robust adaptive optimization algorithm for stability control of electric patrol survey UAV. In: Proceedings of the 2017 International Conference on Computing Intelligence and Information System, pp. 108–115 (2017)
14. Ahn, H.S., Moore, K.L., Chen, Y.: Trajectory-keeping in satellite formation flying via robust periodic learning control. Int. J. Robust Nonlinear Control **20**(14), 1655–1666 (2010)
15. Pasqualetti, F., Dörfler, F., Bullo, F.: Attack detection and identification in cyber-physical systems. IEEE Trans. Autom. Control **58**(11), 2715–2729 (2013)
16. Zhang, Z., Deng, R., Cheng, P., Wei, Q.: On feasibility of coordinated time-delay and false data injection attacks on cyber–physical systems. IEEE Internet Things J. **9**(11), 8720–8736 (2021)
17. Mikhaylenko, D., Zhang, P.: Stealthy local covert attacks on cyber-physical systems. IEEE Trans. Autom. Control **67**(12), 6778–6785 (2022)
18. Pang, Z.H., Ma, B., Liu, G.P., Han, Q.L.: Data-driven adaptive control: An incremental triangular dynamic linearization approach. IEEE Trans. Circuits Syst. II-Express Briefs **69**(12), 4949–4953 (2022)
19. Gao, S., Zhang, H., Wang, Z., Huang, C., Yan, H.: Optimal injection attack for UCPS under vertical depth-keeping task via game approach. IEEE/CAA J. Autom. Sin. **10**(1), 287–289 (2023)
20. Lewis, F.L., Vrabie, D., Syrmos, V.L.: Optimal Control. Wiley (2012)
21. Henderson, H.V., Searle, S.R.: On deriving the inverse of a sum of matrices. SIAM Rev. **23**(1), 53–60 (1981)
22. Binmore, K.G., Davies, J.: Calculus: Concepts and Methods. Cambridge University Press (2001)
23. Verbeke, J., Cools, R.: The newton-Raphson method. Int. J. Math. Educ. Sci. Technol. **26**(2), 177–193 (1995)

Chapter 5
Data-Driven FDI Attack Strategy Against CPSs with Slowly Time-Varying Delays

Abstract This chapter addresses data-driven injection attack against unknown cyber-physical systems with slowly time-varying delays, which is a more general but also more challenging to deal with than the model-based real-time systems. Using the control input and system output measurements, several new data-driven injection attack strategies based on compact form dynamic linearization (CFDL) and incremental triangular dynamic linearization (ITDL) are proposed. The attack strategies are more general than the existing ones, taking into account the unknown model parameters and time-varying delays in control-to-actuator as well as sensor-to-controller data transmission channels. Consequently, the new design attack results are anticipated to have wider applicability. Based on the established attack models of CFDL and ITDL, the optimal data-driven parameter estimation algorithms are employed to overcome the difficulty of the unknown model. Furthermore, with the help of the principle of the linear regression equation, the problem of seeking partial derivatives for the attack inputs with time delays is avoided. Several examples are presented to illustrate the validity of the designed attack strategies.

5.1 Introduction

The injection attacks from the control perspective are primarily investigated in this chapter. The attacker mainly injects elaborate false data into the attacked system to compromise the control performance of the system. The optimal injection attack strategy adopting quadratic control performance is proposed in [1], where the attacker can use minimal energy consumption to maximize deviation from the state of the attacked system. Based on this, the injection attack strategy that considers a form of switching is investigated in [2], where the form of switching makes the attack more stealthy. Further, a spurious injection attack strategy for state estimation is proposed in [3] to achieve maximum deterioration of the state estimation performance for the attacked system. The design of these injection attack strategies is highly dependent on the precise model of the attacked system and has application limitations. For this reason, the data-driven injection attack applying the input and output data was devised in Chap. 4, as well as the stealthy sensor and actuator attack in T-S fussy

S. Gao et al., *Cyberattack Strategies Against Cyber-Physical Systems: From a Control Perspective*, https://doi.org/10.1007/978-981-95-9645-4_5

systems [4], these problems are also not discussed in terms of data transmission delay. Driven by these observations, how to design optimal injection attack strategies for the unknown cyber-physical systems with time-varying delays is one of our main motivations.

Time delay is a frequent phenomenon in data transmission through wireless communications, caused by the number of users accessing the system, or by the environment in which the system is located, e.g., tunnels, underwater, etc. There are many existing works on the control under time delay scenarios, as detailed in [5–9]. Among the relevant studies, the fixed time delay problem [5] is known to be the simplest, which is an idealization of the time delay scenario and is hardly desirable in practical applications. Further, the unilateral time-varying delays are often investigated, e.g., for controller-to-sensor [5] or sensor-to-controller [6, 10] transmission channels. In practice, however, both controller-to-sensor and sensor-to-controller transmission channels are subject to time delays. Unfortunately, to date, there have been fewer results regarding the design of cyberattack strategies from the control viewpoint. Time-varying delays include both slowly and random time-varying delays [7, 8, 11], and this chapter focuses on the case of slowly time-varying delays. Because there is packet disorder when the time-varying delay is generated at any time, this is a huge challenge for both control and attack when no timestamp bit is introduced. When timestamp bits are introduced, the strategy designed in this chapter remains valid and feasible. This motivates us to investigate injection attack strategies against the unknown cyber-physical systems with slowly time-varying delays.

Motivated by the above, the new data-driven attack strategies against the unknown discrete-time cyber-physical systems with slowly time-varying delays are developed in this chapter, which maximizes the deviation of the system output with low energy consumption. The main contributions of this chapter are summarized as follows:

(1) The data-driven attacked discrete-time cyber-physical system models with slowly time-varying delays via CFDL and ITDL are constructed. In contrast to the data-driven models in [12] and Chap. 4, more general data-driven attacked models with slowly time-varying delays are obtained. Meanwhile, comparing to the unilateral time-varying delays studied in [10], bilateral time-varying delays in the controller-to-actuator and sensor-to-controller channels are considered.
(2) A optimal data-driven injection attack strategy based on CFDL is proposed. Compared to existing attack strategies [2, 3], the proposed one breaks away from the dependence on the model of the attacked system. Furthermore, it can guarantee the optimality of the injection attack in the presence of time-varying delays within the data transmission network.
(3) A optimal data-driven injection attack strategy based on ITDL is proposed. Different from the CFDL attack strategy, the designed ITDL attack strategy is more appropriate in the situation where the memory occupation space and the hashrate of the attacker are constrained. Further, the requirements for the devices equipped by attackers have been relaxed.

The remaining sections are organized as follows. The problem formulation about the discrete-time cyber-physical systems with slowly time-varying delays, the data-driven attacked system models via CFDL and ITDL, and the objective function of the attacker are specified in Sect. 5.2. In Sect. 5.3, the attack problem transformation via CFDL and ITDL methods is presented. The strategies of data-driven attack based on CFDL and ITDL are developed in Sect. 5.4. In Sect. 5.5, numerical simulation and experiment based on a leader-following form consisting of five quadcopters illustrate the efficiency of the proposed strategies. Finally, this chapter draws the conclusions in Sect. 5.6.

5.2 Problem Formulation

5.2.1 System Model

Consider a class of discrete-time cyber-physical systems with slowly time-varying delays having repetitive operation characteristics under the iterative situation depicted as

$$\begin{cases} x_k(t+1) = A(t)x_k(t) + B(t)u_k(t-\tau(t)), \\ y_k(t+1) = C(t+1)x_k(t+1), \end{cases} \tag{5.1}$$

where $x_k(t) \in \mathbb{R}^n$, $u_k(t) \in \mathbb{R}^m$, and $y_k(t) \in \mathbb{R}^p$ are the unknown system state, the control input, and the measurable output in the kth iteration at time t. $A(t) \in \mathbb{R}^{n\times n}$, $B(t) \in \mathbb{R}^{n\times m}$, and $C(t) \in \mathbb{R}^{p\times n}$ are the unknown system parameter matrices. $\tau(t) \in \mathbb{N}^+$ is the unknown slowly time-varying delays and $0 < \underline{\tau} \leq \tau(t) \leq \overline{\tau}$, $\overline{\tau}$ and $\underline{\tau}$ are the upper and lower bounds for the time delay as well as the known positive integers. System (5.1) is asymptotically stable, and its initial system state of each iteration is $x_k(0) = x_0$, where x_0 is an unknown constant vector.

The cyber-physical system (5.1) is typically employed in the scenarios such as robotic arm teaching, industrial manufacturing, and artificial satellite motion. For the system (5.1), the following assumptions are given.

Assumption 5.1 ([8, 10, 12]) The norm of the unknown parameter matrices in the system (5.1) is bounded, that is, there are non-negative constants $\mathcal{A}$, $\mathcal{B}$, and C, such that $A(t)$, $B(t)$, and $C(t)$ satisfy $\|A(t)\| \leq \mathcal{A}$, $\|B(t)\| \leq \mathcal{B}$, and $\|C(t)\| \leq C$.

Assumption 5.2 (*Slowly Time-Varying Delays* [7, 8]) There exists a set $H = \{h(j)|h(j+1) = h(j)+1, h(1) = \underline{\tau}, h(v) = \overline{\tau}, h(j) \in \mathbb{N}^+, v \in \mathbb{N}^+, j = 1, 2, ..., v-1\}$, the slowly time-varying delay $\tau(t)$ satisfies the following constraints:

(i) For each $j = 1, 2, ..., v$, there exists $t \in \mathbb{N}$ such that $\tau(t) = h(j)$.
(ii) If $\tau(t) = h(1)$, then $\tau(t+1)$ can be $h(1)$ or $h(2)$; If $\tau(t) = h(j)$, $j = 2, 3, .., v - 1$, then $\tau(t+1)$ can be $h(j-1)$, $h(j)$, or $h(j+1)$; If $\tau(t) = h(v)$, then $\tau(t+1)$ can be $h(v-1)$ or $h(v)$.

Remark 5.1 Assumption 5.1 is a constrained description of the unknown parameter matrices for the discrete-time cyber-physical systems, and it is reasonable for practical physical processes with repetitive operational objects, such as unmanned aerial vehicle (UAV) formation, and manipulator welding. The time-varying delays investigated in this chapter are illustrated in Assumption 5.2, and the comparable description can be found in the literature [7, 8]. Such slowly time-varying delays have been extensively investigated in both discrete-time and continuous-time systems [7, 8, 14, 15], and these types of delays are frequently encountered in practical applications. For instance, as the usage time of a network device increases, its communication transmission module undergoes aging, resulting in the gradual emergence of a slow-growing delay over time.

5.2.2 Attacked Models

The attack framework against discrete-time cyber-physical systems is demonstrated in Fig. 5.1. The remote control center sends the control input signal to the plant by means of wireless network, and the plant utilizes the same way to deliver the system output to the remote control center. Time delays are frequently generated during data transmission due to factors that cannot be ignored, such as external interference and the interaction rate limitations of the network server itself. Time-varying delays are more realistic than the fixed delays. Thus, time-varying delay $\tau(t)$ is taken into account as displayed in Fig. 5.1. Furthermore, this chapter considers how to conduct a false data attack from the attacker's view. Normally, attackers are unable to access the system protection data of the victim system, for instance, the system parameter matrices and the internally transmitted state data. The attacker can steal and tamper with the data transmitted exploiting the wireless channel, as indicated in Fig. 5.1, where the attacker eavesdrops and stores the transmitted control input $u_k(t - \tau(t))$ and the measurable system output $y_k(t - \tau(t) + 1)$. Finally, the attacker generates attack strategy through data eavesdropping and the estimation of the system attribute matrix.

For the attacker, the following assumption is given.

Assumption 5.3 The attacker has a synchronous clock with the remote control center, and the attacker can tamper with the control input as well as inject the designed data to the attacked plant, simultaneously.

Remark 5.2 Assumption 5.3 provides a description of the clock relationship between the attacker and the attacked system as well as the capability of the attacker to compromise a target system. The focus of this chapter is to design the injection

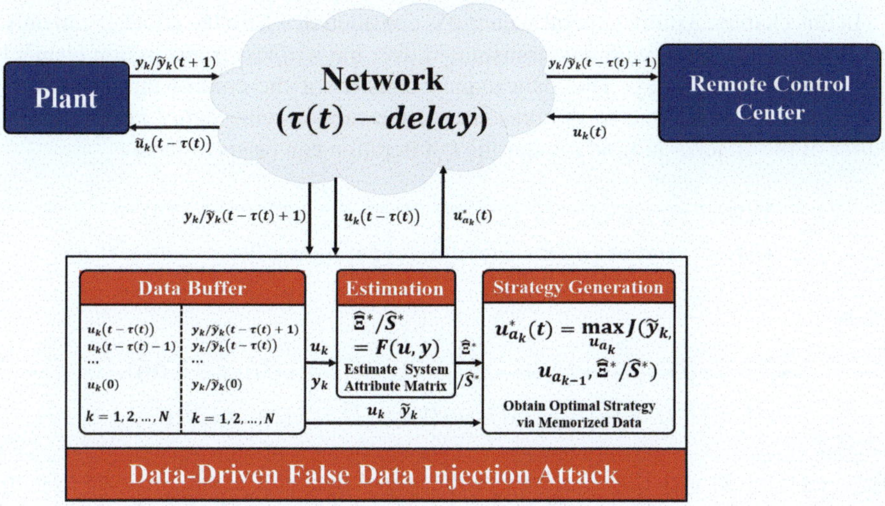

Fig. 5.1 Schematic diagram of discrete-time cyber-physical systems under data-driven injection attack

attack strategy under synchronous clock from the perspective of control. And both the controller-to-actuator and sensor-to-controller connections utilize the same network, which incorporates $\tau(t)$ for data transmission, so the attacker can realize Assumption 5.3 via cover attack technology [16], and similar assumptions can be seen in literature [2, 13].

The control input signal with time-varying delay received by the plant under the injection attack can be expressed as

$$\tilde{u}_k(t-\tau(t)) = u_k(t-\tau(t)) + \Gamma^a(t-\tau(t))u_{a_k}(t-\tau(t)), \tag{5.2}$$

where $\tilde{u}_k(t-\tau(t))$ and $u_{a_k}(t-\tau(t))$, the m-dimension control input after the data injection attack and the q-dimension injection attack input with $\tau(t)$-delay, respectively, are received by the plant at time t under the kth iteration. And the initial value of the attack input is $u_{a_1}(0) = 0$. $\Gamma^a(t) \in \mathbb{R}^{m\times q}$ stands for the attack direction matrix at time t.

Subsequently, the attacked control input (5.2) is brought into the system (5.1), and the attacked model can be formulated as

$$\begin{cases} \tilde{x}_k(t+1) = A(t)\tilde{x}_k(t) + B(t)u_k(t-\tau(t)) + B(t) \\ \qquad\qquad \times\Gamma^a(t-\tau(t))u_{a_k}(t-\tau(t)), \\ \tilde{y}_k(t+1) = C(t+1)\tilde{x}_k(t+1), \end{cases} \tag{5.3}$$

where $\tilde{x}_k(t)$ and $\tilde{y}_k(t)$ are the attacked state and output in the kth iteration at time t, respectively.

In this chapter, a more general scenario is considered, where the attacker can only eavesdrop and tamper with data transmitted over the wireless transmission channel with time-varying delay. This data mainly consists of the control input data and system output data. Due to the presence of the time-varying delay $\tau(t)$, the data buffer of the attacker at time t under the kth iteration can be specified as

$$\begin{aligned}\tilde{\mathbf{Y}}_k(t-\tau(t)+1) &= [\tilde{\mathbf{y}}_k(t-\tau(t)+1); \tilde{\mathbf{y}}_{k-1}(T); ...; \tilde{\mathbf{y}}_1(T)],\\ \mathbf{U}_k(t-\tau(t)) &= [\mathbf{u}_k(t-\tau(t)); \mathbf{u}_{k-1}(T); ...; \mathbf{u}_1(T)],\end{aligned}$$

where

$$\begin{aligned}\tilde{\mathbf{y}}_k(t-\tau(t)+1) &= [\tilde{y}_k(t-\tau(t)+1); \tilde{y}_k(t-\tau(t)); ...; \tilde{y}_k(0)],\\ \mathbf{u}_k(t-\tau(t)) &= [u_k(t-\tau(t)); u_k(t-\tau(t)-1); ...; u_k(0)].\end{aligned}$$

$\tilde{\mathbf{Y}}_k(t-\tau(t)+1)$ and $\mathbf{U}_k(t-\tau(t))$ are the stored data sets of the attacked system output and the control input at time t under the kth iteration. $\tilde{y}_k(t)$ is defined as the attacked system output at time t under the kth iteration. T represents the cycle time for the system to perform repetitive operation.

5.2.3 *Optimal FDI Attack Problem with Slowly Time-Varying Delays*

In this subsection, the research objective and problem of this chapter are elaborated. The main research objective is that the attacker utilizes the accessible input and output data of the attacked system, specifically the communication data of the wireless network with slowly time-varying delays denoted as $\tau(t)$, to design the optimal injection attack strategy against the cyber-physical system with unknown system parameters. In the presence of a time-varying delay $\tau(t)$, the data-driven injection attack strategy should be designed to ensure that the minimum energy consumption of the attacker is employed to cause the maximum offset in the output of the attacked cyber-physical system under maximum time delay. The objective function is specified as

$$\begin{aligned}J_1(\tilde{y}_k(t+1), u_{a_k}(t-\overline{\tau})) =&\frac{1}{2}(\tilde{y}_k^T(t+1)Q\tilde{y}_k(t+1) - \epsilon^2 u_{a_k}^T(t-\overline{\tau})\mathcal{R}\\ &\times u_{a_k}(t-\overline{\tau})),\end{aligned} \tag{5.4}$$

where $Q \in \mathbb{R}^{p\times p}$ and $\mathcal{R} \in \mathbb{R}^{m\times m}$ are semi-positive and positive definite matrices, and $0 \leq \epsilon \leq 1$ stands for the specified constant interference attenuation level.

Subsequently, combining the attacked system (5.3) and the objective function (5.4), the main FDI attack problem with slowly time-varying delays of this chapter can be described as

Problem 5.1

$$\max_{u_{a_k}(t-\overline{\tau})} \quad J_1(\tilde{y}_k(t+1), u_{a_k}(t-\overline{\tau}))$$

$$\text{s.t.} \quad \begin{cases} \tilde{x}_k(t+1) = A(t)\tilde{x}_k(t) + B(t)u_k(t-\tau(t)) + B(t) \\ \qquad\qquad \times\Gamma^a(t-\tau(t))u_{a_k}(t-\tau(t)), \\ \tilde{y}_k(t+1) = C(t+1)\tilde{x}_k(t+1), \\ \mathcal{Q} \succeq 0, \mathcal{R} \succ 0. \end{cases}$$

Remark 5.3 It is worth mentioning that due to the existence of time-varying delays $\tau(t)$, the attacker needs to ensure the performance of the attack under the maximum delay exists. Therefore, when designing the attack value function J_1, the injection attack offsets the measurable output of the attacked cyber-physical system under the maximum time delay $\overline{\tau}$. In addition, similar to Problems 4.1 and 4.2, Problem 5.2 cannot be solved because the attacker does not know the system state $\tilde{x}_k(t)$, time-varying delay $\tau(t)$, and system parameter matrices $A(k)$, $B(k)$, and $C(k)$. To overcome this difficulty, the next section constructs the virtual linear attack models based on CFDL and ITDL forms through I/O data, and uses this data-driven model to transform the unsolvable problems mentioned above.

5.3 Attack Problem Transformation via CFDL and ITDL Methods

This section mainly elaborates on the data-driven virtual linear attacked models via CFDL and ITDL methods, with the aim of solving the Problem 5.1 that cannot be solved due to unknown system states, time-varying delays $\tau(t)$, and system parameter matrices $A(k)$, $B(k)$, and $C(k)$, as well as the transformation of the original attack problem based on these models.

5.3.1 Virtual Linear Attacked Model via CFDL

Joining the data stored by the attacker and the attacked system (5.3), the data-driven attacked discrete-time cyber-physical systems with time-varying delays through CFDL method can be described as Lemma 5.1.

Lemma 5.1 *A data-driven attacked discrete-time cyber-physical system model with $\tau(t)$ via CFDL is constructed as*

$$\begin{aligned} \Delta\tilde{y}_k(t-\tau(t)) =& \Xi(t-\tau(t)-1)\Delta\boldsymbol{u}_k(t-2\tau(t)-1) + \Xi(t-\tau(t)-1) \\ & \times \bar{\Gamma}^a(t-2\tau(t)-1)\Delta\boldsymbol{u}_{a_k}(t-2\tau(t)-1), \end{aligned} \tag{5.5}$$

and the norm of the unknown matrix $\Xi(t)$ is bounded, where

$$\bar{\Gamma}^a(t-2\tau(t)-1) = \mathrm{diag}\{\Gamma^{\mathrm{a}}(\mathrm{t}-2\tau(\mathrm{t})-1), \ldots, \Gamma^{\mathrm{a}}(1), \Gamma^{\mathrm{a}}(0)\},$$
$$\boldsymbol{u}_{a_k}(t-2\tau(t)-1) = [u_{a_k}(t-2\tau(t)-1); \ldots; u_{a_k}(1); u_{a_k}(0)],$$
$$\Xi(t-\tau(t)-1) = [C(t-\tau(t))B(t-\tau(t)-1), C(t-\tau(t))A(t-\tau(t)-1)B(t - \tau(t)-2), \ldots, \prod_{i=0}^{t-\tau(t)-1} C(t+1)A(t-i)B(\tau(t))],$$

$\bar{\Gamma}^a(t-2\tau(t)-1) \in \mathbb{R}^{[m*(t-2\tau(t))]\times[q*(t-2\tau(t))]}$, $\boldsymbol{u}_{a_k}(t-2\tau(t)-1) \in \mathbb{R}^{[q*(t-2\tau(t))]\times 1}$, $\Xi(t-\tau(t)-1) \in \mathbb{R}^{p\times[m*(t-2\tau(t))]}$, *the symbol* Δ *is defined as the difference operator between the kth iteration and the* $(k-1)$*th iteration,* $\Delta\tilde{y}_k(t) = \tilde{y}_k(t) - \tilde{y}_{k-1}(t)$, $\Delta\boldsymbol{u}_k(t) = \boldsymbol{u}_k(t) - \boldsymbol{u}_{k-1}(t)$, *and* $\Delta\boldsymbol{u}_{a_k}(t) = \boldsymbol{u}_{a_k}(t) - \boldsymbol{u}_{a_{k-1}}(t)$.

Proof The proof can refer to [10, 13], thus it is omitted. □

Remark 5.4 Lemma 5.1 is the attacked cyber-physical systems model with a compact form of dynamic linearization, which is built by the attacker utilizing the stored data with time-varying delays. Since the parameter matrices of the system, $A(t)$, $B(t)$, and $C(t)$ are unknown, $\Xi(t)$ characterizing the system properties is uncharted for the attacker. Furthermore, the time-varying delay $\tau(t)$ is also unknown, which prevents attackers from exploiting the data-driven attacked cyber-physical systems (5.5) for attack strategy generation.

As the matrix $\Xi(t)$ is affected by the system itself, in order to overcome the problem posed by uncertain $\Xi(t)$ and $\tau(t)$, the attacker first performs data stealing and estimation for the unknown $\Xi(t)$. At this stage the attacker does not perform the injection attack on the cyber-physical systems. Due to the existence of the time delay, the attacker needs to guarantee the estimation performance of the unknown matrix $\Xi(t)$ under the maximum time delay. Thus, the optimal problem for the estimation of $\Xi(t)$ is designed as

Problem 5.2

$$\begin{aligned}\min_{\hat{\Xi}_k(t)} \quad & J_2(\hat{\Xi}_k(t)) = \|\Delta y_{k-1}(t+1) - \hat{\Xi}_k(t)\Delta\mathbf{u}_{k-1}(t-\overline{\tau})\|^2 \\ & + \alpha\|\hat{\Xi}_k(t) - \hat{\Xi}_{k-1}(t)\|^2, \\ \text{s.t.} \quad & \Delta\hat{y}_k(t+1) = \hat{\Xi}_k(t)\Delta\mathbf{u}_k(t-\overline{\tau}), \alpha > 0,\end{aligned}$$

where $\hat{\Xi}_k(t) \in \mathbb{R}^{p\times[m*(t-\overline{\tau}+1)]}$ is the estimation of $\Xi(t)$ at time t under the kth iteration and α is an adjustment factor.

Lemma 5.2 ([10, 12, 13]) *By solving Problem 5.2, the optimal estimation of* $\Xi(t)$ *is described as*

$$\begin{aligned}\hat{\Xi}_k^*(t) =& \hat{\Xi}_{k-1}^*(t) + (\Delta y_{k-1}(t+1) - \hat{\Xi}_{k-1}^*(t)\Delta\boldsymbol{u}_{k-1}(t-\overline{\tau}))\Delta\boldsymbol{u}_{k-1}^T(t-\overline{\tau}) \\ & \times [(\alpha + \|\Delta\boldsymbol{u}_{k-1}(t-\overline{\tau})\|^2)I - \Delta\boldsymbol{u}_{k-1}(t-\overline{\tau})\Delta\boldsymbol{u}_{k-1}^T(t-\overline{\tau})]/(\alpha^2 \\ & + \alpha\|\Delta\boldsymbol{u}_{k-1}(t-\overline{\tau})\|^2),\end{aligned} \quad (5.6)$$

where the identity matrix with the suitable dimensions is defined as I.

The optimal estimation algorithm for $\Xi(t)$ can be acquired from Lemma 5.2 as follows.

Algorithm 4: The Optimal Estimation Algorithm of $\Xi(t)$ via CFDL

Initialize: Choose arbitrary initial admissible value $\hat{\Xi}_1(t)$. Let $\alpha > 0$ and $\beta > 0$ be the suitable adjustment factor and the sufficiently small computational precision and the maximum number of iterations N.
for $k = 2$ *to* N **do**
    Step 1. Update the matrix $\hat{\Xi}_{k-1}(t)$, $\Delta y_{k-1}(t+1)$, and $\Delta \mathbf{u}_{k-1}(t-\overline{\tau})$;
    Step 2. Compute the matrix $\hat{\Xi}_{k-1}(t)$ using (5.6);
    if $\|\hat{\Xi}_k(t) - \hat{\Xi}_{k-1}(t)\|^2 \leq \beta$ is satisfied **then**
        $\hat{\Xi}^*(t) = \hat{\Xi}_k(t)$;
    else
        Continue;
    Step 3. Store the matrix $\hat{\Xi}^*(t)$.
Return: $\hat{\Xi}^*(t)$.

Remark 5.5 Algorithm 4 is the estimation of $\Xi(t)$ in the developed data-driven model (5.5), which is a critical step before the attack strategy generation, and it releases the attacker from reliance on the model. Furthermore, the estimation of $\Xi(t)$ in Algorithm 4 is performed with the maximum delay, which is advantageous to ensure the estimation performance under the time-varying delay $\tau(t)$.

5.3.2 *Virtual Linear Attacked Model via ITDL*

Combining with the data stored by the attacker and the attacked system (5.3), the data-driven attacked discrete-time cyber-physical systems with slowly time-varying delays through ITDL method can be written as Lemma 5.3.

Lemma 5.3 *A data-driven attacked discrete-time cyber-physical system model with $\tau(t)$-delays via ITDL is developed as*

$$\Delta \tilde{y}_k(t-\tau(t)) = S(t-\tau(t)-1)\Delta\tilde{\varpi}_k(t-2\tau(t)-1), \tag{5.7}$$

and the norm of the unknown matrix $S(t-\tau(t)-1) \in \mathbb{R}^{p\times m}$ is bounded, where $\tilde{\varpi}_k(t-2\tau(t)-1) \triangleq \frac{1}{\gamma+\tau(t)}\sum_{j=1}^{\gamma+\tau(t)} j[u_k(t-2\tau(t)-j)+\Gamma^a(t-2\tau(t)-j)u_{a_k}(t-2\tau(t)-j)] + \frac{1}{\Upsilon-\gamma}\sum_{j=\gamma+\tau(t)+1}^{\Upsilon+\tau(t)}(\Upsilon-j)[u_k(t-2\tau(t)-j)+\Gamma^a(t-2\tau(t)-j)u_{a_k}(t-2\tau(t)-j)] \in \mathbb{R}^m$, $0<\gamma<\Upsilon$, and $\Upsilon \geq \tau(t)+2$.

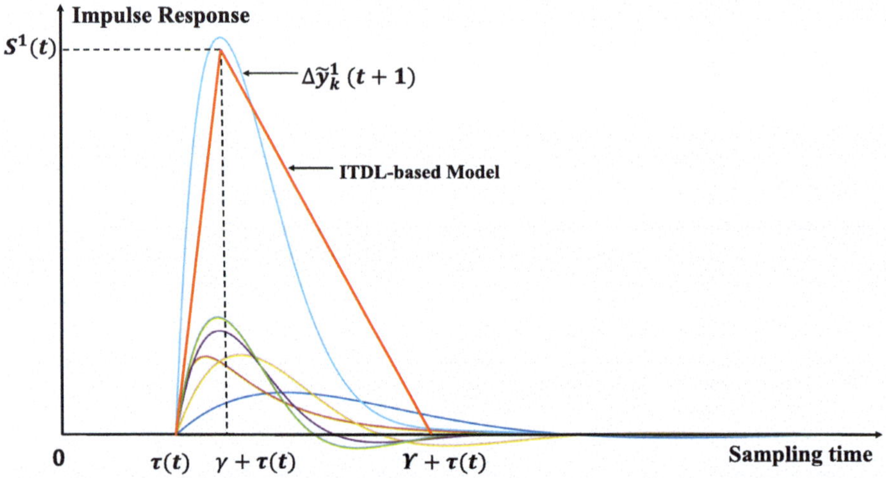

Fig. 5.2 Physical meaning of the data-driven attacked discrete-time cyber-physical systems with $\tau(t)$-delays via ITDL

Proof From [17, 18], it is evident that the compact-based data-driven model can be transformed into a triangular-based data-driven model. When $y_k(t)$ and $u_k(t)$ dimensions are 1-dimensional and no time-varying time delay exists during data transmission, the lemma 5.3 can be degraded to describe as Theorem 1 in [18]. In this chapter, the dimensions of $y_k(t)$ and $u_k(t)$ are $\mathbb{R}^p$ and $\mathbb{R}^m$ respectively, and there is a time-varying delay $\tau(t)$ in the data transmission.

For the $\triangle\tilde{y}_k^1(t-\tau(t))$, with the help of the characteristics of wave superposition and time delay translation as depicted in Fig. 5.2 and following a similar way in [18], one can have

$$\triangle\tilde{y}_k^1(t+1) = S^1(t)\triangle\varpi_k(t-\tau(t)), \tag{5.8}$$

where $\triangle\tilde{y}_k(t+1) = [\triangle\tilde{y}_k^1(t+1); \triangle\tilde{y}_k^2(t+1); \ldots; \triangle\tilde{y}_k^p(t+1)]$ and $S(t) = [S^1(t); S^2(t); \ldots; S^p(t)]$. Then, the equation (5.7) can be obtained via (5.8). Lemma 5.3 can be proved. □

Remark 5.6 As in Lemma 5.1, $S(t)$ is the matrix that embodies the properties of the cyber-physical system and its physical meaning is the peak of the data-driven attacked discrete-time cyber-physical system model with $\tau(t)$ via ITDL. Since the system parameter matrices are unknown, $S(t)$ is also unknown for the attacker. In contrast to the data-driven attacked discrete-time cyber-physical system model with $\tau(t)$ via CFDL, the dimensionality of $S(t)$ is smaller than that of $\Xi(t)$, the ITDL-based model reduces the data storage size requirement for the attacker.

Aiming to cope with the difficulty that $S(t)$ and $\tau(t)$ are unknown, the same estimation approach like Sect. 5.3.1, a detailed description is given in Lemma 5.4.

Lemma 5.4 *By solving the following optimization problem,*

$$\begin{aligned}\min_{\hat{S}_k(t)} \quad & J_3(\hat{S}_k(t)) = \|\Delta y_{k-1}(t+1) - \hat{S}_k(t)\Delta\varpi_{k-1}(t-\overline{\tau})\|^2 \\ & \qquad + \delta\|\hat{S}_k(t) - \hat{S}_{k-1}(t)\|^2, \\ s.t. \quad & \Delta\hat{y}_k(t+1) = \hat{S}_k(t)\Delta\varpi_k(t-\overline{\tau}), \delta > 0,\end{aligned}$$

the optimal estimation of $S(t)$ is obtained as

$$\begin{aligned}\hat{S}_k^*(t) =& \hat{S}_{k-1}^*(t) + (\Delta y_{k-1}(t+1) - \hat{S}_{k-1}^*(t)\Delta\varpi_{k-1}(t-\overline{\tau}))\Delta\varpi_{k-1}^T(t-\overline{\tau}) \\ & \times [(\delta + \|\Delta\varpi_{k-1}(t-\overline{\tau})\|^2)I - \Delta\varpi_{k-1}(t-\overline{\tau})\Delta\varpi_{k-1}^T(t-\overline{\tau})]/(\delta^2 \\ & + \delta\|\Delta\varpi_{k-1}(t-\overline{\tau})\|^2),\end{aligned} \tag{5.9}$$

where $\varpi_k(t-\overline{\tau}) = \frac{1}{\gamma+\overline{\tau}}\sum_{j=1}^{\gamma+\overline{\tau}} j(u_k(t-\overline{\tau}-j+1)) + \frac{1}{\Upsilon-\gamma}\sum_{j=\gamma+\overline{\tau}+1}^{\Upsilon+\overline{\tau}} (\Upsilon-j)u_k(t-\overline{\tau}-j+1)$, *and I is an identity matrix with the suitable dimensions.*

Proof The similar detailed proof process of Lemma 5.4 can also be found in [13, 18]. □

Then, an algorithm similar to Algorithm 4 for estimating $S(t)$ is employed to gain the optimal estimation $\hat{S}^*(t)$.

5.3.3 Optimal Data-Driven FDI Attacked Problems with Delay

Together with $\hat{\Xi}^*(t)$, (5.4), and (5.5), Problem 5.1, in designing optimal data-driven injection attack strategy against cyber-physical systems via CFDL method, can be transformed into Problem 5.3.

Problem 5.3

$$\max_{u_{a_k}(t-\overline{\tau})} \quad J_1(\tilde{y}_k(t+1), u_{a_k}(t-\overline{\tau}))$$

$$\text{s.t.} \quad \begin{cases}\Delta\tilde{y}_k(t+1) = \hat{\Xi}^*(t)\Delta\mathbf{u}_k(t-\overline{\tau}) + \hat{\Xi}^*(t)\bar{\Gamma}^a(t-\overline{\tau})\Delta\mathbf{u}_{a_k}(t-\overline{\tau}), \\ \mathcal{Q} \succeq 0, \mathcal{R} \succ 0.\end{cases}$$

Remark 5.7 From Algorithm 4, derived by solving Problem 5.2, it becomes evident that $\hat{\Xi}_k(t)$ is continuously optimized using the acquired data. After a sufficient number of iterations, $\hat{\Xi}^*(t)$ and $\Xi(t)$ become equal. Consequently, the data-driven cyber-physical system is equivalent to system (5.1). As a result, the relationship between Problems 5.1 and 5.3 is equivalent.

Similarly, united $\hat{S}^*(t)$, (5.7), and (5.4), Problem 5.1, in designing optimal data-driven injection attack strategy against cyber-physical systems via ITDL method, can be transformed into Problem 5.4.

Problem 5.4

$$\max_{u_{a_k}(t-\overline{\tau})} \quad J_1(\tilde{y}_k(t+1), u_{a_k}(t-\overline{\tau}))$$

$$\text{s.t.} \quad \begin{cases} \Delta\tilde{y}_k(t+1) = \hat{S}^*(t)\Delta\tilde{\varpi}_k(t-\overline{\tau}), \\ \tilde{\varpi}_k(t-\overline{\tau}) \triangleq \frac{1}{\gamma+\overline{\tau}} \sum\limits_{j=1}^{\gamma+\overline{\tau}} j[u_k(t-\overline{\tau}-j+1) + \Gamma^a(t-\overline{\tau}-j+1)u_{a_k}(t \\ \qquad -\overline{\tau}-j+1)] + \frac{1}{\Upsilon-\gamma} \sum\limits_{j=\gamma+\overline{\tau}+1}^{\Upsilon+\overline{\tau}} (\Upsilon-j)[u_k(t-\overline{\tau}-j+1) \\ \qquad +\Gamma^a(t-\overline{\tau}-j+1)u_{a_k}(t-\overline{\tau}-j+1)], \\ \mathcal{Q} \succeq 0, \mathcal{R} \succ 0. \end{cases}$$

Remark 5.8 Unlike the data-driven attacked system based on CFDL, which become equivalent to system (5.1) after sufficient iterations of optimization, the data-driven attacked system based on ITDL are approximations of the original system (5.1). However, the data-driven attacked system based on ITDL significantly reduces the computational complexity and storage space requirements of the device. Furthermore, the relationship between Problems 5.1 and 5.4 is approximate, but due to its advantages, studying it is meaningful in terms of reducing the computational device space and computational performance requirements for generating attack strategy.

Remark 5.9 The relationship between the objective functions J_1, J_2, and J_3 is described as follows. Different from J_1, the value functions J_2 and J_3 designed by attackers employing CFDL and ITDL methods in the eavesdropping estimation stage need to minimize the error between the real value and the estimated value. They use the stolen data to estimate the attribute matrix of the attacked cyber-physical system. While the attack target function J_1 is designed when performing the injection attack on the cyber-physical system. At this stage, the attacker wishes to maximize the deterioration of the system performance by utilizing the minimum energy consumption.

5.4 Data-Driven FDI Attack Strategies via CFDL and ITDL Methods

In this section, the optimal data-driven FDI attack strategies and algorithms using CFDL and ITDL methods are presented.

5.4.1 Data-Driven FDI Attack Strategy via CFDL Method

By solving the CFDL-based data-driven attack Problem 5.3, the attackers can obtain the optimal data-driven FDI attack strategy via CFDL method, as described in Theorem 5.1.

Theorem 5.1 *Suppose that Assumptions 5.1–5.3 hold and the control input $u_k(t)$ satisfies $\lim_{k\to\infty} \|\Delta u_k(i) - \Delta u_{k-1}(i)\|^2 = 0, i = 0, 1, \ldots, t$, if $\Theta_k(t) \prec 0$, then the optimal data-driven injection attack input $u^*_{a_k}(t)$ via CFDL method for the Problem 5.3 can be designed as*

$$u^*_{a_k}(t) = \begin{cases} u^{1*}_{a_1}(0), k = 1, t = 0, \\ u^{2*}_{a_k}(0), k > 1, t = 0, \\ u^{3*}_{a_1}(t), k = 1, 0 < t < \overline{\tau}, \\ u^{4*}_{a_k}(t), k > 1, 0 < t < \overline{\tau}, \\ u^{5*}_{a_1}(t), k = 1, t \geq \overline{\tau}, \\ u^{6*}_{a_k}(t), k > 1, t \geq \overline{\tau}, \end{cases} \tag{5.10}$$

where

$$\begin{aligned}
u^{1*}_{a_1}(0) =& -\Theta_1^{-1}(0)[\bar{T}_1^{1^T}(0)\bar{\Gamma}^{a^T}(0)\hat{\Xi}^{*^T}(\overline{\tau})Q(y(\overline{\tau}+1) + \hat{\Xi}^*(\overline{\tau})\bar{\Gamma}^a(0)\bar{T}_2^1(0))],\\
u^{2*}_{a_k}(0) =& u_{a_{k-1}}(0) + \Theta_k^{-1}(0)[\epsilon^2 \mathcal{R} u_{a_{k-1}}(0) - \bar{T}_1^{k^T}(0)\bar{\Gamma}^{a^T}(0)\hat{\Xi}^{*^T}(\overline{\tau})Q(\tilde{y}_{k-1}(\overline{\tau}+1)\\
&+ \hat{\Xi}^*(\overline{\tau})\bar{\Gamma}^a(0)\bar{T}_2^k(0))],\\
u^{3*}_{a_1}(t) =& -\Theta_1^{-1}(t)[\bar{T}_1^{1^T}(t)\bar{\Gamma}^{a^T}(t)\hat{\Xi}^{*^T}(t+\overline{\tau})Q(y(t+\overline{\tau}+1) + \hat{\Xi}^*(t+\overline{\tau})\bar{\Gamma}^a(t)\bar{T}_2^1(t))],\\
u^{4*}_{a_k}(t) =& u_{a_{k-1}}(t) + \Theta_k^{-1}(t)[\epsilon^2 \mathcal{R} u_{a_{k-1}}(t) - \bar{T}_1^{k^T}(t)\bar{\Gamma}^{a^T}(t)\hat{\Xi}^{*^T}(t+\overline{\tau})Q(\tilde{y}_{k-1}(t+\overline{\tau}+1)\\
&+ \hat{\Xi}^*(t+\overline{\tau})\bar{\Gamma}^a(t)\bar{T}_2^k(t))],\\
u^{5*}_{a_1}(t) =& -\Theta_1^{-1}(t)[\bar{T}_1^{1^T}(t)\bar{\Gamma}^{a^T}(t)\hat{\Xi}^{*^T}(t+\overline{\tau})Q(y(t+\overline{\tau}+1) + \hat{\Xi}^*(t+\overline{\tau})\Delta\breve{\mathbf{u}}_1(t)\\
&+ \hat{\Xi}^*(t+\overline{\tau})\bar{\Gamma}^a(t)\bar{T}_2^1(t))],\\
u^{6*}_{a_k}(t) =& u_{a_{k-1}}(t) + \Theta_k^{-1}(t)[\epsilon^2 \mathcal{R} u_{a_{k-1}}(t) - \bar{T}_1^{k^T}(t)\bar{\Gamma}^{a^T}(t)\hat{\Xi}^{*^T}(t+\overline{\tau})Q(\tilde{y}_{k-1}(t+\overline{\tau}+1)\\
&+ \hat{\Xi}^*(t+\overline{\tau})\Delta\breve{\mathbf{u}}_k(t) + \hat{\Xi}^*(t+\overline{\tau})\bar{\Gamma}^a(t)\bar{T}_2^k(t))],\\
\Delta\breve{\mathbf{u}}_k(t) =& [\Delta u_{k-1}(t); \Delta u_{k-1}(t-1); \ldots; \Delta u_{k-1}(t-\overline{\tau}+1); \Delta u_k(t-\overline{\tau}); \Delta u_k(t-\overline{\tau}-1);\\
&\ldots; \Delta u_k(0)],\\
\Theta_k(t) =& (\bar{T}_1^{k^T}(t)\bar{\Gamma}^{a^T}(t)\hat{\Xi}^{*^T}(t+\overline{\tau})Q\hat{\Xi}^*(t+\overline{\tau})\bar{\Gamma}^a(t)\bar{T}_1^k(t) - \epsilon^2\mathcal{R}),\\
\Delta u_{a_k}(h) =& \Phi_k(h)\Delta u_{a_k}(t-\overline{\tau}) + \Lambda_k(h),\ \Phi_k(h) = \mathrm{diag}\{\phi_k^1(h), \phi_k^2(h), \ldots, \phi_k^q(h)\},\\
\bar{\Phi}_k(h) =& \mathrm{diag}\{\bar{\Phi}_k^1(h), \bar{\Phi}_k^2(h), \ldots, \bar{\Phi}_k^q(h)\},\ \Lambda_k(h) = [\lambda_k^1(h); \lambda_k^2(h); \ldots; \lambda_k^q(h)],\\
\bar{\Lambda}_k(h) =& [\bar{\Lambda}_k^1(h); \bar{\Lambda}_k^2(h); \ldots; \bar{\Lambda}_k^q(h)],\ T_1^k(t) = [I_{q\times q}; \Phi_k(t-1); \ldots; \Phi_k(0)],\\
\bar{T}_1^k(t) =& [I_{q\times q}; \bar{\Phi}_k(t-1); \ldots; \bar{\Phi}_k(0)],\ T_2^k(t) = [\mathbf{0}_{q\times 1}; \Lambda_k(t-1); \ldots; \Lambda_k(0)],\\
\bar{T}_2^k(t) =& [\mathbf{0}_{q\times 1}; \bar{\Lambda}_k(t-1); \ldots; \bar{\Lambda}_k(0)],\ 0 < \underline{\phi}_k^v(h) \leq \phi_k^v(h) \leq \overline{\phi}_k^v(h),
\end{aligned}$$

$\underline{\lambda}_k^{\upsilon}(h) \leq \lambda_k^{\upsilon}(h) \leq \overline{\lambda}_k^{\upsilon}(h)$, $\overline{\phi}_k^{\upsilon}(h)$, $\overline{\lambda}_k^{\upsilon}(h)$ are the known positive constants, $h = 0, 1, ...,$ $(t - \overline{\tau} - 1)$, $\upsilon = 1, 2, ..., q$, and $y(t + \overline{\tau} + 1)$ is the healthy output data stored by the attacker at time $(t + \overline{\tau} + 1)$ during the eavesdropping estimation phase.

Proof The attacked system output $\tilde{y}_k(t+1)$ can be expressed as

$$\begin{aligned}\tilde{y}_k(t+1) =& \tilde{y}_{k-1}(t+1) + \Delta\tilde{y}_k(t+1) \\ =& \tilde{y}_{k-1}(t+1) + \hat{\Xi}^*(t)(\Delta\mathbf{u}_k(t-\overline{\tau}) + \bar{\Gamma}^a(t-\overline{\tau})\Delta\mathbf{u}_{a_k}(t-\overline{\tau})),\end{aligned} \tag{5.11}$$

since the equation (5.11) contains $\Delta\mathbf{u}_{a_k}(t-\overline{\tau})$, with the help of linear regression equation $\Delta u_{a_k}(h) = \Phi_k(h)\Delta u_{a_k}(t-\overline{\tau}) + \Lambda_k(h)$, the equation (5.11) can be further described as

$$\begin{aligned}\tilde{y}_k(t+1) =& \tilde{y}_{k-1}(t+1) + \hat{\Xi}^*(t)(\Delta\mathbf{u}_k(t-\overline{\tau}) + \bar{\Gamma}^a(t-\overline{\tau})T_1^k(t-\overline{\tau}) \\ &\times \Delta u_{a_k}(t-\overline{\tau}) + \bar{\Gamma}^a(t-\overline{\tau})T_2^k(t-\overline{\tau})),\end{aligned} \tag{5.12}$$

where $\Phi_k(h)$, $\Lambda_k(h)$, $T_1^k(t)$, and $T_2^k(t)$ are defined in Theorem 5.1.

Through the definition of $\Delta u_{a_k}(t-\overline{\tau})$, the attack input can be rewritten as

$$u_{a_k}(t-\overline{\tau}) = u_{a_{k-1}}(t-\overline{\tau}) + \Delta u_{a_k}(t-\overline{\tau}). \tag{5.13}$$

Then, combining (5.12) and (5.13), the objective function (5.4) can be reformulated as

$$\begin{aligned}J_1 =& \frac{1}{2}\{[\tilde{y}_{k-1}(t+1) + \hat{\Xi}^*(t)(\Delta\mathbf{u}_k(t-\overline{\tau}) + \bar{\Gamma}^a(t-\overline{\tau})T_1^k(t-\overline{\tau})\Delta u_{a_k}(t-\overline{\tau}) \\ &+ \bar{\Gamma}^a(t-\overline{\tau})T_2^k(t-\overline{\tau}))]^T Q[\tilde{y}_{k-1}(t+1) + \hat{\Xi}^*(t)(\Delta\mathbf{u}_k(t-\overline{\tau}) + \bar{\Gamma}^a(t-\overline{\tau}) \\ &\times T_1^k(t-\overline{\tau})\Delta u_{a_k}(t-\overline{\tau}) + \bar{\Gamma}^a(t-\overline{\tau})T_2^k(t-\overline{\tau}))] - \epsilon^2[u_{a_{k-1}}(t-\overline{\tau}) \\ &+ \Delta u_{a_k}(t-\overline{\tau})]^T \mathcal{R}[u_{a_{k-1}}(t-\overline{\tau}) + \Delta u_{a_k}(t-\overline{\tau})]\} \\ \leq& \frac{1}{2}\{[\tilde{y}_{k-1}(t+1) + \hat{\Xi}^*(t)(\Delta\mathbf{u}_k(t-\overline{\tau}) + \bar{\Gamma}^a(t-\overline{\tau})\bar{T}_1^k(t-\overline{\tau})\Delta u_{a_k}(t-\overline{\tau}) \\ &+ \bar{\Gamma}^a(t-\overline{\tau})\bar{T}_2^k(t-\overline{\tau}))]^T Q[\tilde{y}_{k-1}(t+1) + \hat{\Xi}^*(t)(\Delta\mathbf{u}_k(t-\overline{\tau}) + \bar{\Gamma}^a(t-\overline{\tau}) \\ &\times \bar{T}_1^k(t-\overline{\tau})\Delta u_{a_k}(t-\overline{\tau}) + \bar{\Gamma}^a(t-\overline{\tau})\bar{T}_2^k(t-\overline{\tau}))] - \epsilon^2[u_{a_{k-1}}(t-\overline{\tau}) \\ &+ \Delta u_{a_k}(t-\overline{\tau})]^T \mathcal{R}[u_{a_{k-1}}(t-\overline{\tau}) + \Delta u_{a_k}(t-\overline{\tau})]\} \\ \triangleq& \tilde{J}_1.\end{aligned}$$

Consequently, through Pontryagin's maximum principle [20], the following inequality relation can be established

$$\begin{aligned}J_1(\tilde{y}_k(t+1), u_{a_k}(t-\overline{\tau})) &\leq J_1(\tilde{y}_k^*(t+1), u_{a_k}^*(t-\overline{\tau})) \\ &\leq \tilde{J}_1(\tilde{y}_k^*(t+1), u_{a_k}^*(t-\overline{\tau})),\end{aligned} \tag{5.14}$$

where $(\bullet)^*$ stands for the optimal value.

By means of the optimality principle [19] and the Maximum-Minimum Theorem[20], one has

$$\begin{aligned}\frac{\partial \tilde{J}_1}{\partial \Delta u_{a_k}(t-\overline{\tau})} =&\Theta_k(t-\overline{\tau})\Delta u_{a_k}(t-\overline{\tau}) - [\epsilon^2 \mathcal{R} u_{a_{k-1}}(t) - \bar{T}_1^{k^T}(t-\overline{\tau}) \\ &\times \bar{\Gamma}^{a^T}(t-\overline{\tau})\hat{\Xi}^{*^T}(t)Q(y_{k-1}(t+1) + \hat{\Xi}^*(t)\Delta \mathbf{u}_k(t-\overline{\tau}) \\ &+ \hat{\Xi}^*(t)\bar{\Gamma}^a(t-\overline{\tau})\bar{T}_2^k(t-\overline{\tau}))] = \mathbf{0},\end{aligned} \tag{5.15}$$

and the second-order partial derivative condition for $\tilde{J}_1$ should satisfy

$$\begin{aligned}\frac{\partial^2 \tilde{J}_1}{\partial \Delta u_{a_k}^2(t-\overline{\tau})} &= (\bar{T}_1^{k^T}(t)\bar{\Gamma}^{a^T}(t)\hat{\Xi}^{*^T}(t+\overline{\tau})Q\hat{\Xi}^*(t+\overline{\tau})\bar{\Gamma}^a(t)\bar{T}_1^k(t) - \epsilon^2\mathcal{R}) \\ &\prec \mathbf{0}.\end{aligned} \tag{5.16}$$

From Eq. (5.15), one can obtain the optimal $\Delta u_{a_k}^*(t)$ is derived as

$$\begin{aligned}\Delta u_{a_k}^*(t) =&\Theta_k^{-1}(t)[\epsilon^2 \mathcal{R} u_{a_{k-1}}(t) - \bar{T}_1^{k^T}(t)\bar{\Gamma}^{a^T}(t)\hat{\Xi}^{*^T}(t+\overline{\tau})Q(y_{k-1}(t+\overline{\tau}+1) + \hat{\Xi}^*(t \\ &+\overline{\tau})\Delta \mathbf{u}_k(t) + \hat{\Xi}^*(t+\overline{\tau})\bar{\Gamma}^a(t)\bar{T}_2^k(t))],\end{aligned}$$

where $[\Delta u_k(t); \Delta u_k(t-1), \ldots, \Delta u_k(t-\overline{\tau}+1)]$ of $\Delta \mathbf{u}_k(t)$ in (5.15) is unknown for the attacker at time t under the kth iteration. Thus, $[\Delta u_{k-1}(t); \Delta u_{k-1}(t-1), \ldots, \Delta u_{k-1}(t-\overline{\tau}+1)]$ is adopted in Theorem 5.1 to replace the unknown data. Then, the error $E_k^u(t)$ between $\Delta \mathbf{u}_k(t)$ and $\Delta \breve{\mathbf{u}}_k(t)$ is defined as

$$E_k^u(t) = \|\Delta \mathbf{u}_k(t) - \Delta \breve{\mathbf{u}}_k(t)\|^2 = \sum_{i=t-\overline{\tau}+1}^{t} \|\Delta u_k(i) - \Delta u_{k-1}(i)\|^2,$$

when $\lim_{k\to\infty} \|\Delta u_k(i) - \Delta u_{k-1}(i)\|^2 = 0$, the error $E_k^u(t) = 0$, $k \to \infty$. Therefore, the CFDL-based optimal data-driven injection attack input (5.10) can be obtained by classification discussion. Note that (5.16) is the condition that needs to be satisfied in Theorem 5.1. This proof is completed. □

The optimal data-driven injection attack strategy algorithm via CFDL of Theorem 5.1 is as Algorithm 5.

Remark 5.10 The term $(\Delta u_k(i) - \Delta u_{k-1}(i))$, $i = 0, 1, \ldots, t$ stands for the difference in the increment of the control input under two consecutive iterations. Note that $\lim_{k\to\infty} \|\Delta u_k(i) - \Delta u_{k-1}(i)\|^2 = 0$, $i = 0, 1, \ldots, t$ is usually easy to satisfy, especially in the control approaches that employ iterative learning, such as [10, 12]. When the controller applies LQR, the condition can also be satisfied due to Assumption 5.1. It is evident from (5.14) that Theorem 5.1 seeks the upper bound on the optimal

Algorithm 5: The Optimal Data-Driven Injection Attack Strategy via CFDL

```
Initialize: Given the suitable value τ̄, φ̄_k^υ(t), and λ̄_k^υ(t), υ = 1, 2, ..., q. Let 0 ≤ ε ≤ 1 be
    the suitable factor and Q, R be the appropriate weight matrices satisfying the conditions of
    Q ⪰ 0 and R ≻ 0. Set the maximum numb- er of iterations N and the attack duration N_a.
    Store data y and Ξ̂* of the eavesdropping estimation phase.
for k = 1 to N do
    while t ≤ N_a do
        Step 1. Eavesdrop and store ỹ_k(t − τ(t) + 1) and u_k(t − τ(t));
        Step 2. Update the following vectors using stored and historical data;
        if k > 1 then
            Update the vectors ỹ_{k−1}(t + τ̄ + 1), ŭ_k(t), and u_{a_{k−1}}(t);
        else
            Update the vector y(t + τ̄ + 1);
        Step 3. Determine whether condition Θ_k(t) ≺ 0 is satisfied.
        if Θ_k(t) ≺ 0 is satisfied then
            Step 4. Compute the optimal injection attack input u*_{a_k}(t) by (5.10);
            Step 5. Perform injection attack
            ũ_k(t) ← u_k(t) + Γ^a(t)u*_{a_k}(t);
            Step 6. Attack on the system through the network with time-varying delay
            ũ_k(t + τ(t)) ← ũ_k(t);
        else
            Reset Γ_{a_k}(t), and repeat Step 3;
```

strategy when $\Phi_k(t)$ and $\Lambda_k(t)$ take $\bar{\Phi}_k(t)$ and $\bar{\Lambda}_k(t)$. For the attacker, it is natural to want J_3 to be as large as possible, and the closer it is to the upper bound, the better the effect of the attack. And here, $\Delta u_{a_k}(h) = \Phi_k(h)\Delta u_{a_k}(t-\overline{\tau}) + \Lambda_k(h)$ is known to the attacker, and can even be manipulated by himself. Therefore, the upper bound of the optimal strategy in Theorem 5.1 that attackers want to achieve is the significance of solving Problem 5.3. For how to adjust $\Gamma^a(t)$ to satisfy the condition $\Theta_k(t) \prec 0$, a similar approach can be found in Chap. 4.

5.4.2 *Data-Driven FDI Attack Strategy via ITDL Method*

Similarly, by solving the ITDL-based data-driven attack Problem 5.4, attackers can obtain the optimal data-driven FDI attack strategy based on ITDL method, as described in Theorem 5.2.

Theorem 5.2 *Suppose that Assumptions 5.1–5.3 hold and the control input $u_k(t)$ satisfies $\lim_{k\to\infty} \|\Delta u_k(i) - \Delta u_{k-1}(i)\|^2 = 0, i = 0, 1, \ldots, t$, if $\Omega_k(t) \prec 0$, then the optimal data-driven injection attack input $u^*_{a_k}(t)$ via ITDL for the Problem 5.4 can be designed as*

$$u_{a_k}^*(t) = \begin{cases} u_{a_1}^{1*}(0), k = 1, t = 0, \\ u_{a_k}^{2*}(0), k > 1, t = 0, \\ u_{a_1}^{3*}(t), k = 1, 0 < t < \overline{\tau}, \\ u_{a_k}^{4*}(t), k > 1, 0 < t < \overline{\tau}, \\ u_{a_1}^{5*}(t), k = 1, t \geq \overline{\tau}, \\ u_{a_k}^{6*}(t), k > 1, t \geq \overline{\tau}, \end{cases} \tag{5.17}$$

where

$$\begin{aligned}
\Omega_k(t) =& \bar{\Psi}_k^T(t) Q \bar{\Psi}_k(t) - \epsilon^2 \mathcal{R}, \\
u_{a_1}^{1*}(0) =& -\Omega_1^{-1}(0) \bar{\Psi}_1^T(0) Q[y(\overline{\tau}+1) + \bar{H}_1^1(0)], \\
u_{a_k}^{2*}(0) =& u_{a_{k-1}}(0) + \Omega_k^{-1}(0)[\epsilon^2 \mathcal{R} u_{a_{k-1}}(0) - \bar{\Psi}_k^T(0) Q(\tilde{y}_{k-1}(\overline{\tau}+1) + \bar{H}_k^1(0))], \\
u_{a_1}^{3*}(t) =& -\Omega_1^{-1}(t) \bar{\Psi}_1^T(t) Q[y(t+\overline{\tau}+1) + \bar{H}_1^1(t)], \\
u_{a_k}^{4*}(t) =& u_{a_{k-1}}(t) + \Omega_k^{-1}(t)[\epsilon^2 \mathcal{R} u_{a_{k-1}}(t) - \bar{\Psi}_k^T(t) Q(\tilde{y}_{k-1}(t+\overline{\tau}+1) + \bar{H}_k^1(t))], \\
u_{a_1}^{5*}(t) =& -\Omega_1^{-1}(t) \bar{\Psi}_1^T(t) Q[y(t+\overline{\tau}+1) + \bar{H}_1^2(t)], \\
u_{a_k}^{6*}(t) =& u_{a_{k-1}}(t) + \Omega_k^{-1}(t)[\epsilon^2 \mathcal{R} u_{a_{k-1}}(t) - \bar{\Psi}_k^T(t) Q(\tilde{y}_{k-1}(t+\overline{\tau}+1) + \bar{H}_k^2(t))], \\
\bar{\Psi}_k(t) =& \hat{S}^*(t+\overline{\tau})\{\frac{1}{\gamma+\overline{\tau}} \sum_{j=1}^{\gamma+\overline{\tau}} j \Gamma^a(t-j+1) \bar{\Phi}_k(t-j+1) + \frac{1}{\Upsilon-\gamma} \sum_{j=\gamma+\overline{\tau}+1}^{\Upsilon+\overline{\tau}} (\Upsilon \\
& - j) \Gamma^a(t-j+1) \bar{\Phi}_k(t-j+1)\},
\end{aligned}$$

$$\begin{aligned}
\bar{H}_k^1(t) =& \hat{S}^*(t+\overline{\tau})\{\frac{1}{\gamma+\overline{\tau}} \sum_{j=1}^{\gamma+\overline{\tau}} j \Gamma^a(t-j+1) \bar{\Lambda}_k(t-j+1) + \frac{1}{\Upsilon-\gamma} \sum_{j=\gamma+\overline{\tau}+1}^{\Upsilon+\overline{\tau}} (\Upsilon - j) \\
& \cdot \Gamma^a(t-j+1) \bar{\Lambda}_k(t-j+1), \\
\bar{H}_k^2(t) =& \hat{S}^*(t+\overline{\tau})\{\frac{1}{\gamma+\overline{\tau}} \sum_{j=1}^{\gamma+\overline{\tau}} j[\Delta \check{u}_k(t-j+1) + \Gamma^a(t-j+1) \bar{\Lambda}_k(t-j+1)] \\
& + \frac{1}{\Upsilon-\gamma} \sum_{j=\gamma+\overline{\tau}+1}^{\Upsilon+\overline{\tau}} (\Upsilon - j)[\Delta \check{u}_k(t-j+1) + \Gamma^a(t-j+1) \bar{\Lambda}_k(t-j+1)], \\
\Delta \check{u}_k(t-i) =& \begin{cases} \Delta u_{k-1}(t-i), 0 \leq i \leq \overline{\tau}-1, \\ \Delta u_k(t-i), \overline{\tau} \leq i \leq t, \end{cases}
\end{aligned}$$

$\bar{\Phi}_k(t)$, $\bar{\Lambda}_k(t)$, and $y(t+\overline{\tau}+1)$ are defined in Theorem 5.1.

Proof The proof process of Theorem 5.2 is similar to that of Theorem 5.1, and it is omitted here. □

Then, an algorithm similar to Algorithm 5 for Theorem 5.2 is employed to implement the optimal data-driven injection attack strategy via ITDL.

Remark 5.11 The relationship between Problems 5.1 and 5.4 is analogous to the relationship between Problems 5.1 and 5.3, and it is omitted here. By comparison with the CFDL-based optimal injection attack strategy designed in Theorem 5.1, the attack strategy in Theorem 5.2 is more appropriate in the presence of the following two scenarios of restrictions. The first scenario is that the size of memory space occupied by the attacker is limited. In contrast to the CFDL-based attack strategy with dimensional time-varying matrices $\hat{\Xi}^*(t+\overline{\tau}) \in \mathbb{R}^{p\times[m*(t+1)]}$, $\Delta\breve{\mathbf{u}}_k(t) \in \mathbb{R}^{[m*(t+1)]\times 1}$, $\bar{T}_1^k(t), \bar{T}_2^k(t) \in \mathbb{R}^{q\times[q*(t+1)]}$, and $\bar{\Gamma}^a(t) \in \mathbb{R}^{[m*(t+1)]\times[q*(t+1)]}$, the ITDL-based attack strategy, which only involves the dimensional time-invariant matrices $\hat{S}^*(t) \in \mathbb{R}^{p\times m}$, $\Delta\breve{u}_k(t) \in \mathbb{R}^{m\times 1}$, $\bar{\Phi}_k(t) \in \mathbb{R}^{q\times q}$, and $\bar{\Lambda}_k(t) \in \mathbb{R}^{q\times q}$, is more suitable to this scenario. The second scenario is that the hashrate of the attacker is restrained. For the CFDL-based attack strategy, the computational complexity is $O(p*(m*(t+1))*(q*(t+1)))$, while for the ITDL-based attack strategy, the computational complexity mainly is $O(p*m*q)$. Since $O(p*m*q) < O(p*(m*(t+1))*(q*(t+1)))$ is satisfied when $t > 0$, the strategy of Theorem 5.2 is more applicable in this scenario.

5.4.3 Stealthiness Analysis of Attack Strategies

For the stealthiness analysis of the designed optimal data-driven injection-based attack strategies, this subsection focuses on the residual-based detection mechanism, the sum-of-signal-energy-based detection mechanism, and the data-driven framework-based detection mechanism.

(1) The attack strategy designed under the residual detection mechanism [21] is covert for the following reasons. The CFDL-based and ITDL-based attack strategies are based on the value functions J_1 and J_2, where the matrices $\hat{\Xi}^*(t)$ and $\hat{S}^*(t)$ both reflect the properties of the system itself and satisfy $\lim\limits_{k\to\infty} \|\Delta\tilde{y}_k(i) - \Delta\hat{\tilde{y}}_k(i)\|^2 = 0$, i.e., $\lim\limits_{k\to\infty} \|\tilde{y}_k(i) - \hat{\tilde{y}}_k(i)\|^2 = 0 < D^a$, D^a is alarm threshold. Therefore, for the plant equipped with the detection mechanism based on estimated residuals, the attack strategy is stealthy.

(2) For the detection mechanism based on the sum of signal energy [22], the designed attack strategy is stealthy owing to the following reasons. The value function of the attacker in the strategy generation phase is J_1, the energy consumed by the attacking party is characterized as $u_{a_k}^T(t)\mathcal{R}u_{a_k}(t)$, and therefore the total energy consumption term of the attacker at time t under the kth iteration is $\sum\limits_{i=0}^{t} u_{a_k}^T(i)\mathcal{R}u_{a_k}(i)$. The weight matrix $\mathcal{R}$ is there designed by the attacker, and for this reason, the attacker can analyze the traffic data from the eavesdropping estimation phase to obtain the alarm threshold of such detectors, and then, the attacker adjusts the matrix $\mathcal{R}$ to tune the overall energy spent by the attacker to evade such detectors.

(3) For the data-driven framework-based detection mechanism [23], the attack strategy designed with the attacker's knowledge of the operational mechanism is stealthy. Since the detection mechanism adds tags for detection during operation and detects whether the data has been attacked by detecting whether the tags have been altered, the attack strategy devised in this chapter only tampers with the value of the transmitted data and the attacker can achieve stealth, i.e., evade such detectors, by tampering with the data while keeping the tag bits unchanged.

5.5 Simulation and Experiment

In this section, both the numerical simulation and the experiment based on a leader-following form consisting of five quadcopters are presented to demonstrate the validation of the proposed data-driven injection attack strategies based on CFDL and ITDL methods.

5.5.1 Numerical Simulation

In this subsection, the numerical example is considered, and the data employed in the simulation is generated by the following discrete-time error system,

$$\begin{cases} x_k(t+1) = Ax_k(t) + Bu_k(t - \tau(t)), \\ y_k(t+1) = Cx_k(t+1), \end{cases}$$

where system matrices is $\boldsymbol{A} = \begin{bmatrix} 0.1 & 0 \\ 0 & 0.6 \end{bmatrix}$, $\boldsymbol{B} = \begin{bmatrix} 1 & -0.3 \\ 0.1 & -0.5 \end{bmatrix}$, $\boldsymbol{C} = \begin{bmatrix} 0.7 & -0.56 \end{bmatrix}$, $\underline{\tau}$ and $\overline{\tau}$ is 1 and 5, respectively. And the time-varying delays is $\tau(t) = 1, 0 < t \leq N_a/5, \tau(t) = 2, N_a/5 < t \leq 2 * N_a/5, \tau(t) = 3, 2 * N_a/5 < t \leq 3 * N_a/5, \tau(t) = 4, 3 * N_a/5 < t \leq 4 * N_a/5, \tau(t) = 5, 4 * N_a/5 < t.$ The sampling time adopts $\sigma = 0.01s$. The initial state is [1;-0.5]. The number of iterations is $N = 100$ and the attack duration is $N_a = 100$.

The specific description of the parameters adopted in the attack strategies is as follows. The computational precision $\beta = 10^{-3}$, the weight matrix $Q = 1, \mathcal{R} = I_{2\times 2}$, the attenuation factor $\epsilon = 0.93$, the attack direction matrix $\Gamma^a = [1, 0; 0, 1.5]$, $\overline{\phi} = 0.1$, and $\overline{\lambda} = 10$. The parameter of CFDL-based attack strategy is $\alpha = 0.9$. The parameters of ITDL-based attack strategy are $\delta = 0.9$, $\gamma = 3$, and $\Upsilon = 24$. The controllers employed for the system in this subsection of the numerical simulation are two schemes: iterative learning control (ILC) [12] and incremental triangular control (ITC) [18]. Figures 5.3 and 5.4 demonstrate the healthy output and control input of the system under the control schemes at Nth iteration. Figures 5.5, 5.6, 5.7, and 5.8 present the optimal estimations of $\Xi_k(t)$ and $S_k(t)$, respectively. Table 5.1 repre-

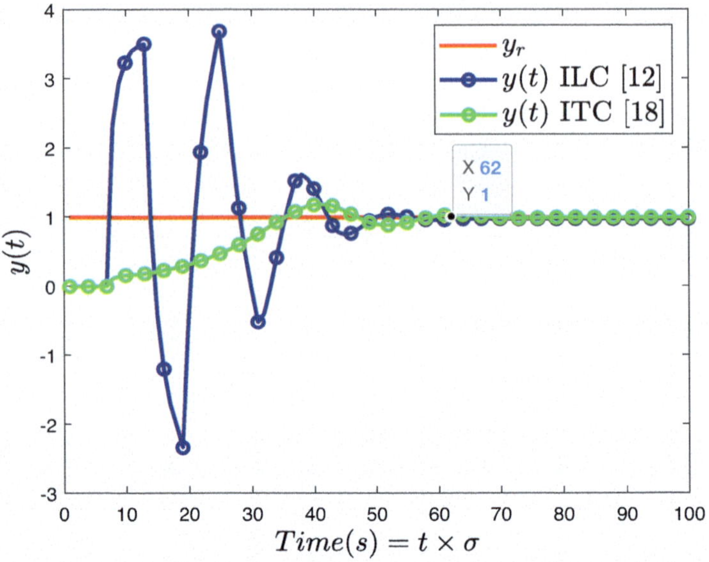

Fig. 5.3 The healthy output of the system under the ILC and ITC schemes at Nth iteration

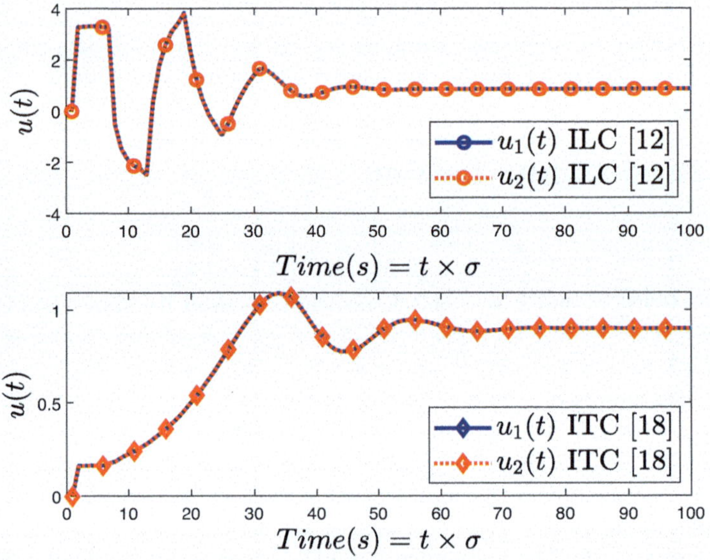

Fig. 5.4 The healthy control input of the system under the ILC and ITC schemes at Nth iteration

sents the comparison of the designed attack strategy with existing attack strategies. Figures 5.9, 5.10, 5.11, 5.12, 5.13, 5.14, 5.15, and 5.16 display the results of the proposed attack strategies under control policies ILC [12] and ITC [18].

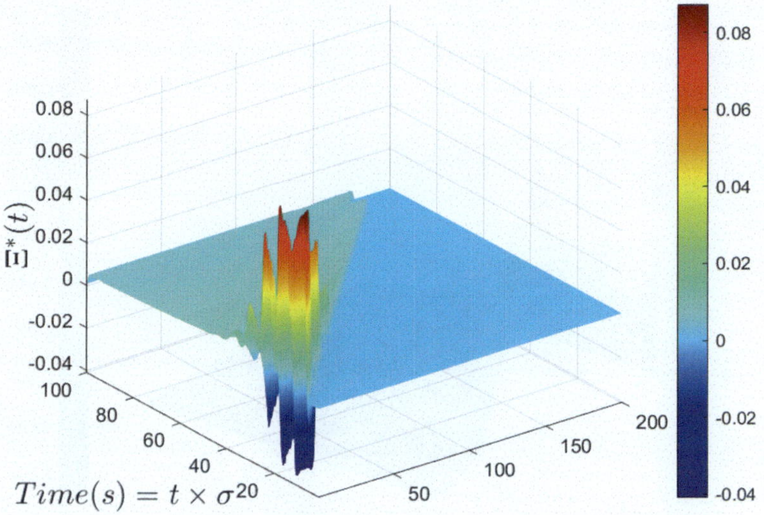

Fig. 5.5 The matrix $\hat{\Xi}^*(t)$ under the ILC scheme at Nth iteration

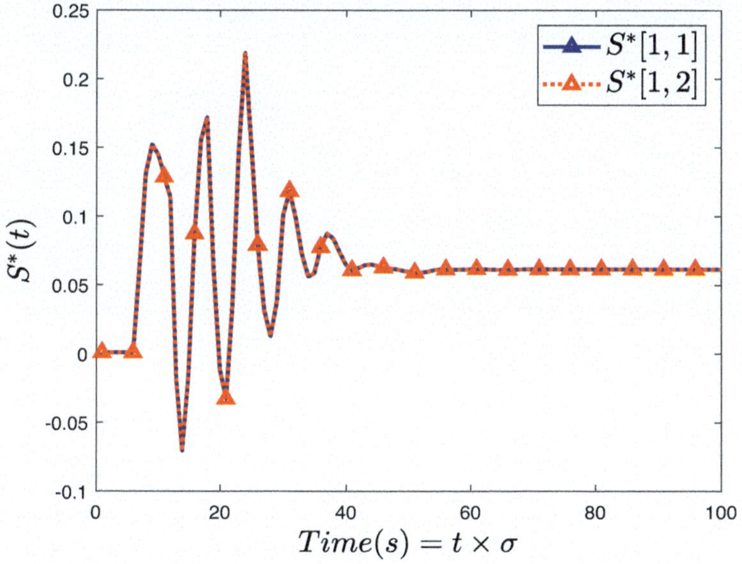

Fig. 5.6 The matrix $\hat{S}^*(t)$ under the ILC scheme at Nth iteration

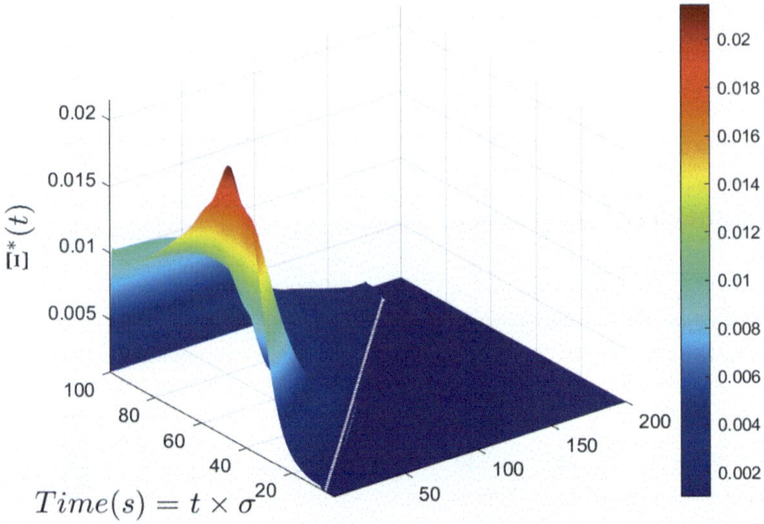

Fig. 5.7 The matrix $\hat{\Xi}^*(t)$ under the ITC scheme at Nth iteration

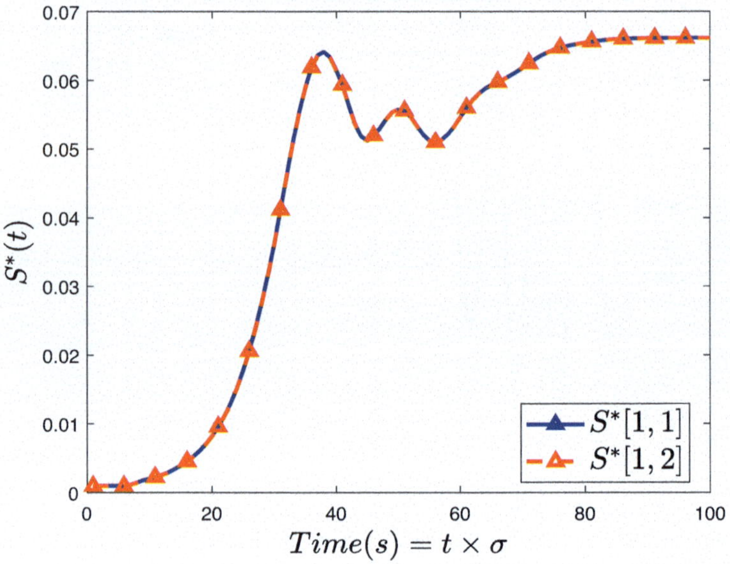

Fig. 5.8 The matrix $\hat{S}^*(t)$ under the ITC scheme at Nth iteration

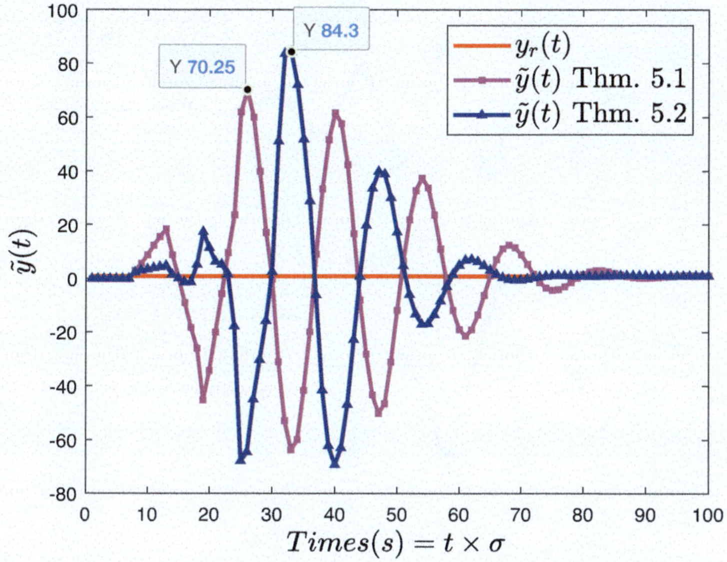

Fig. 5.9 The attacked output of the system under the ILC scheme at Nth iteration

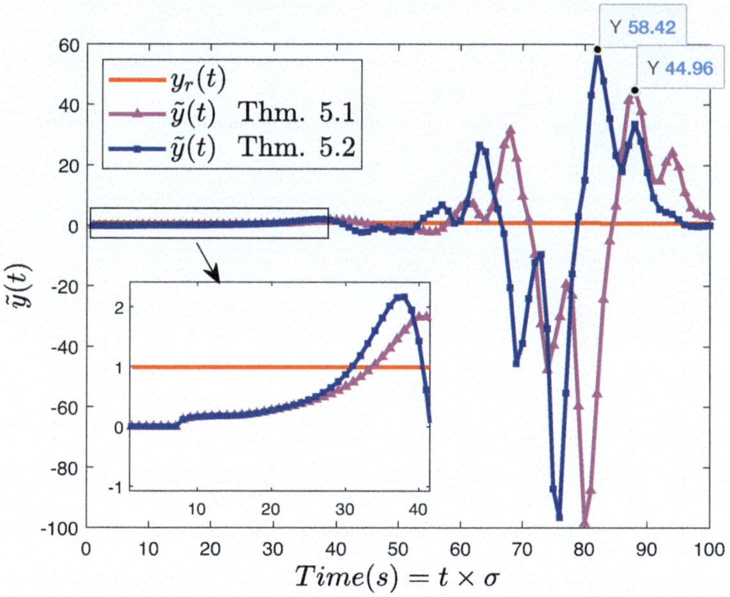

Fig. 5.10 The attacked output of the system under the ITC scheme at Nth iteration

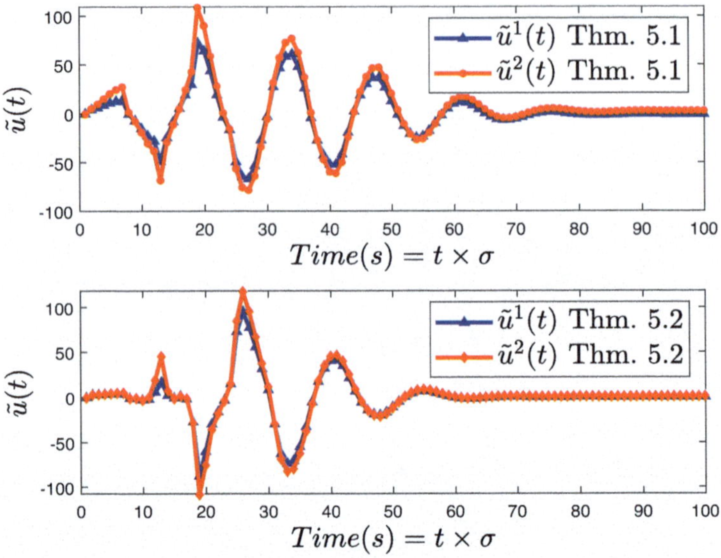

Fig. 5.11 The attacked control input of the system under the ILC scheme at Nth iteration

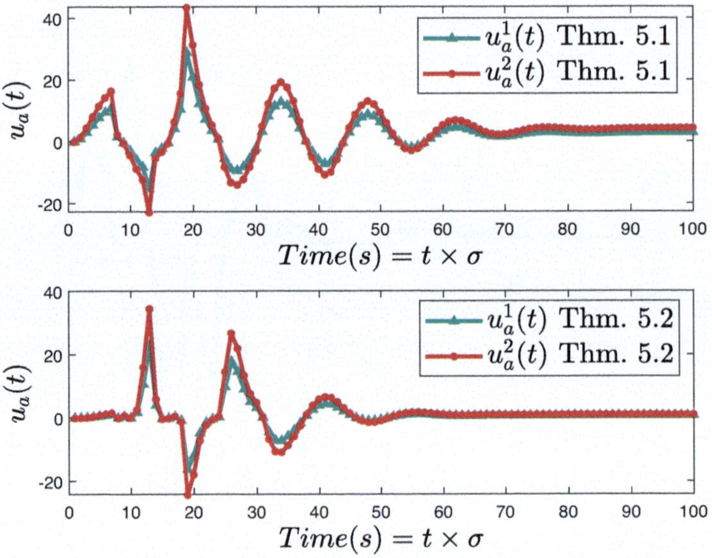

Fig. 5.12 The attack input of the system under the ILC scheme at Nth iteration

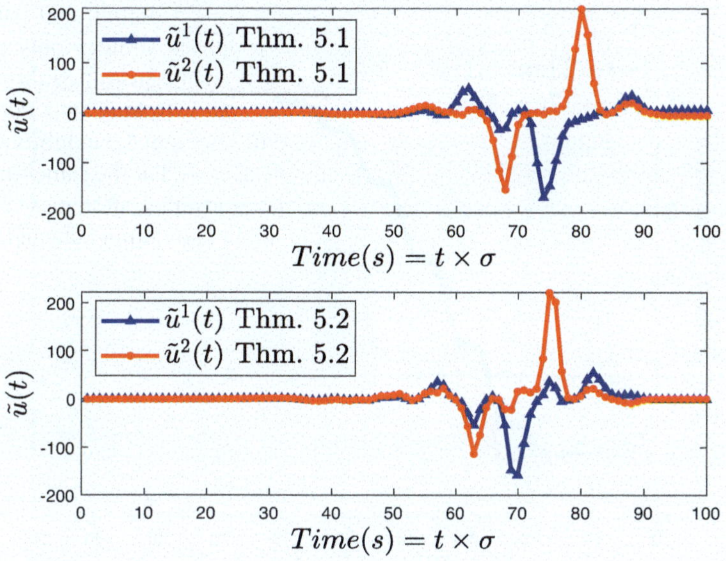

Fig. 5.13 The attacked control input of the system under the ITC scheme at Nth iteration

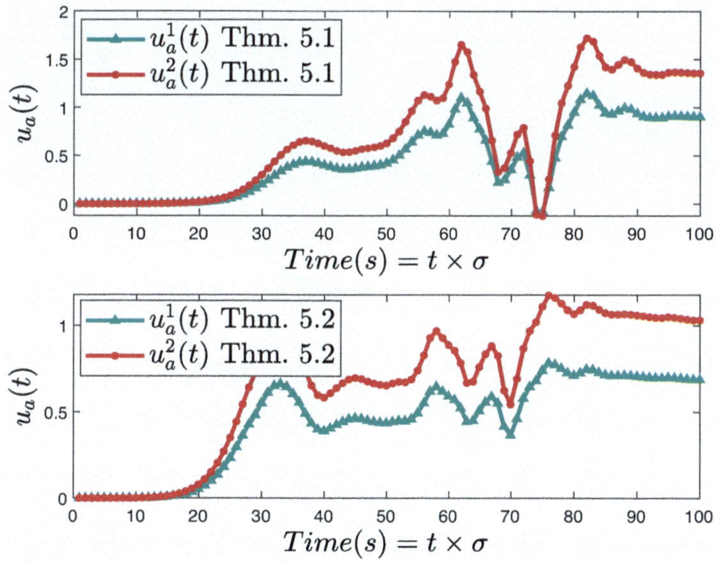

Fig. 5.14 The attack input of the system under the ITC scheme at Nth iteration

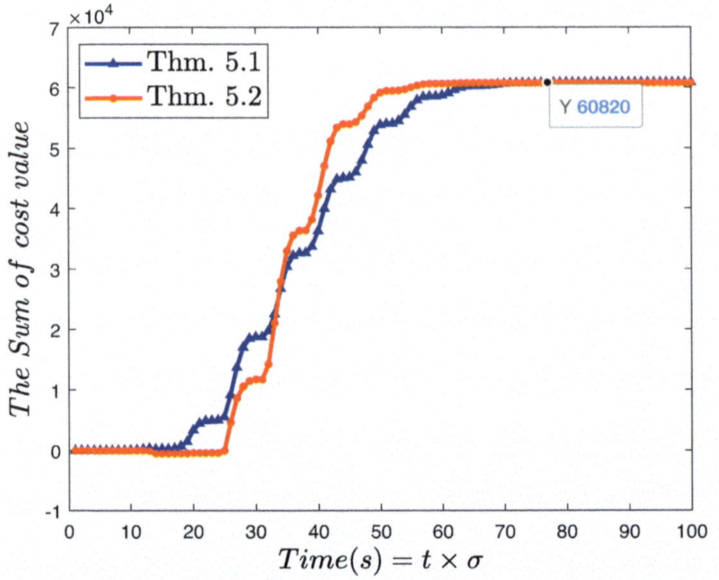

Fig. 5.15 The sum of cost value for the attacker under the ILC scheme at Nth iteration

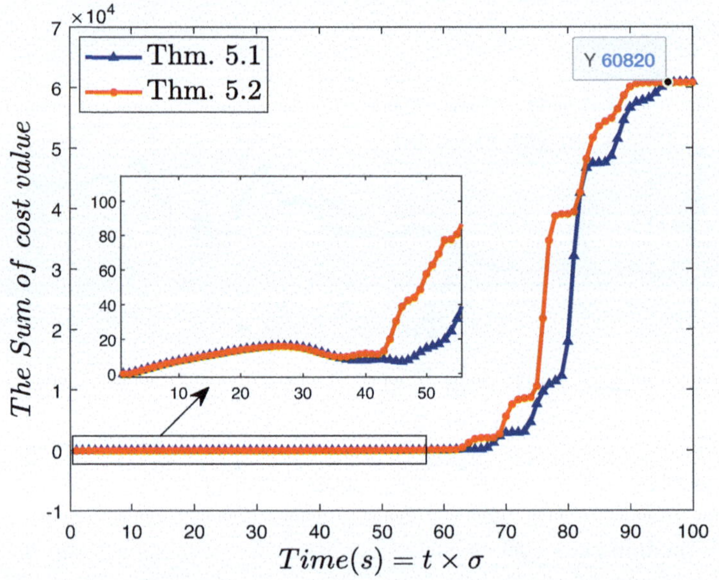

Fig. 5.16 The sum of cost value for the attacker under the ITC scheme at Nth iteration

Table 5.1 A comparison between different injection attack strategies

Strategies	Model-based	Model-free	Fixed time delay	Time-varying delays	Computational complexity
Method in [2], 2021	Yes	No	No	No	–
Method in [3], 2023	Yes	No	No	No	–
Method in [24], 2023	Yes	No	No	No	–
Method in [25], 2013	Yes	No	Yes	No	–
Method in [13], 2023	No	Yes	No	No	$O(p * (m * (t+1)) * (q * (t+1)))$, High
Method in Chap. 4	No	Yes	No	No	$O(p * (m * (t+1)) * (q * (t+1)))$, High
Theorem 5.1 textbf(ours)	No	Yes	No	Yes	$O(p * (m * (t+1)) * (q * (t+1)))$, High
Theorem 5.2 textbf(ours)	No	Yes	No	Yes	$O(p * m * q)$, Low

Combining Figs. 5.3, 5.9, and 5.10, it can be seen that the system outputs resulting from the designed CFDL and ITDL-based data-driven injection attacks deviate far from the desired value $y_r(t)$, i.e., the attack is effective. From Figs. 5.9 and 5.10, it can be viewed that the attacks against ITC scheme mainly affect the later part of the attack period, while the attacks against ILC scheme have a higher immediacy in effect. Combined with Lemma 5.3, it can be known that the attack effectiveness works as the ITDL-based data-driven model needs to spend at least $(\gamma + \Upsilon + \overline{\tau})$ units of time when calculating $\varpi_k(t)$ to obtain the optimal data-driven model. The optimal value of the sum of the attacker's cost values is 60820, which can be seen in Figs. 5.15 and 5.16. Table 5.1 shows the better adaptability of the designed attack strategy and that the ITDL-based attack strategy has low computational complexity and relaxes the memory occupation space for the attacker than the CFDL-based one and [13] and Chap. 4.

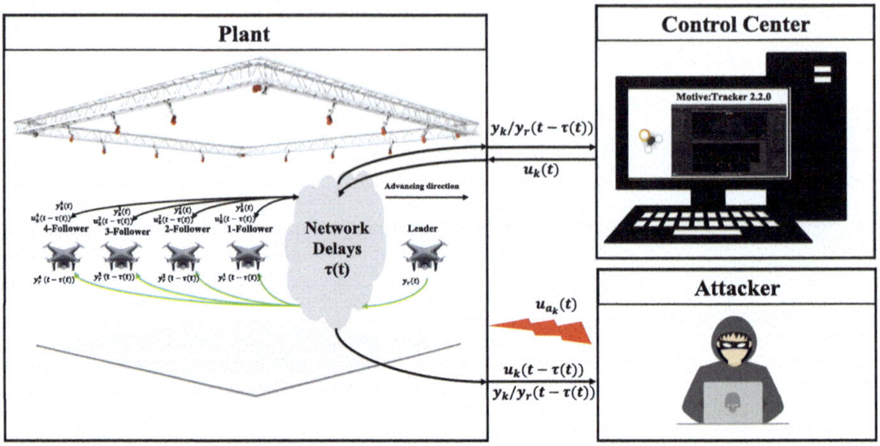

Fig. 5.17 Diagram of the experiment platform based on a leader-follower format consisting of five quadcopters

5.5.2 Experiment

To further validate the effectiveness and feasibility of the proposed attack strategies, the experimental platform was constructed, which consisted mainly of a motion capture system, two Linux-based host computers and five quadcopters in the leader-follower form. The positions of the quadcopters are gained by a motion capture system produced by Optitrack and the data is sent to the host computer via wireless network with time-varying delays. The trajectories of the different quadcopters are calculated on different threads of the host computer and then transmitted to the quadcopters for execution. We implemented altitude control for the quadcopter under attack at different heights compared to the normal quadcopters. This approach not only avoids damage due to collision but also enables the observation of trajectory changes following the attack. The attacker steals the data through the time-delay wireless network and uses the cover attack technique [16] to tamper with the control signals transmitted by the host computer to implement the injection attack, as detailed in the experimental block diagram given in Fig. 5.17.

In the experiment, the five quadcopters are placed around $x = 0\,\text{m}$ and are expected to fly in a circular formation with a speed limit of 0.2 m/s, as shown in Figs. 5.18. After adopting the designed attack strategies on the 2th follower, it can be viewed in Figs. 5.19, 5.20, 5.21, and 5.22, which the strategies successfully causes the quadcopter to collide with other followers when they are at the same altitude or to deviate from the formation.

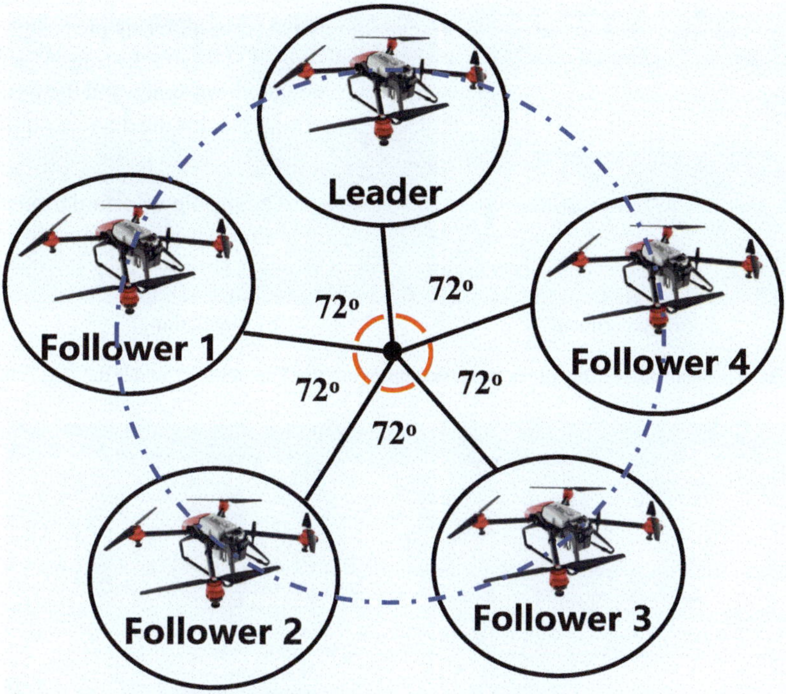

Fig. 5.18 The circular formation of five UAVs flying in a leader-follower mode

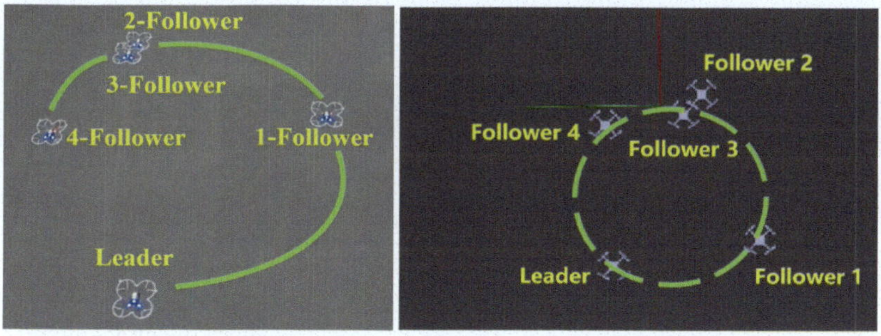

(a) Experiment (b) Simulation

Fig. 5.19 The snapshot of the attack process employing the Theorem 5.1 under the ILC scheme

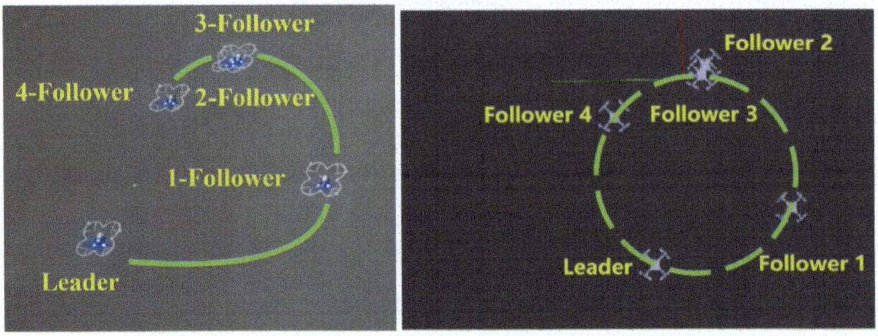

(a) Experiment (b) Simulation

Fig. 5.20 The snapshot of the attack process employing the Theorem 5.2 under the ILC scheme

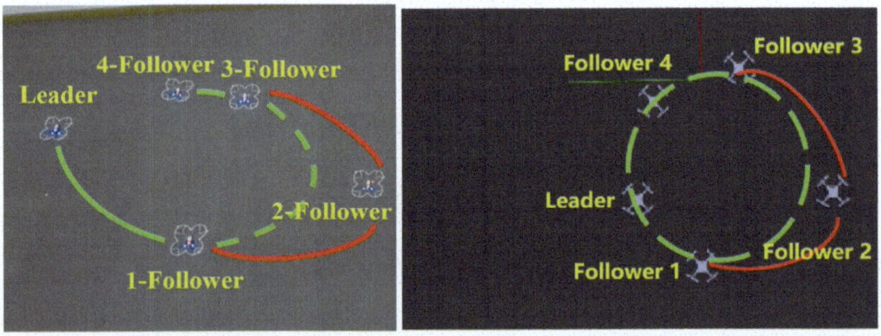

(a) Experiment (b) Simulation

Fig. 5.21 The snapshot of the attack process employing the Theorem 5.1 under the ITC scheme

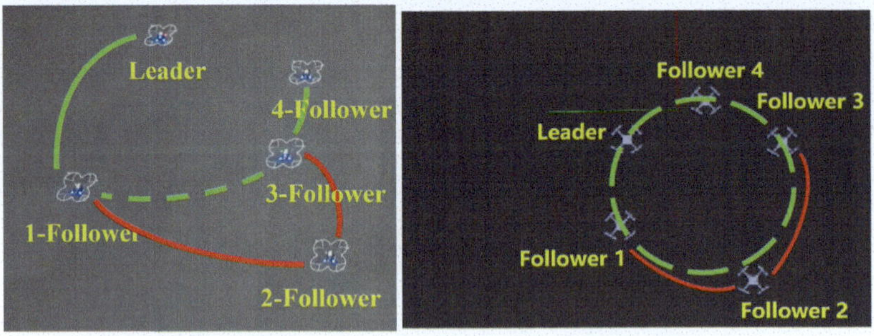

(a) Experiment (b) Simulation

Fig. 5.22 The snapshot of the attack process employing the Theorem 5.2 under the ITC scheme

5.6 Conclusion

The problems of the data-driven injection attacks based on CFDL and ITDL for unknown discrete-time cyber-physical systems with slowly time-varying delays are addressed in this chapter. The obtained data-driven injection attack strategies are not only independent of the system model but also guarantee the effectiveness of the attack in the case of communication networks with slowly time-varying delays. The numerical simulation of the discrete-time error system and the experiment based on a leader-following form consisting of five quadcopters are utilized to verify the effectiveness of the proposed strategies. In the future, we will investigate the optimal attack strategy for unknown heterogeneous cyber-physical systems with random time-varying delays under the zero-sum game framework.

References

1. Chen, Y., Kar, S., Moura, J.M.: Cyber-physical attacks with control objectives. IEEE Trans. Autom. Control **63**(5), 1418–1425 (2018)
2. Wu, G., Wang, G., Sun, J., Xiong, L.: Optimal switching attacks and countermeasures in cyber-physical systems. IEEE Trans. Syst. Man Cybern. -Syst. **51**(8), 4825–4835 (2021)
3. Huo, J.R., Li, X.J.: Stealthy switching attacks on sensors against state estimation in cyber-physical systems. Int. J. Robust Nonlinear Control **33**(2), 1169–1183 (2023)
4. Huang, X., Dong, J.: An adaptive secure control scheme for T-S fuzzy systems against simultaneous stealthy sensor and actuator attacks. IEEE Trans. Fuzzy Syst. **29**(7), 1978–1991 (2021)
5. Zhang, H., Liu, J., Wang, Z., Huang, C., Yan, H.: Adaptive switched control for connected vehicle platoon with unknown input delays. IEEE Trans. Cyber. **53**(3), 1511–1521 (2023)
6. Chen, P., Liu, S., Zhang, D., Yu, L.: Adaptive event-triggered decentralized dynamic output feedback control for load frequency regulation of power systems with communication delays. IEEE Trans. Syst. Man Cybern. -Syst. **52**(9), 5949–5961 (2022)
7. Chen, D., Liu, X., Shi, K., Shao, S., Wang, H.: Exponential stability of discrete-time systems with slowly time-varying delays. Nonlinear Anal.-Hybrid Syst. **46**, 101246 (2022)
8. Chen, D., Liu, X., Song, Y.: Stability Analysis of Discrete-time System with Slowly Time-varying Delays. Procedia Comput. Sci. **199**, 1008–1015 (2022)
9. Li, J., Zhang, L., Li, S., Su, J.: A time delay estimation interpretation of extended state observer-based controller with application to structural vibration suppression. IEEE Trans. Autom. Sci. Eng. **21**(2), 1965–1973 (2024)
10. Li, X., Lu, D., Wang, Y., Zhang, W.: Exponential synchronization of Markovian jump neural networks based on asynchronous delayed-feedback controller with uncertain hidden information. IEEE Trans. Cyber. **54**(4), 2408–2419 (2024)
11. Panda, D.K., Das, S., Townley, S.: Toward a more renewable energy-based LFC under random packet transmissions and delays with stochastic generation and demand. IEEE Trans. Autom. Sci. Eng. **19**(2), 1217–1232 (2022)
12. Chi, R., Lv, Y., Hou, Z.: Compensation-based data-driven ILC with input and output package dropouts. Int. J. Robust Nonlinear Control **30**(3), 950–965 (2020)
13. Gao, S., Zhang, H., Wang, Z., Huang, C., Yan, H.: Optimal injection attack for UCPS under vertical depth-keeping task via game approach. IEEE/CAA J. Autom. Sin. **10**(1), 287–289 (2023)
14. Deng, W., Yao, J., Ma, D.: Time-varying input delay compensation for nonlinear systems with additive disturbance: an output feedback approach. Int. J. Robust Nonlinear Control **28**(1), 31–52 (2018)

15. Wang, Z.P., Wu, H.N., Li, H.X.: Fuzzy control under spatially local averaged measurements for nonlinear distributed parameter systems with time-varying delay. IEEE Trans. Cyber. **51**(3), 1359–1369 (2021)
16. Mikhaylenko, D., Zhang, P.: Stealthy local covert attacks on cyber-physical systems. IEEE Trans. Autom. Control **67**(12), 6778–6785 (2022)
17. Wu, G.H., Xi, Y.G., Zhang, Z.J.: Robust controller design based on simplified triangular model. Automatica **34**(3), 319–325 (1998)
18. Pang, Z.H., Ma, B., Liu, G.P., Han, Q.L.: Data-driven adaptive control: an incremental triangular dynamic linearization approach. IEEE Trans. Circuits Syst. II-Express Briefs **69**(12), 4949–4953 (2022)
19. Lewis, F.L., Vrabie, D., Syrmos, V.L.: Optimal Control. Wiley (2012)
20. Binmore, K.G., Davies, J.: Calculus: Concepts and Methods. Cambridge University Press (2001)
21. Hiranuma, M., Kobayashi, D., Yokota, K., Yamamoto, K.: Chi-square automatic interaction detector decision tree analysis model: predicting cefmetazole response in intra-abdominal infection. J. Infect. Chemother. **29**(1), 7–14 (2023)
22. Ye, D., Zhang, T.Y.: Summation detector for false data-injection attack in cyber-physical systems. IEEE Trans. Cyber. **50**(6), 2338–2345 (2020)
23. Jiang, Y., Wu, S., Yang, H., Luo, H., Chen, Z., Yin, S., Kaynak, O.: Secure data transmission and trustworthiness judgement approaches against cyber-physical attacks in an integrated data-driven framework. IEEE Trans. Syst. Man Cybern. -Syst. **52**(12), 7799–7809 (2022)
24. Huo, J.R., Li, X.J.: Stealthy noise injection attacks against state estimation in interconnected systems. IEEE Trans. Circuits Syst. II-Express Briefs **70**(6), 2042–2046 (2023)
25. Sargolzaei, A., Yen, K.K., Abdelghani, M.N.: Time-delay switch attack on load frequency control in smart grid. Adv. Commun. Technol. **5**, 55–64 (2013)

Chapter 6
Game-Based DoS Attack Strategy Against CPSs via Stackelberg Game

Abstract In this chapter, the optimal DoS attack on cyber-physical systems is studied in Stackelberg game framework, which is manifested by energy allocation on communication channels. To give a realistic picture of DoS attack on a normal user launched by several hackers, a Stackelberg game between one defender and multiple attackers is considered. Compared with the existing literature, which mainly focus on equilibrium in static games, the dynamic process of the Stackelberg game is also reflected in this chapter, which demonstrates the intelligence of attackers switching channels to allocate aggressive energy. In the solution of Stackelberg equilibrium, self-adaptive particle swarm optimization (PSO) algorithm is utilized to cope with the nonlinearity of the reward function. Besides, for the purpose of better performance of both sides in choosing channels to allocate energy, an online computation algorithm is proposed. Finally, numerical examples are equipped to illustrate the theoretical results proved in this work.

6.1 Introduction

Denial-of-service attack is a common type of cyber attack in CPSs, which may jam communication channels with useless or unrelated information, thus reducing the working performance of remote receiver. To fully understand a type of cyber attack, it is necessary to first discuss how to launch the optimal attack in terms of its own modeling. However, as soon as defenders have a good command of attack pattern, they can make appropriate defensive measures, at which point it is no longer effective to design attacks without taking defenders' behaviors into consideration. For this sake, the task of studying the real-time game between attackers and defenders has profound research value. As for modeling of DoS attack, periodic behaviors have been discussed from different perspectives [1, 2]. A pulse-width modulated signal is designed to describe the attacker's periodic behavior in the former work. As soon as the attack signal jumps to a high level, the communication channel between the sensor and the controller will be instantly congested and the control signal input will fall to zero, leading to the system's rapidly entering an unstable state. Different from continuous signal [1], the latter considers the optimal schedule in discrete time

S. Gao et al., *Cyberattack Strategies Against Cyber-Physical Systems: From a Control Perspective*, https://doi.org/10.1007/978-981-95-9645-4_6

situation. An optimal DoS attack schedule is investigated to maximize expected average estimation error. To reveal the interaction between attacker, defender and transmission channels, optimal strategies for launching DoS attacks based on signal-to-noise-ratio (SINR) are proposed [3]. In detail, SINR is used to characterize the Bernoulli distribution of the event of whether there exists packet loss or not. Varied from DoS attack models mentioned above, which demand DoS attack to satisfy periodic condition or probability distribution, recent works [4–6] consider aperiodic DoS attack. Two categories of time intervals are designed to describe the existence of DoS attack, and only a constant M is needed to manifest the maximum continuous number of packet losses induced by DoS attacks. To put in a nutshell, all of these works mentioned above lay emphasis on the possibility of DoS attacks. However, facing with a multi-channel wireless communication network, smart attackers can adjust their choice of channels to allocate energy at any time, which are not discussed thoroughly.

To construct defense mechanism against DoS attacks, quantitative frameworks are required to characterize the attacker's objectives and incentives. Game theory is a suitable mathematical tool to study the behavior of the attacker and defender in the context of cyber security [7], and many researchers have produced recognizable results from a variety of perspectives [8–11]. As to whether this process converges to a stable state, the so-called equilibrium, much existing literature has proved the existence of the equilibrium point for different game models and network models [12, 13], including the optimal DoS attack under Stackelberg game [14, 15]. However, a series of issues such as how the game evolves and how the decisions of both attackers and defenders change in real time have not been studied within the scope of the authors. The main motivation of this chapter is to explore how the attacker and defender evolve dynamically under the Stackelberg game and to see whether the optimal strategy is close to the static equilibrium value.

Inspired by aforementioned discussion, the Stackelberg game between multiple DoS attackers and defender is concerned. The main contributions of this chapter consist of the following.

(1) A multi-stage Stacelberg game, where DoS attackers and defender are limited in energy and defender is uncertain about attackers' information, is constructed. Different from the existing game model [14], not only the static equilibrium point, but online strategies of energy allocation as well are considered.
(2) An improved particle swarm optimization (PSO) algorithm is applied to solve defender's optimal strategies. A Sigmoid-like nonlinear weight function is adopted to better describe the characteristics of particles searching ability than fixed weight. Compared with conventional bio-heuristic algorithms like differential evolution(DE) [14, 16] algorithm, which only preserves good results while iterating, it is more efficient in maintaining the diversity, thus avoiding converging to local optima.

(3) For better application in different network environments, games under different types of network parameters are analyzed. When energy is limited, it is not always a good choice to distribute energy to all channels, let alone channels which are safeguarded. Wiser choices for both sides are to allocate energy on certain channels decided by online computation.

The remainder of the chapter is organized as follows. Section 6.2 introduces some basic concepts related to system models, information transmission and presents formulation of Stackelberg game. Section 6.3 provides analysis of game-based DoS attack strategies with incomplete information from different perspectives, respectively. Numerical examples are given in Sect. 6.4. The conclusion of the chapter is drawn in Sect. 6.5.

6.2 Problem Formulation

6.2.1 System Model

The cyber-physical system is composed of a physical plant, m sensors installed on interference-prone communication channels and a remote estimator. Suppose that there exists n attackers in the wireless environment, which may launch DoS attacks on the vulnerable transmission channels. As for the defender (controller and estimator), m wireless channels can be utilized to transmit information about system states. The system diagram is given in Fig. 6.1, where $x_k \in \mathbb{R}^{n_x}$ is the state vector, $\hat{x}_k^{s,i} \in \mathbb{R}$ is the measurement taken by sensor i at time k, and $\hat{x}_k = [\hat{x}_k^1, \ldots, \hat{x}_k^m]^T \in \mathbb{R}^m$ is the information vector received by remote receiver, which satisfies

$$\hat{x}_k^i = \begin{cases} \hat{x}_k^{s,i} & \text{if } i \text{ th channel is not attacked, } i \in \{1, \ldots, m\}, \\ 0 & \text{otherwise.} \end{cases} \tag{6.1}$$

6.2.2 Communication Model

As Fig. 6.1 indicates, the measurements of the plant are collected by the sensors and conveyed to the remote receiver so that the user can monitor the status of the system. The wireless communication network plays as the medium of transmission. However, fading and interference are inevitable in network transmission, thus resulting in data packet dropout. Assume that independent additive gaussian white noise exist in all channels. In order to evaluate the distortion of data after transmission, the signal-to-noise ratio (SNR) is defined as

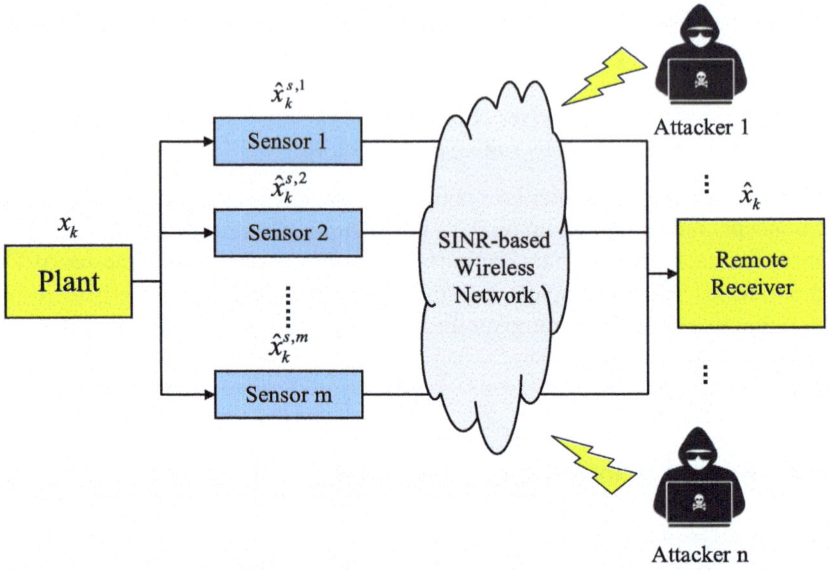

Fig. 6.1 A cyber-physical system under DoS attack launched by multiple attackers

$$\rho_i = \text{SNR}_i = \frac{\alpha_i \lambda_i}{\sigma_{bgn}}, \quad i \in \mathcal{M}, \tag{6.2}$$

where $\mathcal{M} = \{1, 2, \ldots, m\}$, $\alpha_i > 0$ refers to the fading channel gain of the user for the i-th channel. σ_{bgn} describes the level of background noise affecting the wireless communication network. $\lambda_i \geq 0$ denotes the energy allocated by the user to the i-th channel. Once we consider DoS attacks launched by potential attackers in the network, SNR should be revised as signal-to-interference-and-noise ratio (SINR)

$$\pi_i = \text{SINR}_i = \frac{\alpha_i \lambda_i}{\sigma_{bgn} + \beta_i \theta_i}, \quad i \in \mathcal{M}, \tag{6.3}$$

where $\beta_i > 0$ refers to the fading gain of the attacker for the i-th channel. θ_i denotes the energy used by attackers to the i-th channel.

6.2.3 Reward Function and Strategy

In this chapter, a Stackelberg game between multiple attackers and a strong defender with asymmetric information is considered. In detail, the defender is clear about every attacker's energy, but he does not know which attacker exists in the environment. Instead, the defender has a command of the probability that any attacker launches DoS attack.

Assume that the defender is sure that the energy of every attacker is θ_M^j and their attacking probability $\gamma^j \in (0, 1)$, where $j \in \mathcal{N}$, $\mathcal{N} = \{1, 2, \ldots, n\}$, and $\sum_{i=1}^m \gamma_i^j = 1$. Note that the energy levels of attackers are independent and distinct from others.

The reward for the defender is given as follows

$$\begin{aligned}
J_d(\lambda, \theta) &= \gamma_1 \sum_{i=1}^m \frac{\alpha_i \lambda_i}{\sigma_{bgn} + \beta_i \theta_i^1} + \gamma_2 \sum_{i=1}^m \frac{\alpha_i \lambda_i}{\sigma_{bgn} + \beta_i \theta_i^2} \\
&+ \cdots + \gamma_n \sum_{i=1}^m \frac{\alpha_i \lambda_i}{\sigma_{bgn} + \beta_i \theta_i^n} - \eta_d \sum_{i=1}^m \lambda_i, \\
&= \sum_{j=1}^n \sum_{i=1}^m \gamma_j \frac{\alpha_i \lambda_i}{\sigma_{bgn} + \beta_i \theta_i^j} - \eta_d \langle \mathbf{1_m}, \lambda \rangle, \\
&= \langle \gamma, \bar{\pi} \rangle - \eta_d \langle \mathbf{1_m}, \lambda \rangle,
\end{aligned} \tag{6.4}$$

where

$$\begin{aligned}
&\gamma = [\gamma^1, \ldots, \gamma^n]^T, \bar{\pi} = [\bar{\pi}^1, \ldots, \bar{\pi}^n]^T, \bar{\pi}^j = \langle \mathbf{1_m}, \pi^j \rangle, \\
&\pi^j = [\pi_1^j, \ldots, \pi_m^j]^T, \theta = [\theta_1, \ldots, \theta_n]^T.
\end{aligned}$$

The reward for all possible attackers is unique, which is presented as superscript j.

$$\begin{aligned}
J_a^j(\lambda, \theta^j) &= -\sum_{i=1}^m \pi_i^j + \eta_d \sum_{i=1}^m \lambda_i - \eta_a \sum_{i=1}^m \theta_i^j, \\
&= -\langle \mathbf{1_m}, \pi^j \rangle + \eta_d \langle \mathbf{1_m}, \lambda \rangle - \eta_a \langle \mathbf{1_m}, \theta^j \rangle.
\end{aligned} \tag{6.5}$$

The strategy sets for defender and attacker are given as follows, where $\mathcal{R}_d, \mathcal{R}_a^j$ represent the strategy set for defender and j-th attackers, respectively.

$$\mathcal{R}_d = \left\{ (\lambda_1, \ldots, \lambda_m) | \sum_{i=1}^m \lambda_i = \bar{\lambda}_M, \lambda_i \geq 0 \right\}, \tag{6.6}$$

$$\mathcal{R}_a^j = \left\{ (\theta_1^j, \ldots, \theta_m^j) | \sum_{i=1}^m \theta_i^j = \bar{\theta}_M^j, \theta_i^j \geq 0 \right\}. \tag{6.7}$$

Moreover, it should be emphasized that $n > 2$. Otherwise, there is only one DoS attacker in the wireless environment, which means the Stackelberg game is of complete information.

Remark 6.1 The proposed Stackelberg game has profound background in reality. Practically, it is possible for several attackers from one camp to attack a common target with negotiation: In every time interval, only one attacker can launch DoS attack on channels. Corresponding to mathematical expressions, describing the attackers' payoff function in a probability-weighted form reflects the independence of decision making among multiple attackers.

6.2.4 Stackelberg Game

A game is completely determined by four components: decision-makers, options that they can choose, their objectives in choosing an option, and what they know about the structure of the interaction. The components of Stackelberg game in this chapter are listed as follow:

- Player: Defender (user) and DoS Attackers.
- Strategy: $\mathcal{R}_d$, $\mathcal{R}_a$.
- Payoff (Reward): $J_d(\lambda, \theta)$, $J_a^j(\lambda, \theta)$.
- Interaction: A two-stage dynamic game, the time in which is sequential: First the defender makes a decision, then it is the attacker's turn to make a decision, and so on through time.

6.3 Game-Based DoS Attack Strategies

6.3.1 Game-Based DoS Attack Strategy via Static Stackelberg Game

In this subsection, the static equilibrium of the proposed Stackelberg game is investigated. The graphical illustration of the static equilibrium solution is given as Fig. 6.2.

To facilitate the following discussion, the definition of best response for each side is given as follows [17].

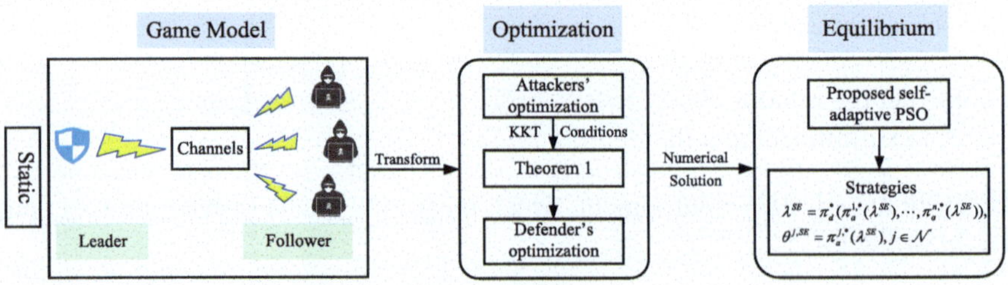

Fig. 6.2 Graphical illustration of static equilibrium solution

Definition 6.1 The best response is the action that produces the maximal payoff for a player while taking other players' actions as given. Specifically, the best response for the attackers and the defender are given as

$$\pi_d^*(\theta) = arg \max_{\lambda \in \mathcal{R}_d} J_d(\lambda, \theta^1, \ldots, \theta^n), \tag{6.8}$$

$$\pi_a^{j,*}(\lambda) = arg \max_{\theta \in \mathcal{R}_a^j} J_a(\lambda, \theta^j), \tag{6.9}$$

respectively.

Lemma 6.1 *The solution to the Stackelberg game can be obtained by solving the following equations sequentially*

$$\lambda^{SE} = \pi_d^*(\pi_a^{1,*}(\lambda^{SE}), \ldots, \pi_a^{n,*}(\lambda^{SE})), \tag{6.10}$$

$$\theta^{j,SE} = \pi_a^{j,*}(\lambda^{SE}), \ j \in \mathcal{N}. \tag{6.11}$$

and $(\lambda^{SE}, \theta^{1,SE}, \ldots, \theta^{n,SE}) \in \mathcal{R}_d \times \mathcal{R}_a^1 \times \cdots \times \mathcal{R}_a^n$ *constructs the Stackelberg equlibrium.*

Proof As the leader in the Stackelberg game, the attacker chooses its strategy according to the defender's previous strategy λ, which is $\theta^j = \pi_a^*(\lambda)$, $j \in \mathcal{N}$. Then, the follower (defender) chooses the corresponding strategy to maximize its reward function $J_d(\lambda, \pi_a^*(\lambda))$

$$\lambda^{SE} = \pi_d(\theta^j) = \pi_d^*(\pi_a^*(\lambda^{SE})),$$

which completes the proof. □

Remark 6.2 "Stackelberg Equilibrium" is a concept different from "Nash Equilibrium", for Nash Equlibrium is based on a simultaneous-action game, while Stackelberg Equilibrium is based on a time sequential game. Besides, the inequalities rewards for each side are different from those in Nash Equilibrium:

$$J_d(\lambda^*, \theta^*) \geq J_d(\lambda, \theta^*), \quad J_a(\lambda^*, \theta^*) \geq J_a(\lambda^*, \theta).$$

In the reward of the defender, the only issue is to consider how to obtain the optimal allocation of defender's total energy. Such explanation is also suitable for the reward of the attacker.

Since it is specified that energy used in the channels for both sides are constant, the optimization problem of the attacker can be simplified as:

Problem 6.1

$$\begin{aligned} &\min_{\theta^j \in \mathcal{R}_a^j} \ \langle \mathbf{1}_m, \pi^j \rangle \\ &\quad s.t. \ \langle \mathbf{1}_m, \theta^j \rangle - \bar{\theta}_M^j = 0, \\ &\qquad -\theta_i^j \leq 0, i \in \mathcal{M}. \end{aligned}$$

For

$$\frac{\partial \langle \mathbf{1}_m, \pi^j \rangle}{\partial \theta_i^j} = -\frac{\alpha_i \beta_i \lambda_i}{(\sigma_{bgn} + \beta_i \theta_i^j)^2}, \tag{6.12}$$

the second order mixed partial derivative is

$$\frac{\partial^2 \langle \mathbf{1}_m, \pi^j \rangle}{\partial \theta_i^j \partial \theta_i^k} = 0, \quad i \in \mathcal{M}, \ j \neq k, \ j, k \in \mathcal{N}. \tag{6.13}$$

Therefore, the explicit form of Hessian matrix can be expressed as

$$\mathcal{H}^j = diag(\frac{\partial^2 \langle \mathbf{1}_m, \pi^j \rangle}{\partial (\theta_1^j)^2}, \ \ldots, \ \frac{\partial^2 \langle \mathbf{1}_m, \pi^j \rangle}{\partial (\theta_m^j)^2}), \tag{6.14}$$

where

$$\frac{\partial^2 \langle \mathbf{1}_m, \pi^j \rangle}{\partial (\theta_i^j)^2} = \frac{2\alpha_i \beta_i^2 \lambda_i}{(\sigma_{bgn} + \beta_i \theta_i^j)^3}, \quad i \in \mathcal{M}. \tag{6.15}$$

Theorem 6.1 *Given the defender's strategy λ, the corresponding optimal strategy for the attacker is*

$$\theta_i^j(\lambda) = \max\left\{\frac{1}{\beta_i}(\sqrt{\frac{\alpha_i \beta_i \lambda_i}{\bar{\mu}^j}} - \sigma_{bgn}), 0\right\}, \ i \in \mathcal{M}, \ j \in \mathcal{N}, \tag{6.16}$$

where $\bar{\mu}$ is the solution of the following equation:

$$\sum_{i=1}^{m} \max\left\{\frac{1}{\beta_i}(\sqrt{\frac{\alpha_i \beta_i \lambda_i}{\bar{\mu}^j}} - \sigma_{bgn}), 0\right\} = \bar{\theta}_M^j. \tag{6.17}$$

Proof Define the Lagrangian function for every single attacker j as follows:

$$L(\hat{\theta}^j, \mu^j) = \langle \mathbf{1_m}, \pi^j \rangle + \mu_0(\langle \mathbf{1_m}, \hat{\theta}^j \rangle - \hat{\theta}_M^j) + \sum_{i=1}^{m} \mu_i^j \theta_i^j, \tag{6.18}$$

where $\mu_0^j > 0$ and $\mu_i^j \geq 0$, $i \in \mathcal{M}$. For this sort of optimization with inequality constraints, KKT conditions are commonly utilized to deal with it. In this problem, KKT conditions are enumerated as follows.

$$\begin{aligned}\frac{\partial L(\hat{\theta}^j, \mu^j)}{\partial \theta_i^j} &= 0, \quad i \in \mathcal{M},\\ \langle \mathbf{1_M}, \hat{\theta}^j \rangle - \bar{\theta}_M &= 0,\\ \mu_0(\langle \mathbf{1_m}, \hat{\theta}^j \rangle - \bar{\theta}_M) &= 0,\\ -\mu_i \theta_i^j &= 0, \quad i \in \mathcal{M},\\ -\theta_i^j &\leq 0, \quad i \in \mathcal{M}.\end{aligned}$$

Calculate the partial derivative in the first equation listed above, whose result is

$$\frac{\partial L(\hat{\theta}^j, \mu^j)}{\partial \theta_i} = -\frac{\alpha_i \beta_i \lambda_i}{(\sigma_{bgn} + \beta_i \theta_i^j)^2} + \mu_0^j + \mu_i^j = 0.$$

Define $\mu^j = \mu_0^j + \mu_i^j > 0$. Then the relationship between θ_i^j and λ can be obtained as (6.16). □

Remark 6.3 It is possible that the defender does not allocate energy on some of the channels. As for cyberattacks and defense in real world, the message that communication channel is not in use cannot be received immediately. There usually exists a time delay. Therefore, in the proposed Stackelberg game, attackers are likely to waste energy in an empty channel.

Since $\langle \mathbf{1_m}, \lambda \rangle$ is a certain value just as assumed, the best response of the defender can be obtained by solving the following optimization problem:

Problem 6.2

$$\begin{aligned}&\max_{\lambda \in \mathcal{R}_d} \ \langle \gamma, \bar{\pi} \rangle\\ &s.t. \ \langle \mathbf{1}_m, \lambda \rangle - \bar{\lambda}_M = 0,\\ &\quad - \lambda_i \leq 0, i \in \mathcal{M}.\end{aligned}$$

The cost function may not be convex, and the relationship between θ_i and λ_i is nonlinear, so traditional analytical methods are not suitable for this problem. Numerical approaches should be taken into consideration.

Inspired by biological evolution and swarm intelligence in nature, differential evolution algorithm and particle swarm optimization algorithm emerged as powerful tools to deal with complex computational problems. Compared with DE, PSO algorithm has more effective memory capability and higher efficiency in maintaining diversity of swarm. Besides, PSO is more likely to converge to the global optimum in various problems, where analytical approaches fail to converge [18]. In this chapter, self-adaptive PSO is selected as a prior algorithm to solve the optimization of the defender, the weight of speed in which is designed as a Sigmoid-like function. Main stages of self-adaptive PSO are shown as Algorithm 6.

Algorithm 6: Self-Adaptive PSO Algorithm

Input: Number of Generations: $G_{\max}$, Quantity of particles: N, Dimensionality of search space: D, possible maximum and minimum values of particles: p_{max}, p_{min}, Fitness function: f, Weight coefficients: c_1, c_2, ω_{ini}, ω_{end}.

Output: Best position of the particles p_{gbest}, Best value J of the fitness function f.

Initialize every particle's position x_{id} and velocity v_{id}.
Initialize J_0. Set iteration number as $k = 1$.
while $G < G_{\max}$ **do**
  for *each particle* $i \in [1, N]$ **do**
    Calculate fitness value $f(\overrightarrow{X_i})$.
    if $f(\overrightarrow{X_i})$ *is better than* $p^k_{id,pbest}$ *in history* **then**
      $p^k_{id,pbest} = f(\overrightarrow{X_i})$.
  Choose the particle having the best fitness value as $p^k_{d,gbest}$.
  Calculate the adaptive coefficient: $c_{ad} = \frac{std(fitness(k))}{std(fitness(k-1))}$.
  for *each particle* $i \in [1, N]$ **do**
    The adaptive weight: $\omega_{ad} = (\omega_{ini} - \omega_{end}) \times \frac{1}{1+exp\{\frac{-1}{c_{ad}} \times \frac{N-i}{N}\}} + \omega_{end}$
    for *each dimension* $d \in [1, D]$ **do**
      Calculate velocity:
      $v^{k+1}_{id} = \omega \times v^k_{id} + c_1 \times rand^1_{[0,1]} \times (p^k_{id,pbest} - x^k_{id}) + c_2 \times rand^2_{[0,1]} \times (p^k_{d,gbest} - x^k_{id})$.
      Update position:
      $x^{k+1}_{id} = x^k_{id} + v^{k+1}_{id}$.
      if *Particle's velocity or position is out of pre-set range* **then**
        Reinitialize velocity v_{id} or position x_{id}.
  $G = G + 1$.

Remark 6.4 To accelerate the convergence of PSO, a Sigmoid-like function is applied to construct the self-adaptive weight in the speed update equation.

$$\omega_{ad} = (\omega_{ini} - \omega_{end}) \times \frac{1}{1 + exp\{\frac{-1}{c_{ad}} \times \frac{N-i}{N}\}} + \omega_{end}.$$

Figure 6.3 shows different curves of ω under different adaptive coefficient values. It can be seen that the smaller c_{ad}, the greater the ω_{ad}. In the sense of the PSO algorithm, this can be explained by the fact that if the standard deviation gap of the previous particles is smaller, the particles will tend to continue exploring along the original search direction. Otherwise, the particles are likely to search in a new direction. In the end, $c_{ad} = 1$ means the algorithm has converged.

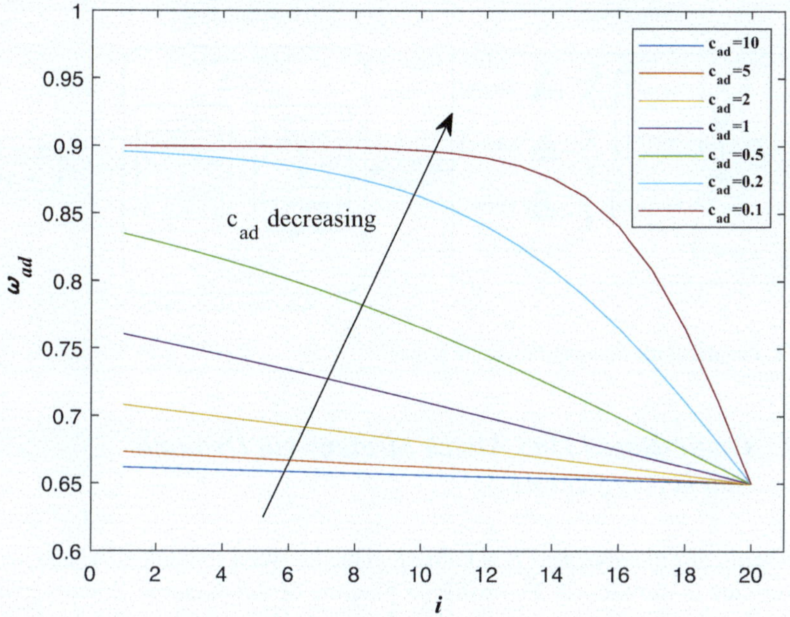

Fig. 6.3 The effect of different adaptive coefficients c_{ad} on adaptive weight ω_{ad}

The proposed PSO algorithm in this chapter consists of several steps: initialization of particles' position and velocity, calculation of fitness value, comparisons to get p_{best}, acquisition of g_{best} and updating of velocity and position information. Note that the last four steps are executed in the same loop.

In the first step, every dimension of each particle's position and velocity ought to be randomly chosen, bringing about the time complexity of $O(ND)$. The second step and the third step are contained in the same loop, the time complexity of which is $O(ND)$. For the worst case, the position information of all particles in all dimensions is covered, which results in the time complexity of $O(ND)$. Likewise, number of comparisons are possible to reach $N \times D$, which indicates all particles of this generation update better fitness values. In the fourth step, numbers of comparisons are smaller than that of p_{best}, whose level is $O(N)$. The last step acts similarly as the first step. The difference lies in that the position and velocity of particles are updated according to certain laws.

Overall, considering the loop times (generations) of last four steps, the time complexity of the proposed PSO is $O(GND)$. It can be concluded that the computation time of the proposed PSO mainly hinges on the population size, the dimension of particles and number of generation iterations.

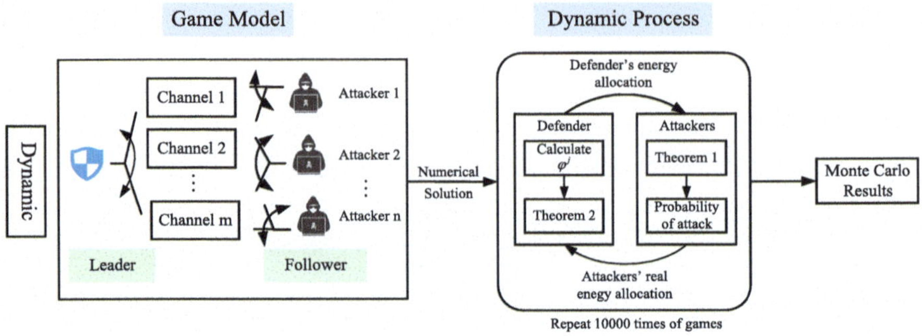

Fig. 6.4 Graphical illustration of dynamic game

6.3.2 Game-Based DoS Attack Strategy via Dynamic Stackelberg Game

In this subsection, a dynamic Stackelberg game between the defender side and the attacker side is introduced. The working pipeline of the dynamic game is given in Fig. 6.4. In the game model, switches beside the attackers and the defender represent their real-time choices to allocate energy on certain channels.

To set the stage for the proposed dynamic Stackelberg game strategy solving algorithm that follows, a theorem related to optimal strategy of the defender needs to be introduced.

First, a vector of weight factors in the reward function of the defender is defined, which is necessary for the theorem proposed next.

Definition 6.2 The weight factors in the reward function of the defender are given by:

$$\varphi^j = [\varphi_1^j, \ldots, \varphi_m^j]^T, \tag{6.19}$$

where

$$\varphi_i^j = \frac{\gamma_j \alpha_i}{\sigma_{bgn} + \beta_i \theta_i^j}, \; i \in \{1, \ldots, m\}. \tag{6.20}$$

Now the reward function of the defender side can be rewritten into the form of the product of vectors: $J_d = \sum_j {\varphi^j}^T \lambda$, which is a linear function respect to λ. Thus, it is natural to obtain the following theorem:

Theorem 6.2 *Given the attacker's strategy θ^j, the corresponding optimal strategy for the defender is to output all of its limited energy to the ith channel, where i is the index of the maximum item in weight factor φ^j. If the index i is not unique, the optimal strategy of the defender is a set*

$$\{(\lambda_{i_1}, \ldots, \lambda_{i_k}) | \Sigma_{r=1}^k \lambda_{i_r} = \bar{\lambda}_M\}, \tag{6.21}$$

where k is the number of the same elements in φ^j.

Proof Suppose φ_p^j, $p \in \mathcal{M}$ is the greatest among all the elements in vector φ^j, and all the energy $\bar{\lambda}_M$ is distributed on the p th channel. If the defender decides to transmit a little bit energy $\Delta\lambda$ onto another channel q, $q \in \mathcal{M}$ and $q \neq p$, then the amount of change in J_d is:

$$\begin{aligned}\Delta J_d = J_{d,new} - J_{d,old} &= \varphi_q^j(\bar{\lambda}_M - \Delta\lambda) + \varphi_p^j \Delta\lambda - \varphi_q^j \bar{\lambda}_M \\ &= (\varphi_q^j - \varphi_p^j)\Delta\lambda < 0,\end{aligned}$$

leading to a loss in reward.

If there are more than one element in φ^j having the greatest value, the defender can alternatively send the energy onto corresponding channels. The proof is similar to the single maximum case, which is omitted here. □

Combining Theorem 6.2 with Theorem 6.1, the online strategy solution algorithm can be constructed as Algorithm 7.

Algorithm 7: Online Strategy via Dynamic Stackelberg Game

Input: Number of rounds: $round_{max}$, Number of channels: m, Number of attackers: N, Network parameters: α_i, β_i.

Output: Defender's optimal strategy: λ_{opt}, Attacker's optimal strategy: θ_{opt}^j, Reward values: J_d, J_a^j.

```
Initialize strategies by averagely allocating energy to every channel.
for round ∈ [1, round_max] do
    if round is odd then
        Using Theorem. 2 to solve the current optimal strategy for defender.
        Remain attackers' strategy in last round.
    if round is even then
        Using Theorem. 1 to solve the current optimal strategy for attackers.
        Remain defender's strategy in last round.
    Store in λ_opt, θ^j_opt the strategies of the defender and attacker calculated at the current
    round.
    Calculate J_d, J^j_a.
```

6.4 Numerical Simulation

In this section, numerical examples of static Stackelberg equilibrium and dynamic Stackelberg game are provided to illustrate the theoretical results given in this chapter. As for the former, the results of two types of Stackelberg game are given through the proposed self-adaptive PSO algorithm. With regard to the latter, three possible scenarios with different wireless channel gains are given for detailed analysis.

6.4.1 Static Stackelberg Game

For the wireless channels in CPS, consider the parameters chosen in [14], where $m = 4$, $\bar{\lambda}_M = 11$, $\eta_a = \eta_d = 0.3$, $\sigma_{bgn} = 0.2$, $\alpha = [0.8, 0.7, 0.6, 0.5]^T$ and $\beta = [0.2, 0.3, 0.4, 0.5]^T$. The parameters in PSO are $N = 200$, $D = 4$, $p_{min} = 0$, $p_{max} = 11$, $c_1 = c_2 = 2$, $\omega_{ini} = 0.9$, $\omega_{end} = 0.4$. Suppose there are $n = 3$ attackers in the wireless environment, possessing energy $\theta = [\theta^1, \theta^2, \theta^3]^T = [4, 3, 5]^T$, and their attacking probability is $\gamma = [0.1, 0.6, 0.3]^T$. Fitness function is $g = \langle \gamma, \bar{\pi} \rangle - X(\lambda)Q(\lambda)$, where $X(\lambda) = 1 + \frac{|\langle \gamma, \bar{\pi} \rangle|}{1+Q(\lambda)}$, $Q(\lambda) = |\langle \lambda, \mathbf{1}_m \rangle - \lambda_M|$.

After 50 individual simulations, the average curve of fitness value is given in Fig. 6.5. Taking advantage of the property that PSO itself preserves the historical global optimum, the best allocation for the defender and attackers are obtained in the 22nd simulation, where

$$\lambda = [4.9558, 3.0484, 1.8283, 1.1676]^T, \ \theta^1 = [1.9443, 1.5131, 0.3384, 0.2031]^T,$$
$$\theta^2 = [1.4962, 1.1814, 0.2108, 0.1113]^T, \ \theta^3 = [2.3927, 1.8451, 0.4661, 0.2949]^T.$$

The corresponding reward values of both sides are exhibited in Fig. 6.6, where $J_d = 10.3070$, $J_a^1 = -10.8987$, $J_a^2 = -13.1559$, $J_a^3 = -9.3440$. It can be observed

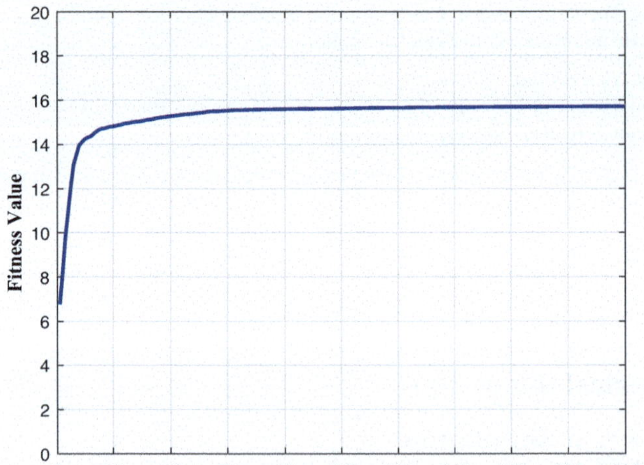

Fig. 6.5 Fitness value of game type II

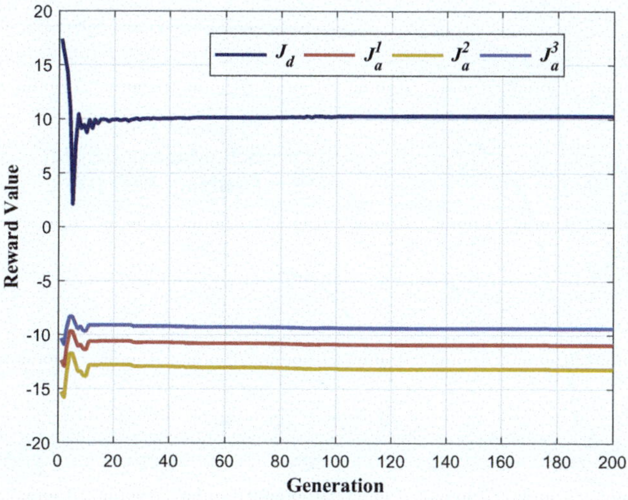

Fig. 6.6 Reward values of both sides

that the less probability the attacker launches DoS attacks, the more loss he suffers in his reward.

To show the advantages of the PSO algorithm proposed in this chapter, a set of comparative experiments are considered. With the same network parameters, population size and generations, the proposed PSO, the self-adaptive DE algorithm [14] and the conventional PSO algorithm [18] are operated respectively, and their computational time and the final results obtained are shown in Table 6.1.

It can be observed from Table 6.1 that although the computation time of the adaptive DE is significantly lower than the other two algorithms for the same spatial complexity, the optimal policy obtained violates the energy constraint and the mean fitness value is smaller than those of PSOs. Besides, in this simulation, the two PSO algorithms do not show significant differences, both from the perspective of com-

Table 6.1 Comparison between different optimization algorithms

Algorithm	Average computational time (s)	Optimal strategy for the defender	Mean fitness value of 50 repeated simulations
Proposed PSO	5.15	$[4.9654,2.9088,1.8856,1.2402]^T$	15.7267
Self-adaptive DE [14]	1.03	$[6.3706,0.0547,3.2665,2.1871]^T$	14.5355
Ordinary PSO [18]	4.94	$[4.9768,2.9053,1.8714,1.2464]^T$	15.7252

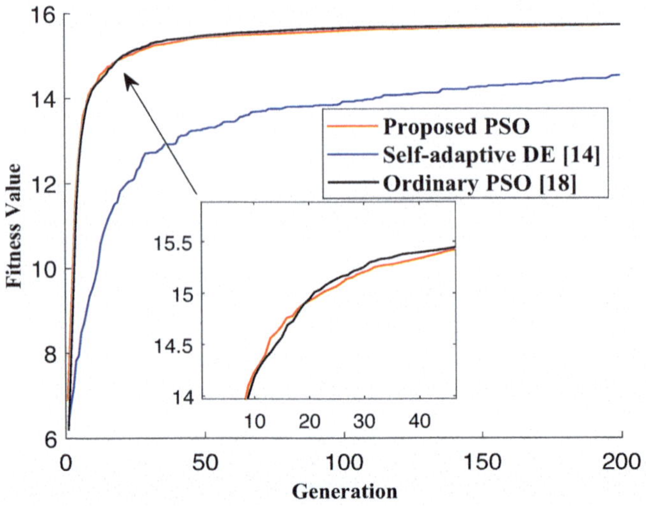

Fig. 6.7 Comparison of different optimization algorithms

Table 6.2 Comparison between PSOs

Algorithm	Average computational time (s)	Optimal strategy for the defender	Mean fitness value of 50 repeated simulations
Proposed PSO	5.06	$[7.3098, 3.1070, 0, 1.4393, 4.6495, 1.4734]^T$	14.0786
Ordinary PSO [18]	5.40	$[1.3249, 3.1096, 6.9165, 1.1726, 3.2874, 2.3281]^T$	13.6112

putation time and convergence. From Fig. 6.7, it can be seen that there only exists slight difference between PSOs in evolution process.

In order to contrast more clearly the superiority of the proposed PSO algorithm, a six-channel CPS system is considered and the number of attackers is increased to five. The parameters of the channels are $m = 6, \bar{\lambda}_M = 18, \eta_a = \eta_d = 0.3, \sigma_{bgn} = 0.2, \alpha = [0.8, 0.7, 0.6, 0.5, 0.4, 0.3]^T$ and $\beta = [0.2, 0.3, 0.4, 0.5, 0.6, 0.7]^T$. The attackers' energy is $\theta = [\theta^1, \theta^2, \theta^3, \theta^4, \theta^5]^T = [4, 6, 3, 2, 3]^T$, and their attacking probability is $\gamma = [0.1, 0.2, 0.3, 0.3, 0.1]^T$. Other settings remain the same. The computational time and mean fitness value are displayed in Table 6.2.

From Tables 6.1 and 6.2, it can be discovered that as the complexity of the CPS and the number of attackers increase, the processing time of the ordinary PSO starts to grow, while the computation time of the proposed PSO remains stable. The main reason is that the search particles are more autonomous and can adjust their own search direction in time according to the convergence extent of the previous particles. What's more, Fig. 6.8 also confirms the statement of autonomous particles. The mean fitness value derived by proposed PSO is better than that of ordinary PSO,

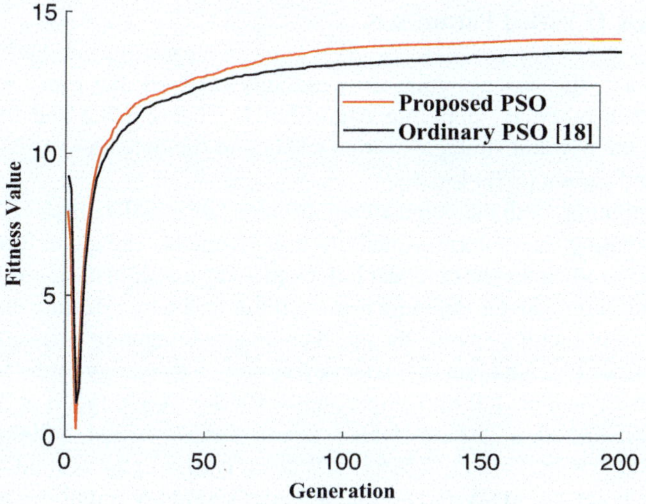

Fig. 6.8 Comparison of PSOs

which implies that proposed PSO is more promising for obtaining static equilibrium solutions of complex Stackelberg games.

6.4.2 Dynamic Stackelberg Game

In this subsection, multi-stage game under different groups of network parameters and two types of games are discussed in detail. All the parameters except channel gains are the same as the previous subsection. Three possible groups of channel gains for both sides are listed in Table 6.3.

Table 6.3 Channel gains of both sides on different channels

Channel number i		1	2	3	4
Case 1	α_i	0.8	0.7	0.6	0.5
	β_i	0.2	0.3	0.4	0.5
Case 2	α_i	0.7	0.7	0.7	0.1
	β_i	0.4	0.4	0.4	0.1
Case 3	α_i	0.7	0.7	0.7	0.7
	β_i	0.4	0.4	0.4	0.4

6.4.2.1 Case 1: Varied Parameters

In the first case, the channel gains are randomly selected, but every two of them vary from each other. By applying Algorithm 7, Theorem 6.2 and Monte Carlo simulations, the optimal strategies for attackers and the defender on every channel are solved and shown in Table 6.4.

At the beginning, both the defender and the attacker initialize their energy allocation by distributing their energy equally to four channels. The game formally starts from the first round. Since the defender is defined as the leader in the game framework, the first round starts with the defender making the decision by selecting the maximum item from weight factor φ^j while the attackers maintaining their original strategy. It can be seen from Fig. 6.8a that defender devotes all of its energy to the first channel while the other channels fall into empty status, for the weight factor at this round is $\varphi^1 = [5.5000, 3.8500, 2.7500, 2.7500]^T$, $\varphi^2 = [6.2857, 4.5294, 3.3000, 2.3913]^T$, $\varphi^3 = [4.8889, 3.3478, 2.3571, 1.6667]^T$. When it comes to the second round, decision making is left to attackers and the defender's strategy remains unchanged. A randomly selected attacker chooses its optimal strategy according to Theorem 6.1. As the game goes on through time, the defender and attackers will take turns to make decisions, forming defender's cyclical switching strategy presented between channel 1 and channel 2. However, the attackers are constrained by the probability distribution, which means they will not always have the opportunity to launch attacks, leading to the expected energy allocation lower than the results obtained in static equilibrium. An example of the multi-stage game is given in Figs. 6.9, 6.10, 6.11 and 6.12.

Besides, compared with the results of static equilibrium, although the exact value of energy distributed on channels and reward values are different, the attacker and the defender are likely to devote most of their energy to channel 1 and channel 2 in both scenarios.

6.4.2.2 Case 2: Partly Identical Parameters

In this case, some of the elements the weight vector φ^j are set to be the same, which means the defender can choose several possible channels to distribute energy. For simplicity, some of the channel gains for the attackers and the defender are set to be the same, as are given in Table 6.3.

Different from Case 1, the defender and attacker no longer choose their strategies in a periodic order. According to Theorem 6.2, when the attacker's strategy is fixed, the optimal solution to defender's optimization problem is a set:

$$\{(\lambda_1, \lambda_2, \lambda_3, \lambda_4) | \lambda_1 + \lambda_2 + \lambda_3 = \bar{\lambda}_M, \lambda_1, \lambda_2, \lambda_3 \geq 0, \lambda_4 = 0\}.$$

The defender can make relatively free decisions in this situation. A simple idea is to separate the energy equally to three channels, but it is too simple to be missed by the attacker. Here a discrete uniform distribution is introduced to characterize

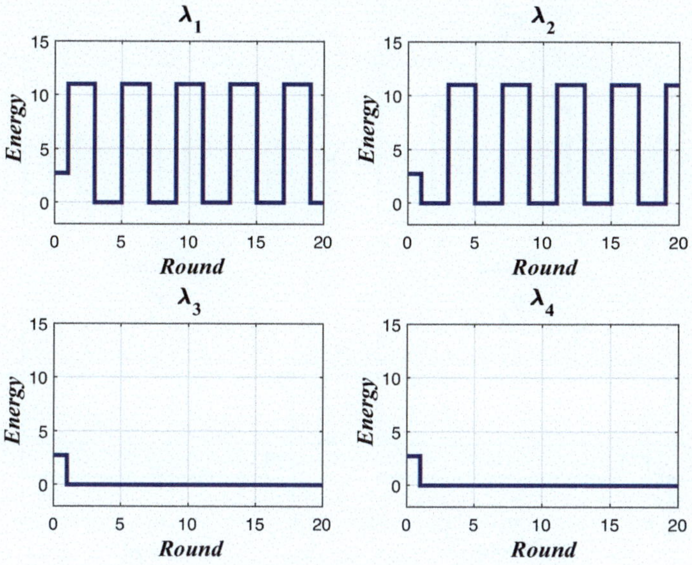

Fig. 6.9 Optimal dynamic strategies and related rewards of the defender in Case 1

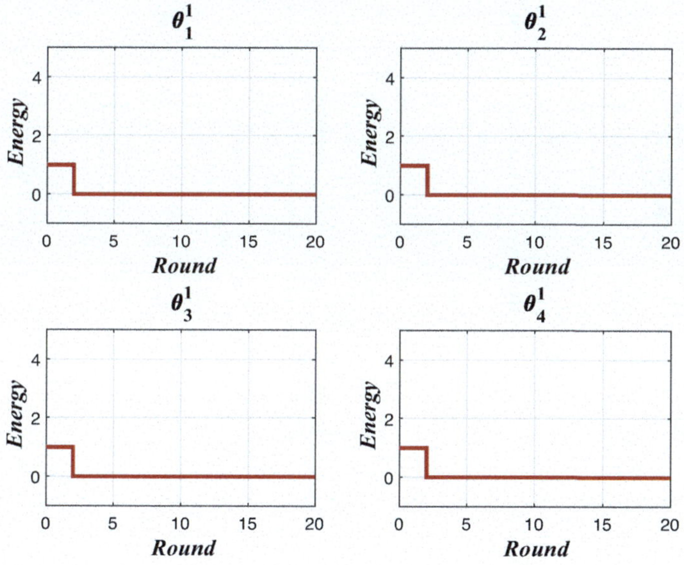

Fig. 6.10 Optimal dynamic strategies and related rewards of the attacker 1 in Case 1

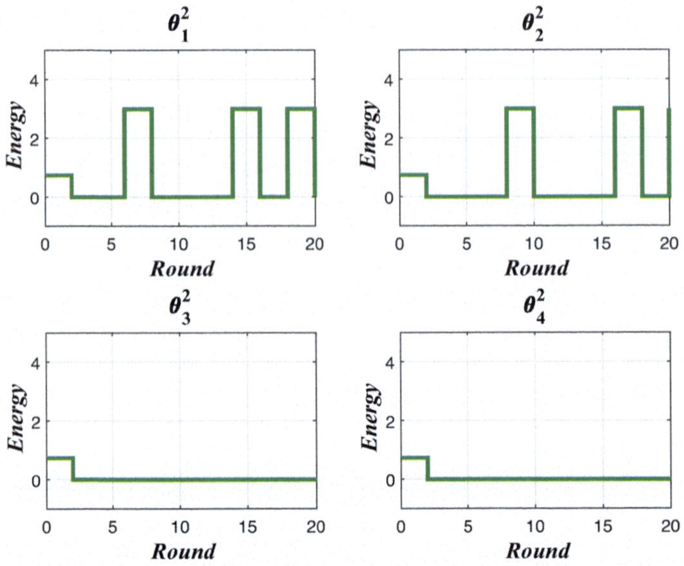

Fig. 6.11 Optimal dynamic strategies and related rewards of the attacker 2 in Case 1

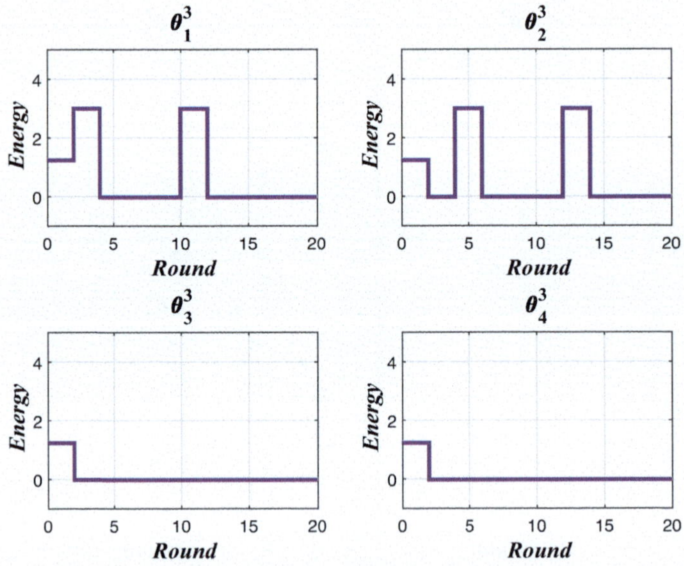

Fig. 6.12 Optimal dynamic strategies and related rewards of the attacker 3 in Case 1

defender's random choice on channels, thus making extra trouble to the attacker. In other words, when it turns to defender's round,

$$P(\mathbf{I_{D1}} = \mathbf{1}) = P(\mathbf{I_{D2}} = \mathbf{1}) = P(\mathbf{I_{D3}} = \mathbf{1}) = \frac{1}{3},$$

where $\mathbf{I_{Di}} = 0 \ or \ 1$ indicates the defender's decision to distribute all of his energy on the ith channel.

The corresponding simulation results are given in Table 6.4. In this case, the strategy is more random than the one obtained in Case 1, where the attackers and defenders switch between only two certain channels. However, in practice, the numerical characteristics (expectation) of the switching strategies of the attackers and defenders are determined due to the uniform distribution. After 10,000 times of Monte Carlo simulations, it can be found that the expected strategy for both attackers and defenders is still the simple average distribution.

6.4.2.3 Case 3: Identical Parameters

In this case, all the channels are set to be numerically same. There is no need for both sides to play chasing games, just like the example in Case 1. It's fair for both sides to averagely allocate energy to all channels. However, on account of the probability, the expected optimal energy allocation of the attacker is lower than theoretic results computed by PSO, as is shown in Table 6.4.

6.5 Conclusion

In this chapter, optimal DoS attack strategy in an SINR-based wireless network under the Stackelberg game framework has been investigated. The influence attackers make on wireless channels are characterized by signal-to-noise ratio. A Stackelberg game framework has been proposed to analyze both off-line and online decision making. For off-line decision making, KKT conditions have been utilized to determine the relationship between defender's optimal strategies and those of attackers'. To obtain the value of static equilibrium, a self-adaptive PSO algorithm has been applied to select the best particle to act as the optimal strategies for the defender, thus to calculate the attackers' optimal strategies. For online decision making, a detailed algorithm has been provided to describe the process of a multi-stage Stackelberg game and players' interaction. Future work includes designing a Stackelberg game framework in the background of cyber attacks in continuous time model.

Table 6.4 Optimal energy allocation of static and dynamic game

Cases	Methods	Players	Reward value	Energy allocation on channels			
				1	2	3	4
Case 1	Theoretic approach	λ	10.3070	4.9558	3.0484	1.8283	1.1676
		θ^1	−10.8987	1.9443	1.5131	0.3384	0.2031
		θ^2	−13.1559	1.4962	1.1814	0.2108	0.1113
		θ^3	−9.3440	2.3927	1.8451	0.4661	0.2949
	Monte Carlo method	λ	17.8445	5.8465	5.1535	0	0
		θ^1	−28.0379	0.2120	0.1872	0	0
		θ^2	−15.0925	0.9576	0.8511	0	0
		θ^3	−21.8078	0.4779	0.4140	0	0
Case 2	Theoretic approach	λ	5.9160	3.7132	3.6668	3.6200	0
		θ^1	−6.6020	1.3445	1.3329	1.3212	0
		θ^2	−8.5336	1.0094	1.0000	0.9904	0
		θ^3	−5.3849	1.6803	1.6666	1.6527	0
	Monte Carlo method	λ	15.2581	3.6179	3.6443	3.7378	0
		θ^1	−25.0331	0.1324	0.1400	0.1292	0
		θ^2	−12.6603	0.5958	0.5898	0.6267	0
		θ^3	−19.0532	0.2916	0.2991	0.2958	0
Case 3	Theoretic approach	λ	7.8561	2.665	2.6497	2.9262	2.7592
		θ^1	−8.1920	0.9770	0.9727	1.0476	1.0028
		θ^2	−10.2112	0.7307	0.7272	0.7896	0.7523
		θ^3	−6.8610	1.2230	1.2180	1.3054	1.2532
	Monte Carlo method	λ	17.8378	2.7500	2.7500	2.7500	2.7500
		θ^1	−26.7487	0.0999	0.0999	0.0999	0.0999
		θ^2	−15.5485	0.4467	0.4467	0.4467	0.4467
		θ^3	−21.2941	0.2284	0.2284	0.2284	0.2284

References

1. Shisheh Foroush, H., Martínez, S.: On triggering control of single-input linear systems under pulse-width modulated DoS signals. SIAM J. Control. Optim. **54**(6), 3084–3105 (2016)
2. Zhang, H., Cheng, P., Shi, L., Chen, J.: Optimal denial-of-service attack scheduling with energy constraint. IEEE Trans. Autom. Control **60**(11), 3023–3028 (2015)
3. Peng, L., Shi, L., Cao, X., Sun, C.: Optimal attack energy allocation against remote state estimation. IEEE Trans. Autom. Control **63**(7), 2199–2205 (2018)
4. Peng, C., Sun, H.: Switching-like event-triggered control for networked control systems under malicious denial of service attacks. IEEE Trans. Autom. Control **65**(9), 3943–3949 (2020)
5. Ma, Y.S., Che, W.W., Deng, C.: Dynamic event-triggered model-free adaptive control for nonlinear CPSs under aperiodic DoS attacks. Inf. Sci. **589**, 790–801 (2022)
6. Li, X., Wen, C., Wang, J., Chen, C., Deng, C.: Resilient leader tracking for networked Lagrangian systems under DoS attacks. Inf. Sci. **577**, 622–637 (2021)
7. Kumar, B., Bhuyan, B.: Using game theory to model DoS attack and defence. Sādhanā **44**(12), 1–12 (2019)
8. Agah, A., Das, S.K., Basu, K.: A game theory based approach for security in wireless sensor networks. In: Procceedings of the IEEE International Conference on Performance, Computing, and Communications, pp. 259–263 (2004)
9. Vamvoudakis, K.G., Hespanha, J.P., Sinopoli, B., Mo, Y.: Detection in adversarial environments. IEEE Trans. Autom. Control **59**(12), 3209–3223 (2014)
10. Yuan, Y., Yuan, H., Guo, L., Yang, H., Sun, S.: Resilient control of networked control system under DoS attacks: a unified game approach. IEEE Trans. Ind. Inform. **12**(5), 1786–1794 (2016)
11. Li, Y., Mehr, A.S., Chen, T.: Multi-sensor transmission power control for remote estimation through a SINR-based communication channel. Automatica **101**, 78–86 (2019)
12. Li, Y., Shi, D., Chen, T.: False data injection attacks on networked control systems: a Stackelberg game analysis. IEEE Trans. Autom. Control **63**(10), 3503–3509 (2018)
13. Liu, X., Yang, G.H.: Optimal encryption strategy for cyber-physical systems against stealthy attacks with energy constraints: a Stackelberg game approach. Inf. Sci. **610**, 674–693 (2022)
14. Liu, H.: SINR-based multi-channel power schedule under DoS attacks: a Stackelberg game approach with incomplete information. Automatica **100**, 274–280 (2019)
15. Zhang, Y., Xu, Y., Xu, Y., Yang, Y., Luo, Y., Wu, Q., Liu, X.: A multi-leader one-follower Stackelberg game approach for cooperative anti-jamming: no pains, no gains. IEEE Commun. Lett. **22**(8), 1680–1683 (2018)
16. Das, S., Suganthan, P.N.: Differential evolution: a survey of the state-of-the-art. IEEE Trans. Evol. Comput. **15**(1), 4–31 (2010)
17. Gibbons, R.: A Primer in Game Theory. Harvester Wheatsheaf New York (1992)
18. Del Valle, Y., Venayagamoorthy, G.K., Mohagheghi, S., Hernandez, J.C., Harley, R.G.: Particle swarm optimization: basic concepts, variants and applications in power systems. IEEE Trans. Evol. Comput. **12**(2), 171–195 (2008)

Chapter 7
Game-Based DoS Attack Strategy Against CPSs via Zero-Sum Game

Abstract This chapter studies the optimal distributed denial-of-service attack strategy against cyber-physical systems with multiple attackers and multiple defenders. An advanced attack strategy is proposed to cause the great damage to system in a multi-attacker-defender form. Firstly, a novel model of signal-to-interference-to-noise ratio for the multi-attacker and multi-defender is built. Taking the energy constraints into consideration, the objective of defenders is to minimize the system performance, while the attackers tend to deteriorate it by emitting interference energy. Thus, the optimal channel selection and optimal energy allocation strategies are proposed to answer which channel both of them should choose and how much power both of them should allocate to each channel in a finite time horizon. Secondly, a two-player zero-sum matrix game is formulated to solve the optimal problem by linear programming and obtain the Nash equilibrium. When the channel parameters are time-varying, a dynamic optimal channel selection problem is considered and a multi-stage game algorithm is proposed to find the Nash equilibrium. Additionally, the designed optimal strategies of both players are demonstrated and analyzed. Finally, a numerical simulation is provided to illustrate the effectiveness of the proposed approach.

7.1 Introduction

A variety of works seem to address the game issue as an interaction between the attacker and the defender. Thinking about attack and defense strategies from a game perspective helps us understand our opponent's actions and formulate our next move. The problem that maximizes the profit of an attacker and a defender to decide when to attack and when to send is transformed into a zero-sum game problem [1]. By designing the infinite time quadratic cost function, an optimal false data injection attack strategy is studied in the zero-sum game framework [2]. The work [3] investigates an online attacker and defender zero-sum game in infinite time horizon. The node occupation game between attackers and defenders in a multi-node system is studied, the transfer relationship between nodes in different states is considered, and the cost function under the zero-sum game is designed in [4]. Nevertheless, the

S. Gao et al., *Cyberattack Strategies Against Cyber-Physical Systems: From a Control Perspective*, https://doi.org/10.1007/978-981-95-9645-4_7

works mentioned above explore only an attacker existing situation. Unfortunately, until now, the issue of multi-attacker and multi-defender games has received few amount of attention, which is one of the motivations for this chapter.

Compared to DoS attacks, distributed denial-of-service attacks are more threatening. These DoS attacks have stronger concealment, more flexibility, and greater destructiveness [5]. Distributed DoS attacks are that the attacker first obtains control of the puppet host in some way, such as worms, backdoors, or Trojan horses [5]. Then, the attacker manipulates the puppet host to send a large number of attack signals to the victim, such that the victim is unable to respond to normal service requests [6]. Some actual cases of distributed DoS attacks are exemplified in [7]. The challenges of distributed DoS attacks can be presented in three parts. First, in terms of scale, there is coupling between multiple attackers. Therefore, a more appropriate channel model is needed. Single attacker is a special case of multiple attackers. Second, the channel number encoding model should be considered to determine which attacker chooses which defender to attack. Third, the attack strategies and constraints are more complicated and challenging than the scenario of an attacker. The widespread use of distributed DoS attacks has prompted us to further analyze the behavior of the attackers and the destructive nature of this kind of attacks.

In summary, this chapter is motivated by the greater system destructiveness caused by multiple attackers as opposed to a single attacker under the specific counterattack methods by the defender. Thus, the optimal distributed DoS attack strategy for cyber-physical systems under the framework of multi-attacker-defender game is studied in this chapter. The multi-process systems are considered. Multiple defenders choose which channel to send their local estimate to a remote estimator through a package-dropping wireless network, which reflects the ability of the defenders as intelligent agents to participate in the game and ensures the security of the system. The objective of attackers is to select the channel that causes the greatest damage in the sense of Nash equilibrium. The channel selection problem is discussed in both static and dynamic situations, and solved by one stage game algorithm and multi-stage game algorithm. Apart from channel selection, the energy allocation problem is also presented in this chapter, and the energy allocation theorem is proposed to solve the problem in the same framework as optimal static channel selection case. The main contributions of this chapter are summarized as follows:

(1) A novel model of SINR for the multi-attacker is built. Some existing models either only consider single attacker and single defender [8] or do not consider the channel gain [9]. This chapter considers multiple attackers and defenders with different channel gains, expanding the SINR model to a more general scenario. Relying on the rationality of the design, the model is suitable for both homogeneous and heterogeneous systems.
(2) The static two-player zero-sum matrix game problems between the attackers and the defenders are studied. Different from one attacker, there is coupling between multiple attackers. Thus the strategies and constraints design and analysis are more complex than an attacker. The optimal rational channel selection and energy allocation strategies are then obtained by settling linear programming.

(3) One stage and multi-stage game algorithm for static and dynamic optimal channel selection are proposed. In contrast to an attacker, a channel number encoding model should be established to decide which attacker to attack with defender. In addition, the energy allocation algorithm has a same framework as optimal channel selection, which considers a more general situation including the channel resource limit and energy resource limit.

The remainder of this chapter is organized as follows. Section 7.2 provides the problem formulation. The multi-sensor and multi-attacker game via channel selection and the energy allocation are presented in Sect. 7.3. Numerical examples are presented in Sect. 7.4, followed by the conclusion in Sect. 7.5.

7.2 Problem Formulation

7.2.1 System Model

Consider N independent discrete linear time-invariant dynamic cyber-physical systems with N sensors as follows:

$$\begin{cases} x_i(k+1) = A_i x_i(k) + w_i(k), \\ y_i(k+1) = C_i x_i(k) + v_i(k), \end{cases} \tag{7.1}$$

where $i = 1, 2, \ldots, N$ is the i-th plant of N plants. $k \in \mathbb{Z}$ denotes the indices of discrete time steps. $x_i(k) \in \mathbb{R}^{n_i}$ and n_i refer to the state and state dimension of the i-th system, respectively. $y_i(k) \in \mathbb{R}^{m_i}$ and m_i stand for the i-th system measurements and measurements' dimension, respectively. $w_i(k) \in \mathbb{R}^{n_i}$ and $v_i(k) \in \mathbb{R}^{m_i}$ are process noise and measurement noise, respectively. $w_i(k)$ is assumed to be independent and identically distributed white Gaussian with zero mean, and its covariance matrix is given by $Q_i \geqslant 0$, i.e., $w_i(k) \sim N(0, Q_i)$, $Q_i = E(w_i(k)w_i(k)^T)$. $v_i(k)$ is assumed to be independent and identically distributed white Gaussian with zero mean, and its covariance matrix is given by $R_i > 0$, i.e., $v_i(k) \sim N(0, R_i)$, $R_i = E(v_i(k)v_i(k)^T)$. A_i and C_i are matrices with appropriate dimensions. For simplicity, it is assumed that the pair $(A_i, \sqrt{Q_i})$ is stabilizable, and the pair (A_i, C_i) is observable.

Remark 7.1 The system matrix A_i and the output matrix C_i can differ, as well as the state dimension n_i of each plant, according to system (7.1). This research is applicable to both heterogeneous and homogeneous systems since the total performance metrics are afterward generated to be the sum of the performance of individual intelligences and the proposed algorithm later in this chapter being unaffected by the structure of each node.

7.2.2 Kalman Filter

Kalman filtering is performed separately at the sensor and remote estimator ends to obtain an estimate of the true state of system. Each sensor first runs a local Kalman Filter as shown in (7.2) to obtain the local state estimate of its monitored system. The i-th sensors local state estimate is defined as $\hat{x}_i^s(k) \triangleq E[x_i(k)|y_i(1), y_i(2), \ldots, y_i(k)]$, and its corresponding estimation error covariance matrix is defined as $P_i^s(k) \triangleq E[(x_i(k) - \hat{x}_i^s(k))(x_i(k) - \hat{x}_i^s(k))^T|y_i(1), y_i(2), \ldots, y_i(k)]$.

$$\begin{cases} \hat{x}_i^{s-}(k) = A_i\hat{x}_i^s(k-1), \\ P_i^-(k) = A_iP_i(k-1)A_i^T + Q_i, \\ K_i(k) = P_i^-(k)C_i^T(C_iP_i^-(k)C_i^T + R_i)^{-1}, \\ \hat{x}_i^s(k) = \hat{x}_i^{s-}(k) + K_i(k)(y_i(k) - C\hat{x}_i^{s-}(k)), \\ P_i(k) = (I - K_i(k)C)P_i^-(k), \end{cases} \tag{7.2}$$

where $x_i^{s-}(k)$ is the priori estimate, $P_i^-(k)$ denotes the priori error covariance, and $K_i(k)$ refers to Kalman gain. Similar to [8, 9], without loss of generality, $P_i^s(k)$ converges to a steady-state value when time approaches infinity, i.e., $\lim_{k\to\infty} P_i^s(k) = \bar{P}_i$.

The remote estimator runs intermittent Kalman Filters. N sensors independently send data packet $\hat{x}_i^s(k)$ through the wireless communication channel at the time step k. Considering the existence of the attacker and the characteristics of the wireless channels, the data packet might be lost. Define random variable $\gamma_i(k)$ with Bernoulli distribution, which indicates whether data packet in i-th channel is arrived or not. $\gamma_i(k) = 1$ refers to $\hat{x}_i^s$ arrives at the time step k, while $\gamma_i(k) = 0$ indicates the loss of data packet at the time step k. The minimum mean-square error (MMSE) estimate of the systems states is defined as $\hat{x}_i(k) \triangleq E[x_i(k)|y_i(1), y_i(2), \ldots, y_i(k), \gamma_i(1), \gamma_i(2), \ldots, \gamma_i(k)]$, and its corresponding estimation error covariance matrix is defined as $P_i(k) \triangleq E[(x_i(k) - \hat{x}_i^s(k))(x_i(k) - \hat{x}_i^s(k))^T|y_i(1), y_i(2), \ldots, y_i(k), \gamma_i(1), \gamma_i(2), \ldots, \gamma_i(k)]$. For convenience, define the Lyapunov operators $h_i : \mathbb{S}_+ \to \mathbb{S}_+$ as $h_i(X) = A_iXA_i^T + Q_i$. Thus, the iteration relationship between time step $k-1$ and k is obtained as follows:

$$\hat{x}_i(k) = \begin{cases} \hat{x}_i^s(k), & \gamma_i(k) = 1, \\ A_i\hat{x}_i(k-1), & otherwise, \end{cases} \tag{7.3}$$

$$P_i(k) = \begin{cases} \bar{P}_i, & \gamma_i(k) = 1, \\ h_i(P_i(k-1)), & otherwise. \end{cases} \tag{7.4}$$

7.2.3 Communication Model

The sensors transmit data packet to the remote estimator through an N-channel communication network. All N channels are subject to several independent additive white Gaussian noises. The interference of attacker is regarded as additive Gaussian white noise and the gain of the channels is considered. The signal-to-interference-to-noise ratio (SINR) measures the transmission quality of the channel, and SINR of the i-th sensor is defined as follows:

$$\mathrm{SINR}_i = \frac{u_i^s}{\sigma_i^2 + u_i^a}, \tag{7.5}$$

where σ_i^2 is the noise power of channel i. u_i^s denotes the received power of i-th channel. u_i^a denotes the total received power of M attackers in the i-th channel. With respect to sensors, $u_i^s = G_i^s q_i^s(e_i)$, where $q_i^s(e_i)$ denotes the transmit power, which is a function of sensors actions. G_i^s means the i-th channel gain. As for attackers, since multiple attackers attack simultaneously, there exists $u_i^a = \sum_{j=1}^{M} G_{ij}^a a_{ij} q_j^a$, where q_j^a refers to the transmit power of the j-th attacker. G_{ij}^a means the channel gain from j-th attacker to i-th channel.

Remark 7.2 Attackers and defenders in this game share a common spectrum of N orthogonal tones and transmit Gaussian signals [10]. From (7.5), we observe that for players to obtain the SINR of the i-th sensor, it not only needs to measure the noise-plus-interference level on corresponding tone, but also needs to know the corresponding received power. It can be seen that the transmit power is not exactly equivalent to the received power, but is related to the channel gain (the gain of the channel is usually caused by the multipath propagation of the channel, the transmission distance or the obstacles in the environment, which reflected as channel fading). In practical applications, both the attackers and the sensors can obtain the channel parameters G by pilot-aided channel estimation techniques [11].

The symbol error rate (SER) of channel is related to the modulation method and the signal-to-noise ratio, and generally SER is a non-increasing function of SINR [11]. Similar to literature [3, 9], the quadrature amplitude modulation (QAM) method is adopted for coding, so the relationship of the SER, SINR, and i-th channel parameters is as follows:

$$\mathrm{SER}_i = 2Q(\sqrt{\eta \mathrm{SINR}_i}), \tag{7.6}$$

where $Q(x) \triangleq \frac{1}{\sqrt{2\pi}} \int_x^\infty e^{-\frac{t^2}{2}} dt$ and $\eta > 0$ is a parameter related to i-th channel.

The packet arrival rate of i-th channel is as follows:

$$\alpha_i(k) = 1 - \mathrm{SER}_i. \tag{7.7}$$

Similar to [9, 12], the remote estimator is assumed to be able to detect the arrival of packets by cyclic redundancy check (CRC). Thus, the probability distribution of data arrival is obtained as $Pr(\gamma_i(k) = 1) = \alpha_i(k)$, $Pr(\gamma_i(k) = 0) = 1 - \alpha_i(k)$.

Above all, according to the iterative formula of the error covariance (7.4) and the probability distribution of data arrival, the expected error covariance can be obtained as

$$E[P_i(k)] = \alpha_i(k)\bar{P}_i + (1 - \alpha_i(k))h_i(E[P_i(k-1)]). \tag{7.8}$$

7.2.4 The Actions of Multi-attacker and Multi-defender

The framework of the multi-defender and multi-attacker structure depicted in Fig. 7.1 is introduced, which is composed of physical plants, sensors, attackers, estimation, and communication parts.

As Fig. 7.1 described, there are N processes with noise in CPSs, the i-th sensor measuring the i-th physical process. As defenders, sensors send the Kalman filtered data to the remote estimator through a packet-dropping channel for data fusion. In the process of data transmission, M attackers can launch attacks on all channels, thereby destroying the estimation performance of the remote estimator. The sensors side can realize the existence of the attackers and participate in the process of resisting the attack, thus forming a game between them. In the diagram, the yellow lightning indicates which channels the attackers attack and the soft black light indicates which channels the defenders defend.

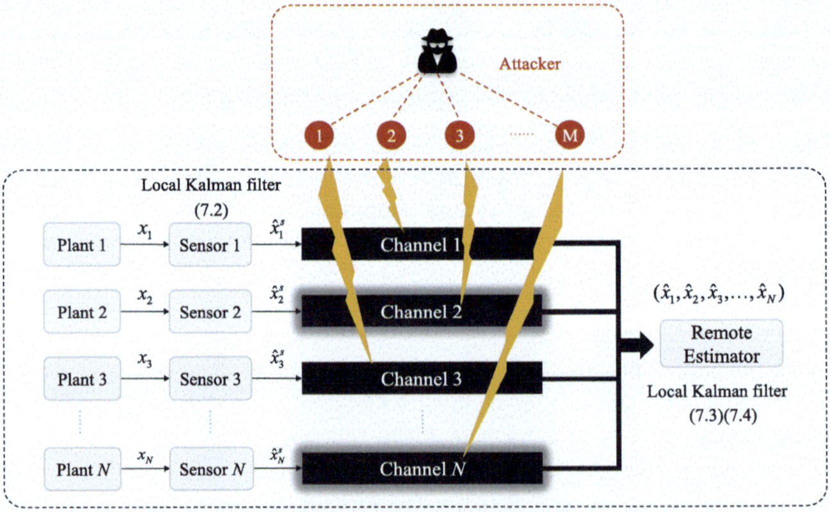

Fig. 7.1 The framework of multi-attacker-defender game for cyber-physical systems

Assumption 7.1 The attackers have complete information, including system information A_i, C_i, Q_i, R_i, environment information $M, N, G^a_{ij}, G^s_i, \sigma_i$ and energy information q^s_i, q^a_j, N_a, N_e.

Remark 7.3 The attackers can obtain system information through eavesdropping the input and output data for a sufficient time [13]. In practice, the sensors can adopt pilot-aided channel estimation techniques to obtain channel gain. Because of the Kerckhoffs principle, that is, the security of a system should not rely on its obscurity, the complete knowledge of the channel parameters and energy information are often assumed to be known by the attackers [11, 14].

In the open network, multiple attackers are trying to compromise the estimation performance of CPSs, as demonstrated in Fig. 7.1. Distributed DoS attackers are hypothesized to attack only for a limited period of time T. It is assumed that the attackers attack with a fixed energy, while the sensor transmits alternately with high and low energy to achieve a balance between saving energy and improving estimation performance [11]. At the time step k, M attackers decide which channels to attack, and N sensors decide which channels to strengthen defense.

The attacker-side actions and sensor-side actions are defined as ϕ_k and θ_k:

$$\begin{aligned} \phi_k &\in \mathbb{A}^a = \{(a_{11}(k), a_{12}(k), \dots, a_{NM}(k))\}, \\ \theta_k &\in \mathbb{A}^s = \{(e_1(k), e_2(k), \dots, e_N(k))\}, \end{aligned} \tag{7.9}$$

where $\phi_k \in \mathbb{R}^{NM}$ indicates the action of attackers at the time step k. $a_{ij}(k) \in \{0, 1\}$ denotes whether the j-th attacker jams the i-th channel. $a_{ij}(k) = 1$ refers to j-th attacker jams the i-th channel at the time step k. Otherwise, $a_{ij}(k) = 0$ means j-th attacker does not jam the i-th channel at the time step k. $\theta_k \in \mathbb{R}^N$ refers to the action decision of the sensors side at time step k. $e_i(k) \in \{0, 1\}$ signifies whether the i-th sensor applies a larger energy to transmit at time step k or not. $e_i(k) = 1$ means i-th sensor applies a larger energy to transmit at time step k. $\mathbb{A}^a$ and $\mathbb{A}^s$ are strategy set for attackers and sensors, respectively. Therefore, over the finite time horizon T, the attackers and sensors actions are $\phi = \{\phi_1, \phi_2, \dots, \phi_T\}$, $\theta = \{\theta_1, \theta_2, \dots, \theta_T\}$.

Since sensors are usually battery powered and are sparsely deployed, it is difficult to replace batteries, so the energy loss should be minimized. The attacker also hopes to achieve the maximum attack effect with as few energy as possible. So this chapter considers the energy constraints of the sensors and the attackers.

For the sensors, $\sum_{i=1}^{N} e_i(k) \leqslant N_e$, where N_e represents the maximum number of channels that M attackers attack simultaneously at time step k.

For the attackers, $\sum_{j=1}^{M} \sum_{i=1}^{N} a_{ij}(k) \leqslant N_a$, where N_a indicates the number of channels that N sensors at time k can utilize a larger energy to transmit data packets at the same time.

7.2.5 Optimal DoS Attack Problem via Zero-Sum Game

In the theory of Kalman Filter, the estimation error covariance of the system reflects the quality of estimation. As a result, it is reasonable to apply the total estimation error covariance of the system as the objective function (Similar settings in [8, 9]).

It is presumed that all the attackers and the sensors are intelligent. They have certain perception, computing and communication capabilities, and the ability to play the game. The performance index of system is as follows:

$$J(\phi, \theta) = \sum_{i=1}^{N} \sum_{k=1}^{T} tr\left(E[P_i(k)]\right). \tag{7.10}$$

The goal of attackers is to maximize this performance index, the goal of sensors is to minimize this performance index, and the goals of the two sides are just opposite. Therefore, the following optimization problem is considered.

Problem 7.1 For the attackers,

$$\begin{aligned} \max_{\phi} \min_{\theta} \quad & J(\phi, \theta), \\ s.t. \quad & \sum_{j=1}^{M} \sum_{i=1}^{N} a_{ij}(k) \leqslant N_a, a_{ij}(k) \in \{0, 1\}, \\ & \sum_{i=1}^{N} e_i(k) \leqslant N_e, e_i(k) \in \{0, 1\}. \end{aligned}$$

For the sensors,

$$\begin{aligned} \min_{\theta} \max_{\phi} \quad & J(\phi, \theta), \\ s.t. \quad & \sum_{j=1}^{M} \sum_{i=1}^{N} a_{ij}(k) \leqslant N_a, a_{ij}(k) \in \{0, 1\}, \\ & \sum_{i=1}^{N} e_i(k) \leqslant N_e, e_i(k) \in \{0, 1\}. \end{aligned}$$

Based on the objective functions of both sides, the sensors and the attackers are regarded as the two players of a zero-sum game. The game framework $\mathcal{G}$ is defined as follows:

- Players: N sensors and M attackers.
- Action Space: $\phi = \{\phi_1, \phi_2, \ldots, \phi_T\}$ and $\theta = \{\theta_1, \theta_2, \ldots, \theta_T\}$ for attackers and sensors, respectively.
- Payoff: $J^a = J$ and $J^s = -J$ for attackers and sensors, respectively.

To construct the relationship between Problem 7.1 and the zero-sum game, the concept of maxminimizer and the Lemma 7.1 should be introduced.

Definition 7.1 ([15]) Let $\mathcal{G}$ be a zero-sum strategic game. The action ϕ^* is a maxminimizer for attackers if for $\forall\phi$,

$$\min_{\theta} J^a(\phi^*, \theta) \geq \min_{\theta} J^a(\phi, \theta).$$

Similarly, the action θ^* is a maxminimizer for sensors if for $\forall\theta$,

$$\min_{\phi} J^s(\phi, \theta^*) \geq \min_{\phi} J^s(\phi, \theta).$$

Remark 7.4 Due to the fact that for $\forall\theta$, $J^a(\phi^*, \theta) \geq \max_{\phi} \min_{\theta} J^a(\phi, \theta) = \min_{\theta} J^a(\phi^*, \theta)$, when attackers choose the maxminimizer strategy ϕ^*, at least the attackers can get payoff of $\min_{\theta} J^a(\phi^*, \theta)$, no matter how the sensors select. This is the best outcome in the worst-case scenario.

Lemma 7.1 ([15]) *Let $\mathcal{G}$ be a zero-sum strategic game.*

(1) If (ϕ^*, θ^*) is a Nash equilibrium of $\mathcal{G}$, ϕ^* is a maxminimizer for attackers and θ^* is a maxminimizer for sensors.
(2) If $\max_{\phi} \min_{\theta} J^a(\phi, \theta) = \min_{\theta} \max_{\phi} J^a(\phi, \theta)$, $\mathcal{G}$ has a Nash equilibrium, ϕ^* is a maxminnimizer for attackers, and θ^* is a maxminimizer for sensors, then (ϕ^*, θ^*) is a Nash equilibrium of $\mathcal{G}$.
(3) If (ϕ^*, θ^*) is a Nash equilibrium of $\mathcal{G}$, then we have $\max_{\phi} \min_{\theta} J^a(\phi, \theta) = \min_{\theta} \max_{\phi} J^a(\phi, \theta) = J^a(\phi^*, \theta^*)$, and thus all Nash equilibrias of $\mathcal{G}$ yield the same payoff, i.e., Nash equilibrium value.

According to Definition 7.1, ϕ^* and θ^* are the maxminimizer for attackers and sensors, which are the solutions of Problem 7.1. When $\max_{\phi} \min_{\theta} J^a(\phi, \theta) = \min_{\theta} \max_{\phi} J^a(\phi, \theta)$, (ϕ^*, θ^*) is a Nash equilibrium of $\mathcal{G}$. Till now, the framework of multi-attacker-defender game has been established.

Remark 7.5 Lemma 7.1 shows conclusions of the pure strategy Nash equilibrium. The result can be extended to mixed Nash equilibrium, which will be leveraged in the following part. In a zero-sum game, increasing utility and action space will improve the Nash equilibrium value [15]. Therefore both sides of the game will maximize their utility as much as possible and the effect will be illustrated in the simulation part.

In the framework of game theory, if all players reach the Nash equilibrium, they will not change their decisions, because this is the best response to other players. For this reason, the Nash equilibrium of the game will be seeking in the following sections.

7.3 Game-Based DoS Attack Strategies

In this section, game-based DoS attack strategies via channel selection and energy allocation are presented.

7.3.1 Game-Based DoS Attack Strategy via Channel Selection

7.3.1.1 Game-Based DoS Attack Strategy via Static Channel Selection

To solve Problem 7.1, the solution of a one stage game between M attackers and N sensors in a finite time is first taken into consideration. In order to reduce the computation amount, suppose that both the attackers and the sensors make the same decision at each moment, and the optimization objective is still J, so as to obtain an optimal channel selection strategy. This assumption is reasonable, because this chapter does not limit the total energy in the whole time horizon T, but limits the energy at each time instant. For the purpose of maximizing their own interests, both sides of the game choose to implement a fixed strategy after comprehensive evaluation of the whole time period T. Similar to [8, 11], when the energy of both sides is infinite, they only need to use as much energy as possible to achieve their goals, because the performance index J is respectively increasing and decreasing in the attack energy and transmission energy. The optimal value is obtained at the point where both sides adopt the maximum energy. Then the energy constraint can be transformed to $\sum_{j=1}^{M}\sum_{i=1}^{N} a_{ij}(k) = N_a, \sum_{i=1}^{N} e_i(k) = N_e$. Thus, an optimal channel selection problem is defined as the following.

Problem 7.2 For the attackers,

$$
\begin{aligned}
&\max_{\phi}\min_{\theta} \quad J(\phi, \theta), \\
&s.t. \quad \sum_{j=1}^{M}\sum_{i=1}^{N} a_{ij}(k) = N_a, a_{ij}(k) \in \{0, 1\}, a_{ij}(k) = a_{ij}, \\
&\qquad\quad \sum_{i=1}^{N} e_i(k) = N_e, e_i(k) \in \{0, 1\}, e_i(k) = e_i.
\end{aligned}
$$

For the sensors,

$$\min_{\theta}\max_{\phi} \quad J(\phi,\theta),$$

$$s.t. \quad \sum_{j=1}^{M}\sum_{i=1}^{N} a_{ij}(k) = N_a, a_{ij}(k) \in \{0, 1\}, a_{ij}(k) = a_{ij},$$

$$\sum_{i=1}^{N} e_i(k) = N_e, e_i(k) \in \{0, 1\}, e_i(k) = e_i.$$

In order to solve this optimization problem, the specific expression of J should be given. According to (7.4) and [8], $P_i(k)$ takes its value in $\{\bar{P}_i, h(\bar{P}_i), h^2(\bar{P}_i), \ldots, h^T(\bar{P}_i)\}$, i.e., $P_i(k) = h^m(\bar{P}_i)$, $m = 0, 1, 2, \ldots, T$, where m represents the number of consecutive packet losses. Therefore,

$$J(\phi,\theta) = \sum_{i=1}^{N}\sum_{k=1}^{T} tr(E[P_i(k)]) = tr(\sum_{i=1}^{N}\sum_{k=1}^{T}\sum_{m=0}^{T} p^i_{m,k} h^m(\bar{P}_i)), \tag{7.11}$$

where $p^i_{m,k}$ means the i-th channel drops packet from time step $k-m+1$ to k, and receives packet successfully at the time step $k-m$. It is supposed that packet will arrive at time step 0, i.e., $\alpha_i(0) = 1$. When $k < m$, $p^i_{m,k} = 0$. When $k \geqslant m$, $p^i_{m,k} = \alpha_i \beta_i^m$, where $\beta_i = 1 - \alpha_i$ indicates the packet drop rate. As for i-th sensor, we can get a probability matrix $T_i = [p^i_{m,k}]_{(T+1)\times T}$, $m = 0, 1, \ldots, T$, $k = 1, 2, \ldots, T$, $i = 1, 2, \ldots, N$,

$$T_i = \begin{bmatrix} \alpha_i & \alpha_i & \cdots\cdots & \alpha_i \\ \beta_i & \alpha_i\beta_i & & \alpha_i\beta_i \\ 0 & \beta_i^2 & & \alpha_i\beta_i^2 \\ \vdots & & \ddots & \vdots \\ \vdots & & \ddots & \vdots \\ 0 & 0 & \cdots\cdots & \beta_i^T \end{bmatrix}, \tag{7.12}$$

where $\sum_{k=1}^{T} p^i_{m,k} = (T_i\mathbf{1})_{m+1}$, and the subscript $m+1$ represents the $(m+1)$-th element of the vector. Thus,

$$\begin{aligned} J(\phi,\theta) &= tr\left(\sum_{i=1}^{N}\sum_{m=0}^{T} h^m(\bar{P}_i)\sum_{k=1}^{T} p^i_{m,k}\right), \\ &= \sum_{i=1}^{N}\sum_{m=0}^{T} tr(h^m(\bar{P}_i)(T_i\mathbf{1})_{m+1}), \\ &= tr\left(\sum_{i=1}^{N}(1-\beta_i)\sum_{m=0}^{T-1}[\beta_i^m h^m(\bar{P}_i)(T-m)]\right) \end{aligned}$$

$$+tr\left(\sum_{i=1}^{N}\sum_{m=1}^{T}[\beta_i^m h^m(\bar{P}_i)]\right). \tag{7.13}$$

When $\beta_i \to 1$, $J(\phi,\theta) = Tr\left(\sum_{i=1}^{N}\sum_{m=1}^{T} h^m(\bar{P}_i)\right)$, which signifies when attack energy is infinite, the larger T and N, the more the attack will damage the estimated performance of the system. In order to find the optimal response of both sides of the game to seek the Nash equilibrium, the following Definition and Lemmas are needed.

Definition 7.2 ([16]) The Nash equilibrium $((\pi^a)^*, (\pi^s)^*)$ in the two-player matrix game refers to the set of strategies that satisfy the following formula:

$$\begin{aligned} &V^i((\pi^i)^*, (\pi^{-i})^*) \geqslant V^i(\pi^i, (\pi^{-i})^*), \\ &\forall \pi^i \in PD(\mathbb{A}^i), i = a, s, \end{aligned} \tag{7.14}$$

where $-i$ refers to the opponent of player i's strategy, $V^i(\cdot)$ means expected reward value function of the player i, π^i represents the probability that player i chooses a pure strategy in the mixed strategy space and $PD(\mathbb{A}^i)$ denotes the discrete probability distribution of strategy set $\mathbb{A}^i$.

Lemma 7.2 ([17]) *A game with a finite set of strategies has at least one mixed Nash equilibrium solution.*

Lemma 7.3 ([16, 18]) *Solving a Nash equilibrium in a two-player zero-sum matrix game is equivalent to finding the minimum solution to the following equation:*

$$\max_{\pi^i \in PD(\mathbb{A}^i)} \min_{d_j^{-i} \in \mathbb{A}^{-i}} \sum_{d_j^i \in \mathbb{A}^i} R^i_{d_j^{-i}, d_j^i} \pi^i(d_j^i), \tag{7.15}$$

where $i = a, s$, $d_j^i \in \mathbb{A}^i$. $\pi^i(d_j^i)$ denotes the probability of choose strategy d_j^i in $\mathbb{A}^i$. d_j^{-i} represents the strategy chosen by player i's opponent. R^i is the reward matrix of player i and $R^i_{d_j^{-i}, d_j^i}$ is the reward to the player i for choosing action d_j^i while its opponent chooses action d_j^{-i}.

For Problem 7.2, the objective function of attacker side is $J^a = J$, the objective function for the defender side is $J^s = -J$, which constitute a zero-sum game. Under the pure strategy determined by both sides, specific payoff are generated. So this problem can be transformed into the problem of matrix game. Redefine finite policy sets for attackers and sensors as $\mathbb{A}^a = \{(a_{11}, a_{12}, \ldots, a_{NM}) | \sum_{j=1}^{M}\sum_{i=1}^{N} a_{ij} = N_a, a_{ij} \in \{0, 1\}\}$ and $\mathbb{A}^s = \{(e_1, e_2, \ldots, e_N) | \sum_{i=1}^{N} e_i = N_e, e_i \in \{0, 1\}\}$. For the attackers' side, the number of possible pure strategies is $C_{MN}^{N_a}$, and for the sensors' side, the number of possible pure strategies is $C_N^{N_e}$. Define the pure strategy of the attackers and sensors as: d_j^a and d_i^s so that $\mathbb{A}^a = \{d_j^a | j = 1, 2, \ldots, C_{MN}^{N_a}\}$ and

$\mathbb{A}^s = \{d_i^s | i = 1, 2, \ldots, C_N^{N_e}\}$. Define the attacker-side and sensor-side reward matrix as R^a and R^s, respectively.

To sum up, the solutions of the two max-min problems can be obtained by solving the corresponding two linear programming problems.

Theorem 7.1 *Under Assumption 7.1, mixed Nash equilibrium* $((\pi^a)^*, (\pi^s)^*)$ *to Problem 7.2 can be obtained by solving the following two linear programming.*

For attackers,

$$\begin{aligned} \max_{\pi^a} \ & v, \\ s.t. \ & (\pi^a)^T R \geqslant v\mathbf{1}^T, \\ & \pi^a = (\pi^a(d_1^a), \pi^a(d_2^a), \ldots, \pi^a(d^a_{C_{MN}^{N_a}}))^T, \\ & \sum_{j=1}^{C_{MN}^{N_a}} \pi^a(d_j^a) = 1, \\ & \pi^a \geqslant 0. \end{aligned}$$

For sensors,

$$\begin{aligned} \min_{\pi^s} \ & v, \\ s.t. \ & R\pi^s \leqslant v\mathbf{1}, \\ & \pi^s = (\pi^s(d_1^s), \pi^s(d_2^s), \ldots, \pi^s(d^s_{C_N^{N_e}}))^T, \\ & \sum_{j=1}^{C_N^{N_p}} \pi^s(d_j^s) = 1, \\ & \pi^s \geqslant 0, \end{aligned}$$

where

$$R = \begin{bmatrix} J(d_1^a, d_1^s) & J(d_1^a, d_2^s) & \cdots & J(d_1^a, d^s_{C_N^{N_e}}) \\ J(d_2^a, d_1^s) & \ddots & & \\ \vdots & & \ddots & \vdots \\ J(d^a_{C_{MN}^{N_a}}, d_1^s) & \cdots & & J(d^a_{C_{MN}^{N_a}}, d^s_{C_N^{N_e}}) \end{bmatrix}, \tag{7.16}$$

v is the payoff of the attacker under the Nash equilibrium solution, i.e., J^*_{nash}.

Proof Since the two parties are in a zero-sum game relationship, $R^a = -R^s = R$, $R \in C_{MN}^{N_a} \times C_N^{N_e}$. Correspondingly, the reward matrix (7.16) is gained.

In order to find the mixed Nash equilibrium solution, the payoff function needs to be written as the expectation of the pure policy payoff function. The expected value functions of the attackers and sensors are $V^a(\pi^a, \pi^s) = (\pi^a)^T R\pi^s$ and

$V^s(\pi^a, \pi^s) = -(\pi^a)^T R\pi^s$, respectively. Then the problem becomes $\max_{\pi^a} \min_{\pi^s} (\pi^a)^T R\pi^s$ and $\min_{\pi^s} \max_{\pi^a} (\pi^a)^T R\pi^s$ for attackers and sensors, respectively. According to well-known von Neumann's minimax theorem, the function $(\pi^a)^T R\pi^s$ is bilinear for π^s and π^a, so that $\max_{\pi^a} \min_{\pi^s} (\pi^a)^T R\pi^s = \min_{\pi^s} \max_{\pi^a} (\pi^a)^T R\pi^s$. Based on Lemma 7.1, which could be extended to mixed Nash equilibrium case, the mixed Nash equilibrium $((\pi^a)^*, (\pi^s)^*)$ for two-player zero-sum games is exactly the solution. By Lemma 7.3, this problem is equal to solve

$$\max_{\pi^i \in PD(A^i)} \min_{d_j^{-i} \in A^{-i}} \sum_{d_j^i \in A^i} R^i_{d_j^{-i}, d_j^i} \pi^i(d_j^i).$$

Let $v = \min_{d_j^{-i} \in A^{-i}} \sum_{d_j^i \in A^i} R^i_{d_j^{-i}, d_j^i} \pi^i(d_j^i)$, the problem can be transformed to $\max_{\pi^i \in PD(A^i)} v$, and solved by settling two linear programming problems.

In addition to the game framework, the single part decision scheme can be deduced by Theorem 7.1. As the sensor or attacker side applies a fixed strategy, there is $V^i((\pi^i)^*) \geqslant V^i(\pi^i)$, by the definition of Nash equilibrium $V^i((\pi^i)^*, (\pi^{-i})^*) \geqslant V^i(\pi^i, (\pi^{-i})^*), \forall \pi^i \in PD(A^i), i = a, s$. Hence the solution tends to be the optimal solution for unilateral decision.

Corollary 7.1 *When the sensor or attacker side uses a fixed strategy, the result of Theorem 7.1 is the optimal solution for unilateral decision.*

7.3.1.2 Game-Based DoS Attack Strategy via Dynamic Channel Selection

Assumption 7.2 Suppose that the attackers can obtain the arrival of data packets $\gamma_i(k)$ by intercepting the feedback channel from the estimator to the corresponding sensor.

Remark 7.6 It is a general assumption when the attackers can eavesdrop the information of ACKs from the feedback channel of the sensors to remote estimator (e.g., transmission control protocol). Moreover, the same assumption can be found in [11].

The game in Sect. 7.3.1.1 is an offline process, that is, the strategy is obtained at time step $k = 0$. The one stage game is applicable to the time-invariant system. While there are time-varying channel parameters in the system, a continuous game is considered from the perspective of attackers, that is, the game is played again at every moment, lasting till the time step T, to realize the real-time game between the attacker and defender sides.

Analyze the meaning of T in Sect. 7.3.1.1, which means that the attacker attacks only in the time period T. From another perspective, the meaning of T can also be the prediction of T steps in the future, which is similar to the idea of model predictive

Algorithm 8: Game-Based DoS Attack Strateg via Dynamic Channel Selection.

Initialize: $M, N, N_a, N_e, A_i, C_i, Q_i, R_i, G_{ij}^a, G_i^s, q_i^s, q_j^a, \eta, \sigma_i, T, t = 1$.
Calculate $C_{MN}^{N_a}, C_N^{N_e}$.
Calculate $\bar{P}_i$, $J(\phi, \theta) = \sum_{i=1}^{N} \sum_{k=t}^{T} tr\,(E[P_i(k)])$ by (7.13).
Implement Theorem 7.1 to calculate R and get the Nash equilibrium strategy $((\pi^a)^*, (\pi^s)^*)$.
while $t < T$ **do**
    Randomly sampling a Nash equilibrium action to use;
    $t = t + 1$;
    Observe $P_i(t)$, $i = 1, 2, \ldots, N$;
    if *The parameters are dynamic* **then**
        Observe channel gain G_{ij} and energy constraints;
    Adopt (7.8) to update performance index $J(\phi, \theta) = \sum_{i=1}^{N} \sum_{k=t}^{T} tr(E[P_i(k)])$;
    Utilize Theorem 7.1 to calculate R and get the new Nash equilibrium strategy $((\pi^a)^*, (\pi^s)^*)$;

control. The optimal channel of each prediction is only applied to the current moment. In this way, a dynamic channel selection game can be proposed. According to the current state, current channel parameters, new energy constraints, and packet arrivals for all channels (i.e., $\gamma_i(k)$), the performance index from the current time to time step T can be predicted again, and the corresponding attack strategy and defense strategy can be obtained. Compared with the one stage game, this multi-stage game takes the actual state at each moment into account in the process of proceeding and reduces the uncertainty. For example, when the game reaches time $T - 1$, only the state at time T is uncertain. The algorithm is summarized in Algorithm 8.

7.3.2 *Game-Based DoS Attack Strategy via Energy Allocation*

Section 7.3.1 analyzes game-based DoS attack strategies via optimal channel selection under the framework of zero-sum game. In the actual implementation of an attack, the above problems can be transformed into a problem of energy allocation between the attackers and the sensors on the communication channel. The main difference between them is the meaning of both sides' action and energy constraints.

Redefine the action of attackers as $\tilde{a}_{ij}(k) \in \{0, 1, \ldots, M_j\}$, which means the j-th attacker choose how much energy to launch attack to i-th channel, where M_j is the energy level of the j-th attacker. q_j^a refers to a unit energy. $\tilde{a}_{ij}(k) = 0, 1, \ldots, M_j$ denotes that the attacker j attacks the channel i with energy level $0, 1, \ldots, M_j$, that is, adopts the energy $0, q_j^a, \ldots, M_j q_j^a$ to attack. Similarly, redefine the action of sensors as $\tilde{e}_i(k) \in \{0, 1, 2, \ldots, N_i\}$. The transmit power of sensors $q_i^s(\tilde{e}_i)$ is a function of action, and each action is mapped to a specific energy level, i.e., $\tilde{e}_i(k) = 0, 1, 2, \ldots, N_i$. When $q_i^s(\tilde{e}_i) \in \{q_i^s(\tilde{e}_i) | \tilde{e}_i(k) \in \{0, 1, 2, \ldots, N_i\}\}$, i-th sensor applies the energy $q_i^s(0), q_i^s(1) \ldots, q_i^s(N_i)$ to send the data packet. To set this energy allocation problem in the same framework as optimal channel selection, a piece of energy can be send through one channel. For instance, the energy level of j-th attacker is M_j, which means that the j-th attacker has M_j channels to choose. The details are illustrated in Section V.

For the attackers, the energy constraint is $\tilde{N}_a = \sum_{j=1}^{M}\sum_{i=1}^{N}\tilde{a}_{ij}(k)$, where $\tilde{N}_a$ represents M attackers limit of total energy, and pure strategy of the attackers turned into $C_{\tilde{M}}^{\tilde{N}_a}$. $\tilde{M} = N\sum_{j=1}^{M} M_j$ denotes the total possible channel to choose.

For the sensors, the energy constraint is $\tilde{N}_e = \sum_{i=1}^{N}\tilde{e}_i(k)$, where $\tilde{N}_e$ refers to total energy limits of N sensors. And then the pure strategy of the sensors are $C_{\tilde{N}}^{\tilde{N}_e}$, where $\tilde{N} = \sum_{i=1}^{N} N_i$ indicates the total possible channel to choose.

Thus, the energy allocation problem and its corresponding solution are shown in Problem 7.3 and Theorem 7.2.

Problem 7.3 For the attackers,

$$
\begin{aligned}
&\max_{\phi}\min_{\theta} \quad J(\phi,\theta),\\
&\quad s.t. \quad \sum_{j=1}^{M}\sum_{i=1}^{N}\tilde{a}_{ij}(k) = \tilde{N}_a,\\
&\qquad\quad \tilde{a}_{ij}(k) \in \{0, 1, \ldots, M_j\}, \tilde{a}_{ij}(k) = \tilde{a}_{ij},\\
&\qquad\quad \sum_{i=1}^{N}\tilde{e}_i(k) = \tilde{N}_e,\\
&\qquad\quad \tilde{e}_i(k) \in \{0, 1, 2, \ldots, N_i\}, \tilde{e}_i(k) = \tilde{e}_i.
\end{aligned}
$$

For the sensors,

$$
\begin{aligned}
&\min_{\theta}\max_{\phi} \quad J(\phi,\theta),\\
&\quad s.t. \quad \sum_{j=1}^{M}\sum_{i=1}^{N}\tilde{a}_{ij}(k) = \tilde{N}_a,\\
&\qquad\quad \tilde{a}_{ij}(k) \in \{0, 1, \ldots, M_j\}, \tilde{a}_{ij}(k) = \tilde{a}_{ij},\\
&\qquad\quad \sum_{i=1}^{N}\tilde{e}_i(k) = \tilde{N}_e,\\
&\qquad\quad \tilde{e}_i(k) \in \{0, 1, 2, \ldots, N_i\}, \tilde{e}_i(k) = \tilde{e}_i.
\end{aligned}
$$

Theorem 7.2 *Under Assumption 7.1 and substituting the pure strategy of the attackers and the sensors in Theorem 7.1 by $C_{\tilde{M}}^{\tilde{N}_a}$ and $C_{\tilde{N}}^{\tilde{N}_e}$, the mixed Nash equilibrium $((\pi^a)^*, (\pi^s)^*)$ of Problem 7.3 can be obtained.*

Proof The proof of Theorem 7.2 is similar to Theorem 7.1, omitted here.

7.4 Numerical Simulation

In this section, the effectiveness of optimal attack and defense channel selection with energy constraint for different scenarios are illustrated.

7.4.1 The Channel Selection Case

The cyber-physical systems (7.1) with two plants (i.e., $N = 2$) are considered as

$$A_1 = 1.1, C_1 = 1.2, Q_1 = 0.8, R_1 = 0.6,$$

$$A_2 = 1.1, C_2 = 1.0, Q_2 = 0.9, R_2 = 0.9.$$

As previously hypothesized, the local Kalman Filter runs for a enough long time, and we can get the steady error covariance at the sensors side $\bar{P}_1 = 0.3074$, $\bar{P}_2 = 0.5755$.

Let $G_i^s = 1$, $G_{ij}^a = 1$, for $i = 1, 2$, $j = 1, 2$. The duration is $T = 10$. Two attackers (i.e., $M = 2$) are considered. The noise powers are $\sigma_1^2 = 0.8$, $\sigma_2^2 = 1.0$. The channel parameter is $\eta = 3$.

We solve Problem 7.2 through Theorem 7.1 and compare the corresponding error covariance under different power levels of attackers and sensors. The energy constraints and power levels are as the following three cases:

(1) $N_e = 1$, $N_a = 1$, $q_i^s(0) = 1$, $q_i^s(1) = 5$, for $i = 1, 2$ and $q_j^a = 5$, for $j = 1, 2$;
(2) $N_e = 1$, $N_a = 2$, $q_i^s(0) = 1$, $q_i^s(1) = 2$, for $i = 1, 2$ and $q_j^a = 10$, for $j = 1, 2$;
(3) $N_e = 2$, $N_a = 1$, $q_i^s(0) = 1$, $q_i^s(1) = 10$, for $i = 1, 2$ and $q_j^a = 2$, for $j = 1, 2$.

It is assumed that the power q is same for both sides, which is reasonable when all sensors are of the same type, and so do the attackers. In detail, $q_i^s(1)$ indicates the high energy, and $q_i^s(0)$ means the low energy. Case (1) demonstrates that both sensors and attackers sides have considerable energy. Case (2) shows that the attackers have more energy than the sensors, while Case (3) leads to the opposite situation.

Figure 7.2 presents the channel number encoding model for optimal channel selection simulation. The symbol of black stars signifies the location of two sensors, the black triangle delegates remote estimator and the red rhombus are two attackers. The numbers in the yellow and black squares pointed by the arrow indicate the channel number of attackers and defenders, respectively. Because $MN = 2 \times 2 = 4$, there are 4 channels of attackers in Fig. 7.2. Then the channel number coding method is that the first N channel refers to the 1-st attacker to attack the 1-st $\sim$ N-th sensor, and so on. In this specific case, the 1-st channel refers to the first attacker to attack the 1-st sensor, the second channel refers to the first attacker to attack the 2-nd sensor, and so on. Similarly, because $N = 2$, there are 2 channels of sensors in Fig. 7.2. Sensors transmit data through the corresponding channel.

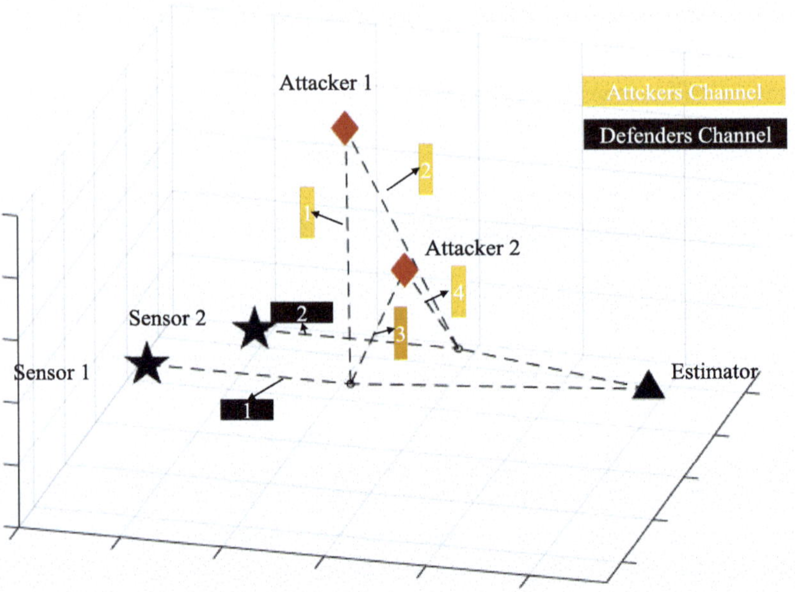

Fig. 7.2 Channel number encoding model for optimal channel selection simulation

From Theorem 7.1, the Nash channel selection strategy and Nash trace of expected error covariance for the above three cases are as follows:

(1) $(\pi^a)^* = (0, 0, 0.57, 0.43)$,$(\pi^s)^* = (0.44, 0.56)$, $J^*_{nash} = 14.30$.
(2) $(\pi^a)^* = (0, 0, 0, 0, 0, 1)$, $(\pi^s)^* = (0, 1)$, $J^*_{nash} = 32.11$.
(3) $(\pi^a)^* = (0, 0, 0, 1)$, $(\pi^s)^* = (1)$, $J^*_{nash} = 8.85$.

As defined before, π^a is the mixed Nash equilibrium. In Case (1), there are $C^{N_a}_{MN} = 4$ possible pure strategies for attackers and $C^{N_e}_{N} = 2$ possible pure strategies for sensors. In terms of Case (2), $C^{N_a}_{MN} = 6$ and $C^{N_e}_{N} = 2$ possible pure strategies for attackers and sensors are set. According to Case (3), it has $C^{N_a}_{MN} = 4$ and $C^{N_e}_{N} = 1$ possible pure strategies for attackers and sensors. Table 7.1 shows the details of channel chosen in each strategy for those different cases. The result of Case (1) is a mixed Nash equilibrium, while the consequences of Cases (2) and 3) are pure Nash equilibriums.

In accordance with the Nash equilibrium and Nash value, the relevant explanation is given. The trace of expected error covariance without being attacked is 10.25, where the sensors always adopt the low energy $q^s_i(0) = 1$ to keep fundamental function. Figures 7.3 and 7.4 depict the processes of Kalman Filter for the case where there is no attacker, and Figs. 7.5, 7.6, 7.7, 7.8, 7.9, and 7.10 demonstrate the corresponding Kalman Filter curves for three cases, respectively. The black line presents the actual state of plant, and the blue line refers to estimate at the sensor side, which is not attacked. The dashed line in red stands for estimate at remote estimator

Table 7.1 The channel selection strategies and corresponding selected channels for different cases

Strategies and selected channels	Case (1)		Case (2)		Case (3)	
Players	Attackers	Sensors	Attackers	Sensors	Attackers	Sensors
Strategies	d_1^a	d_1^s	d_1^a	d_1^s	d_1^a	d_1^s
Selected channels	Ch_1^a	Ch_1^s	Ch_1^a, Ch_2^a	Ch_1^s	Ch_1^a	Ch_1^s, Ch_2^s
Strategies	d_2^a	d_2^s	d_2^a	d_1^s	d_2^a	
Selected channels	Ch_2^a	Ch_2^s	Ch_1^a, Ch_3^a	Ch_2^s	Ch_2^a	
Strategies	d_3^a		d_3^a		d_3^a	
Selected channels	Ch_3^a		Ch_1^a, Ch_4^a		Ch_3^a	
Strategies	d_4^a		d_4^a		d_4^a	
Selected channels	Ch_4^a		Ch_2^a, Ch_3^a		Ch_4^a	
Strategies			d_5^a			
Selected channels			Ch_2^a, Ch_4^a			
Strategies			d_6^a			
Selected channels			Ch_3^a, Ch_4^a			

[1] $d_j^a, j = 1, 2, \ldots, C_{MN}^{N_a}$ refers to the j-th strategy of attackers, $d_i^s, i = 1, 2, \ldots, C_N^{N_e}$ means the i-th strategy of sensors

[2] $Ch_j^a, j = 1, 2, \ldots, MN$ refers to the j-th channel of attackers, $Ch_i^s, i = 1, 2, \ldots, N$ means the i-th channel of sensors

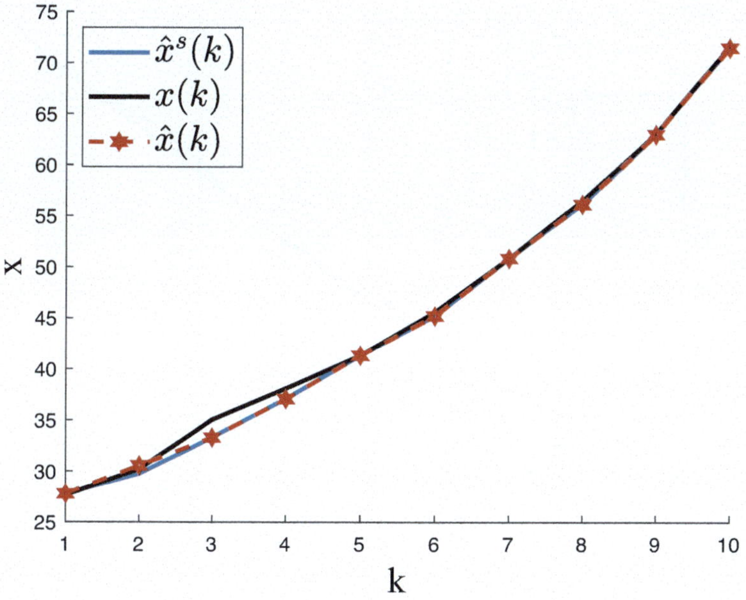

Fig. 7.3 State and estimate of plant 1

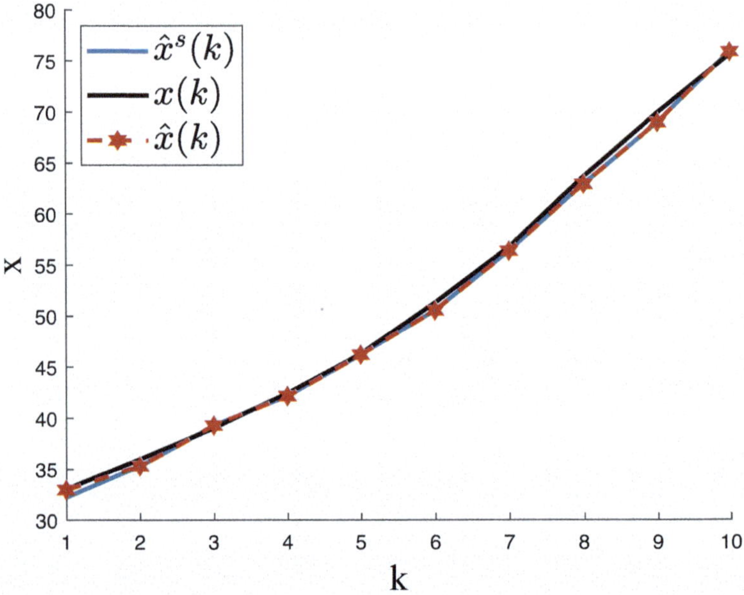

Fig. 7.4 State and estimate of plant 2

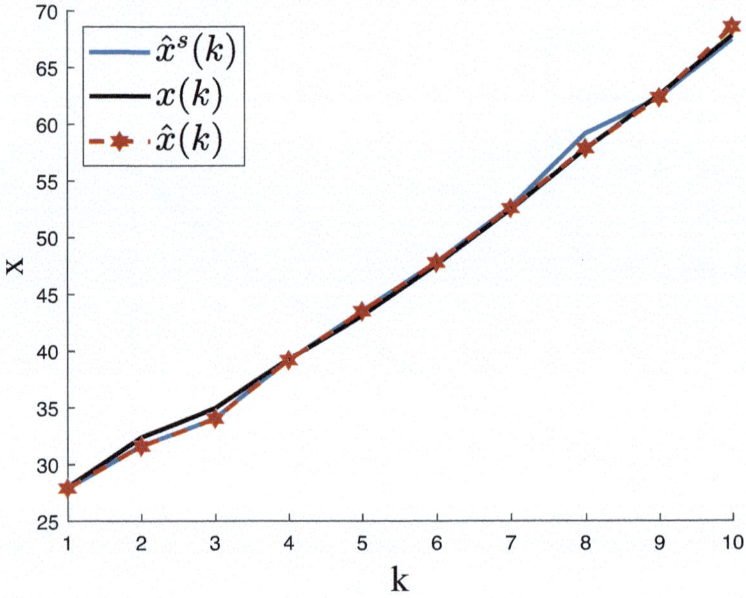

Fig. 7.5 State and estimate of plant 1 under Case (1)

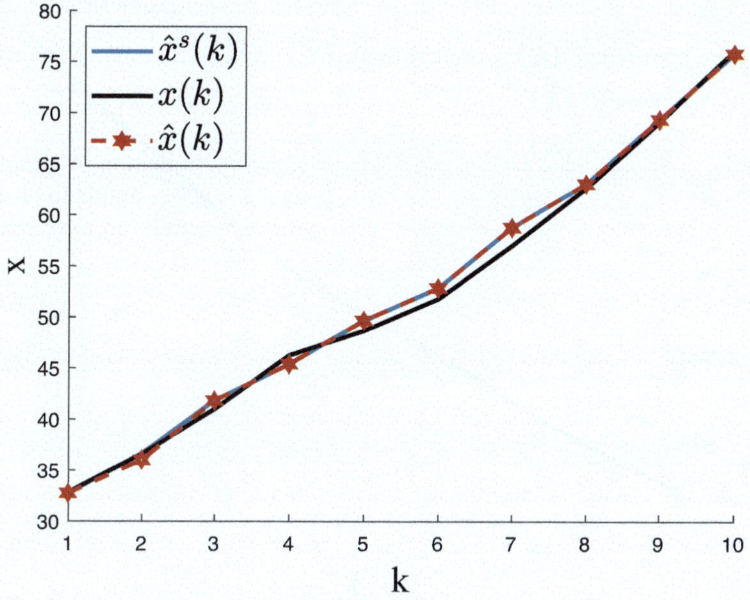

Fig. 7.6 State and estimate of plant 2 under Case (1)

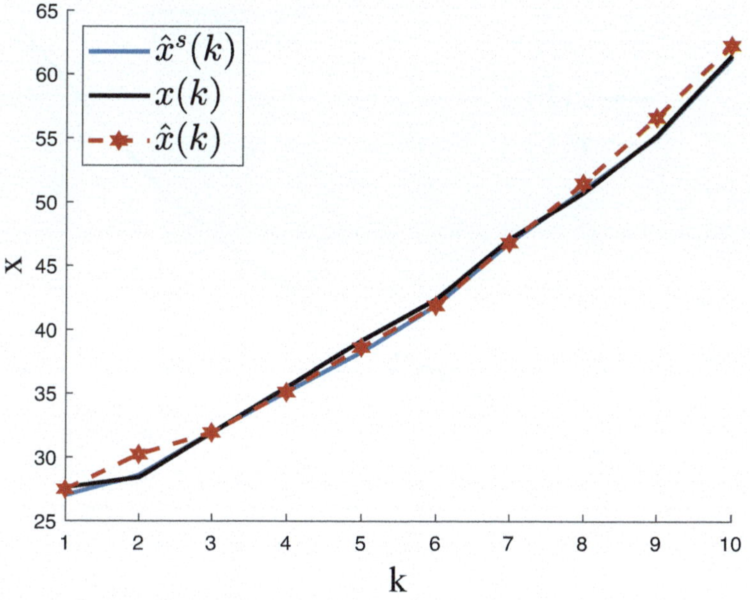

Fig. 7.7 State and estimate of plant 1 under Case (2)

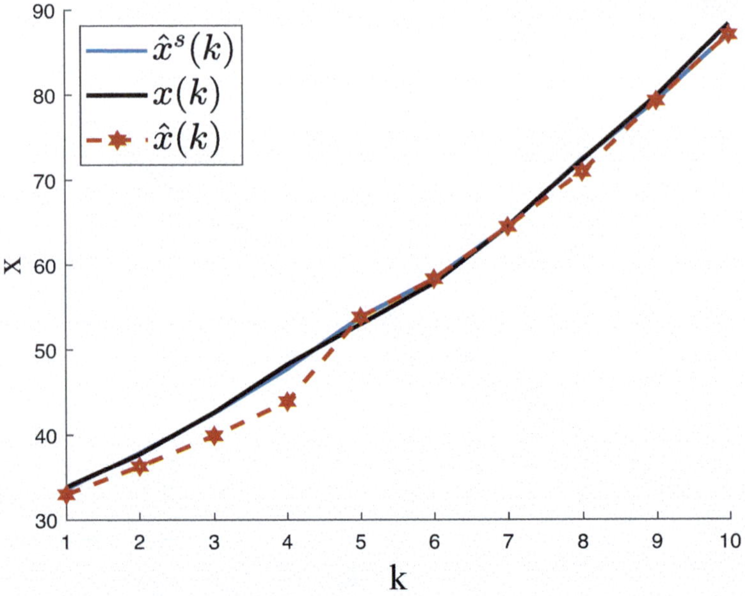

Fig. 7.8 State and estimate of plant 2 under Case (2)

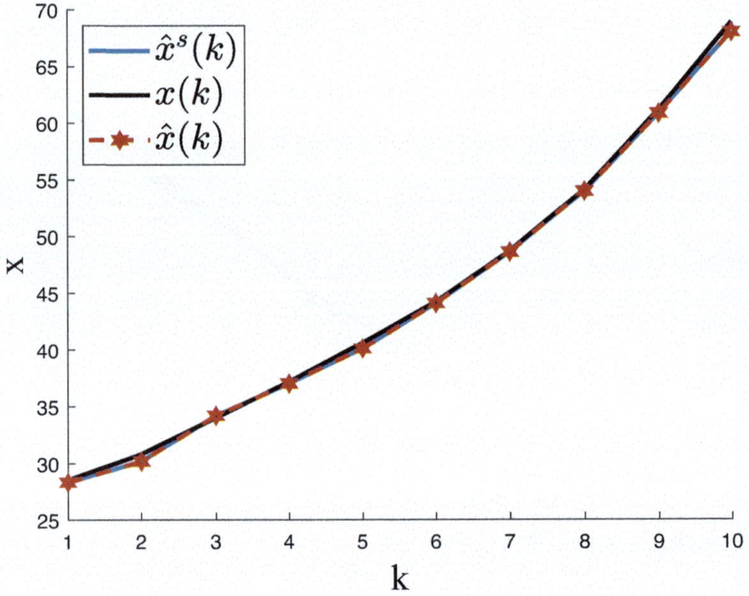

Fig. 7.9 State and estimate of plant 1 under Case (3)

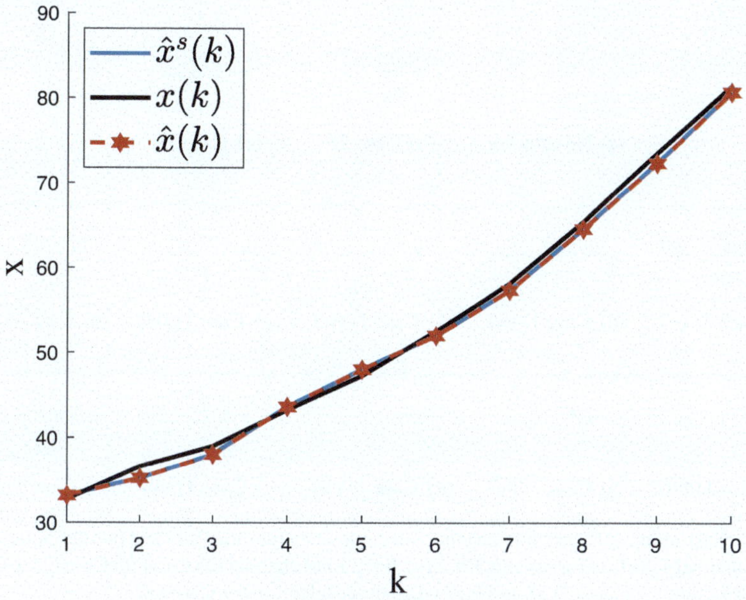

Fig. 7.10 State and estimate of plant 2 under Case (3)

side. Figure 7.11 presents the Nash strategy for optimal channel selection when two attackers and two defenders exist. We utilize the width of line to represent the energy level. The channel numbers of attackers 1, 2 and sensors 1, 2 are shown in Fig. 7.2. Because the results of Cases (2) and (3) are pure Nash equilibrium strategies, the same channel is chosen each time. Figure 7.11d, m demonstrate a channel selection result under a mixed Nash equilibrium strategy in finite time T for Case (1). As Table 7.1 shows, the 3-th and 4-th strategies for attackers mean channels 3 and 4 are selected. By decoding the channel number we can get that the second attacker attacks the first sensor or the second sensor with the energy of $q_2^a = 5$. Figure 7.12 plots the trace of expected error covariance $\sum_{i=1}^{N} \sum_{k=1}^{t} tr(E[P_i(k)])$. When Case (1) occurs (the yellow line), the sensors are able to resist the attackers to some extent while the attackers also cause some degree of damage to the system, because they are of comparable energy. In terms of Case (2) (the red line), the attackers possess more energy than the sensors so that the systems suffer from more dramatic damage. In Case (3) (the black line), one can find that the system performance is protected by the sensors which own large energy, comparing to the case there is no attacker (the blue line).

We then investigate the dynamic optimal channel selection toward the time-varying channel parameters in the time horizon $T = 10$. Take the same setup of attackers and sensors for the static optimal channel selection case. Applying the Algorithm 8, the Nash trace of expected error covariance J_{nash}^* for the above three cases are in Table 7.2, which reveals that greater the energy, the more dominant.

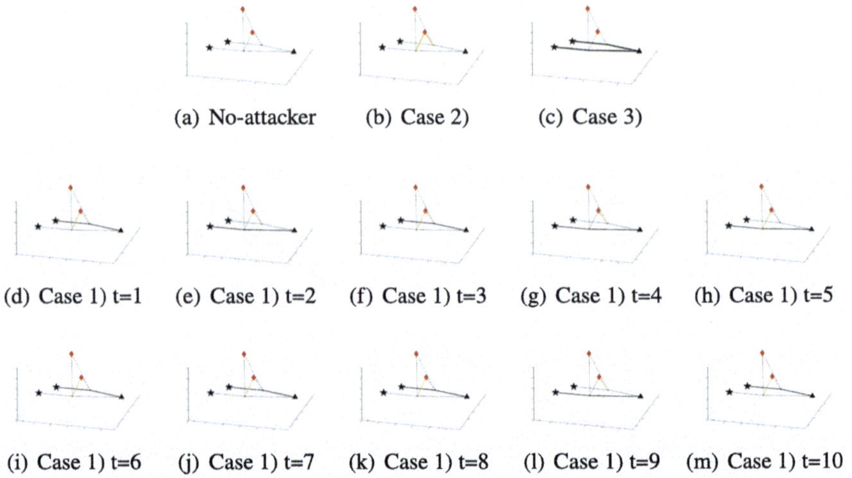

Fig. 7.11 Nash strategy for optimal channel selection when two attackers and two defenders exist. **a** Optimal channel selection for no-attacker case. **b** Optimal channel selection for Case (2). **c** Optimal channel selection for Case (3). **d–m** Optimal channel selection for Case (1)

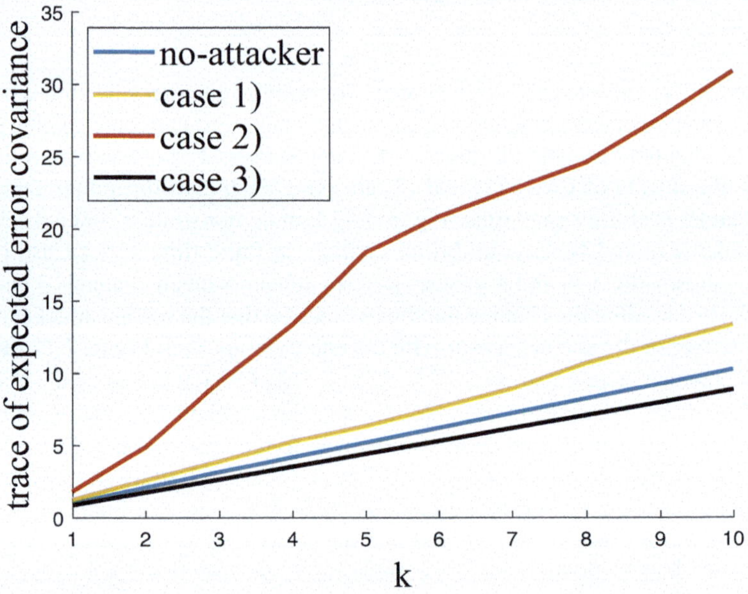

Fig. 7.12 Expected error covariance under static channel selection for different cases

Table 7.2 Nash equilibrium J^*_{nash} at each time step for Cases (1), (2), (3) under dynamic optimal channel selection

T	1	2	3	4	5	6	7	8	9	10
Case (1)	14.30	13.18	12.06	9.81	8.60	6.83	5.36	4.19	2.52	1.31
Case (2)	32.11	29.65	27.47	21.00	20.58	16.50	11.66	11.05	6.08	2.44
Case (3)	8.85	7.96	7.08	6.19	5.31	4.42	3.54	2.65	1.77	0.88

7.4.2 The Energy Allocation Case

In this subsection, Theorem 7.2 is implemented to show the result of optimal energy allocation under zero-sum game, i.e., Problem 7.3.

The same system in Sect. 7.4.1 is used in this part. We solve Problem 7.3 through Theorem 7.2 and compare the corresponding error covariance under different power levels of attackers and sensors as the following three cases:

(1) $\tilde{N}_e = 2$, $\tilde{N}_a = 2$. $N_i = 2$, $q_i^s(0) = 1, q_i^s(1) = 5, q_i^s(2) = 10$, for $i = 1, 2$ and $M_j = 2$, $q_j^a = 5$, for $j = 1, 2$.
(2) $\tilde{N}_e = 2$, $\tilde{N}_a = 2$. $N_i = 2$, $q_i^s(0) = 1, q_i^s(1) = 2, q_i^s(2) = 3$, for $i = 1, 2$ and $M_j = 2$, $q_j^a = 5$, for $j = 1, 2$.
(3) $\tilde{N}_e = 2$, $\tilde{N}_a = 2$. $N_i = 2$, $q_i^s(0) = 1, q_i^s(1) = 5, q_i^s(2) = 10$, for $i = 1, 2$ and $M_j = 2$, $q_j^a = 2$, for $j = 1, 2$.

Similar to Sect. 7.4.1, Case (1) represents both sides have equal energy, Case (2) means the attackers own more energy and Case (3) shows the defenders are superior in energy. Taking Case (1) for example, energy constraints are set as $\tilde{N}_a = 2$, $\tilde{N}_e = 2$. It is assumed that each attacker has 2 ($M_j = 2$, for $j = 1, 2$) energy to choose, and the specific possible attack power for attackers is $0, q_j^a, 2q_j^a$, where $q_j^a = 5$. Sensors also have 2 ($N_i = 2$, for $i = 1, 2$) energy level to choose, and the specific possible transmit power for sensors is $q_i^s(\tilde{e}_i) \in \{1, 5, 10\}$.

As the Sect. 7.3.2 mentioned, in order to set this energy allocation problem in the same framework as optimal channel selection case, a piece of energy can be send through one channel. For example, the energy level of j-th attacker is M_j. It is equivalent to the j-th attacker having M_j channels to choose. Figure 7.13 depicts the channel number encoding model for energy allocation simulation. The numbers in the yellow and black squares pointed by the arrow indicate the channel number of attackers and defenders, respectively. Because $\tilde{M} = 2 \times (2 + 2) = 8$, there are 8 channels of attackers in Fig. 7.13. Then the channel number coding method is that the first channel refers to the 1-st energy level of the first attacker to attack the 1-st sensor, the second channel refers to the 1-st energy level of the first attacker to attack the 2-nd sensor, and so on. Similarly, because $\tilde{N} = 2 + 2 = 4$, there are 4 channels of sensors in Fig. 7.13. The first N_1 channel means the 1-st—N_1-th energy of the first sensor to transmit data through the corresponding channel and the following N_2 channel means the 1-st—N_2-th energy of the second sensor to transmit data

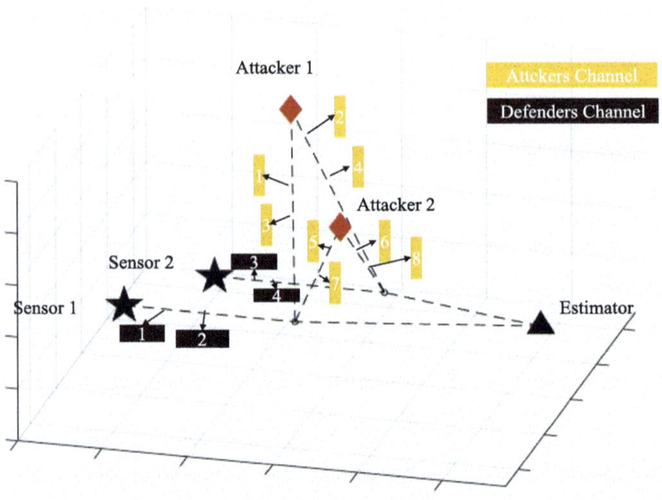

Fig. 7.13 Channel number encoding model for optimal energy allocation simulation

through the corresponding channel. There are $C_{\tilde{M}}^{\tilde{N}_a} = 28$ possible pure strategies for attackers. There are $C_{\tilde{N}}^{\tilde{N}_e} = 6$ possible pure strategies for sensors. Table 7.3 illustrates the energy allocation strategies and corresponding selected channels of different cases for attackers and sensors. From Theorem 7.2, the Nash energy allocation strategy and Nash trace of expected error covariance for the above three cases are as follows:

(1) $(\pi^a)^* = d_{24}^a$, $(\pi^s)^* = d_1^s$, $J_{nash}^* = 10.91$;
(2) $(\pi^a)^* = d_{24}^a$, $(\pi^s)^* = d_1^s$, $J_{nash}^* = 20.22$;
(3) $(\pi^a)^* = d_{24}^a$, $(\pi^s)^* = d_1^s$, $J_{nash}^* = 9.10$.

The result is pure strategy Nash equilibrium, which means both attackers and sensors choose to leverage same strategy within the time horizon T. Figure 7.14 presents the Nash strategy simulation effect for no-attacker case and Cases (1), (2), (3). The symbol of black stars signifies the location of two sensors, the black triangle delegates remote estimator and the red rhombus are two attackers. We utilize the width of line to indicate the energy level. It is supposed that the number of attacker 1, 2 and sensor 1, 2 is same with Fig. 7.13. Taking Fig. 7.14b for example, the 24-th strategy and 1-st strategy are selected by attackers and defenders, respectively. As Table 7.3 shows, the 24-th strategy for attackers means that the channel 5 and 7 are allocated. By decoding the channel number we can get that the second attacker attacks the first sensor with the energy of $2q_2^a = 10$. As Table 7.3 shows, the 1-st strategy for sensors states that the 1-st and 2-nd channels are selected. By decoding the channel number, the first sensor defends the attackers with the energy of $q_1^s(2) = 10$. Figure 7.15 shows the expected error covariance for energy allocation. Then blue line implies the results under not being attacked, while the yellow, red, black lines refer to Cases (1), (2), (3), respectively. Similar consequences can be obtained with the channel selection

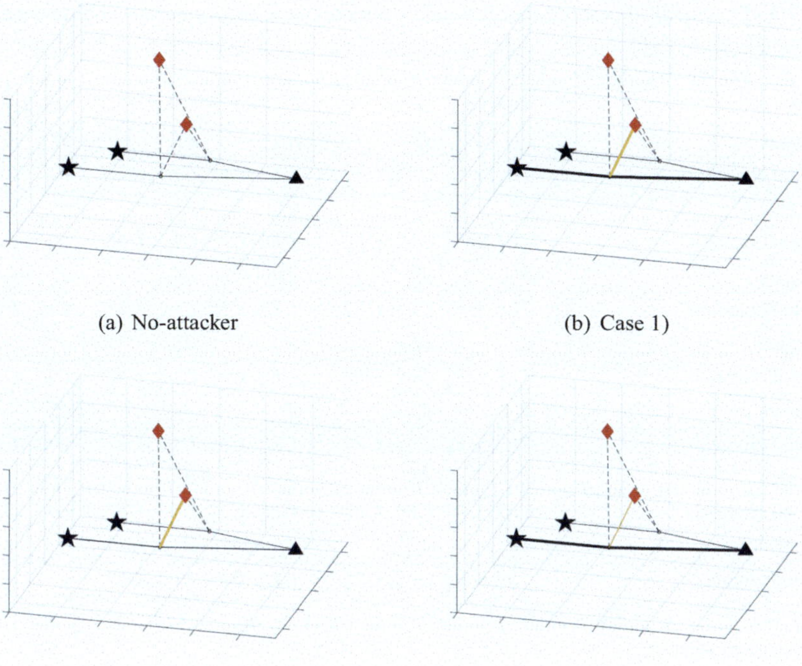

(a) No-attacker (b) Case 1)

(c) Case 2) (d) Case 3)

Fig. 7.14 Nash strategy for optimal energy allocation when two attackers and two defenders exist

case. Thus, the simulation between sensors and attackers in case of energy allocation shows that this framework of multi-attacker-defender game is effective.

7.5 Conclusion

In this chapter, a framework of multi-attacker-defender game is proposed to seek the Nash distributed DoS attack strategy and defend strategy in a packet-dropping network. The multi-attacker and multi-defender model of SINR is first set up. The attackers with limited energy incline to maximize the cumulative expected estimation error covariance by jamming the channels through cooperation. The objective of defenders is to minimize the cumulative expected estimation error covariance. Therefore, a zero-sum game is composed. The static channel selection problem is taken into consideration, and the Nash equilibrium is found by linear programming. The dynamic channel selection problem is further investigated when it comes to time-varying channel parameters. Based on channel selection problem, there comes energy allocation problem which is probably more practical. Eventually, kinds of numerical examples are given to show the results with different energy levels. In the

Table 7.3 The energy allocation strategies and corresponding selected channels of different cases for attackers and sensors

Strategies and selected channels	Cases (1), (2), and (3)				
Players	Attackers				Sensors
Strategies	d_1^a	d_8^a	d_{15}^a	d_{22}^a	d_1^s
Selected channels	Ch_1^a, Ch_2^a	Ch_2^a, Ch_3^a	Ch_3^a, Ch_5^a	Ch_4^a, Ch_8^a	Ch_1^s, Ch_2^s
Strategies	d_2^a	d_9^a	d_{16}^a	d_{23}^a	d_2^s
Selected channels	Ch_1^a, Ch_3^a	Ch_2^a, Ch_4^a	Ch_3^a, Ch_6^a	Ch_5^a, Ch_6^a	Ch_1^s, Ch_3^s
Strategies	d_3^a	d_{10}^a	d_{17}^a	d_{24}^a	d_3^s
Selected channels	Ch_1^a, Ch_4^a	Ch_2^a, Ch_5^a	Ch_3^a, Ch_7^a	Ch_5^a, Ch_7^a	Ch_1^s, Ch_4^s
Strategies	d_4^a	d_{11}^a	d_{18}^a	d_{25}^a	d_4^s
Selected channels	Ch_1^a, Ch_5^a	Ch_2^a, Ch_6^a	Ch_3^a, Ch_8^a	Ch_5^a, Ch_8^a	Ch_2^s, Ch_3^s
Strategies	d_5^a	d_{12}^a	d_{19}^a	d_{26}^a	d_5^s
Selected channels	Ch_1^a, Ch_6^a	Ch_2^a, Ch_7^a	Ch_4^a, Ch_5^a	Ch_6^a, Ch_7^a	Ch_2^s, Ch_4^s
Strategies	d_6^a	d_{13}^a	d_{20}^a	d_{27}^a	d_6^s
Selected channels	Ch_1^a, Ch_7^a	Ch_2^a, Ch_8^a	Ch_4^a, Ch_6^a	Ch_6^a, Ch_8^a	Ch_3^s, Ch_4^s
Strategies	d_7^a	d_{14}^a	d_{21}^a	d_{28}^a	
Selected channels	Ch_1^a, Ch_8^a	Ch_3^a, Ch_4^a	Ch_4^a, Ch_7^a	Ch_7^a, Ch_8^a	

[1] $d_j^a, j = 1, 2, \ldots, C_{\tilde{M}}^{\tilde{N}_a}$ refers to the j-th strategy of attackers, $d_i^s, i = 1, 2, \ldots, C_{\tilde{N}}^{\tilde{N}_e}$ means the i-th strategy of sensors

[2] $Ch_j^a, j = 1, 2, \ldots, \tilde{M}$ refers to the j-th channel of attackers, $Ch_i^s, i = 1, 2, \ldots, \tilde{N}$ means the i-th channel of sensors

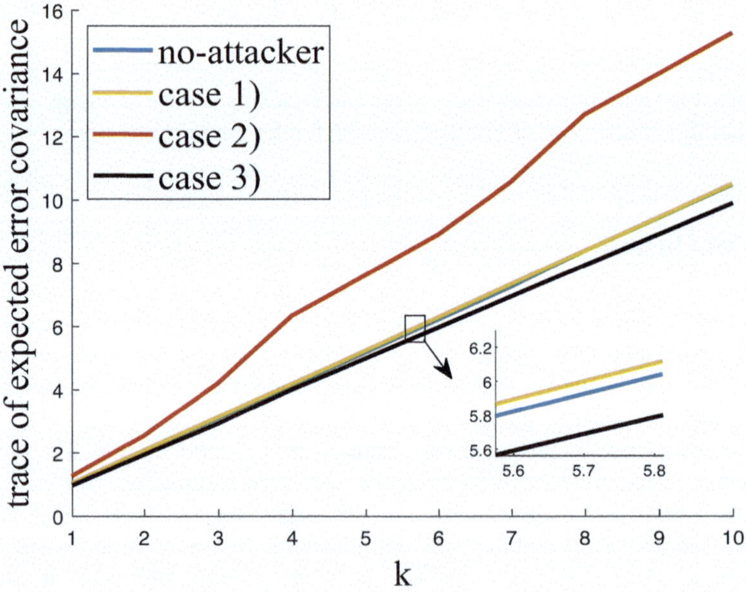

Fig. 7.15 Expected error covariance under static energy allocation with or without being attacked

future, we are tending to study the situation when the system parameter and channel information are not able to obtain, that is, the attackers have incomplete information when launching attack.

References

1. Li, Y., Shi, L., Cheng, P., Chen, J., Quevedo, D.E.: Jamming attacks on remote state estimation in cyber-physical systems: a game-theoretic approach. IEEE Trans. Autom. Control **60**(10), 2831–2836 (2015)
2. Wu, C., Li, X., Pan, W., Liu, J., Wu, L.: Zero-sum game-based optimal secure control under actuator attacks. IEEE Trans. Autom. Control **66**(8), 3773–3780 (2021)
3. Li, Y., Quevedo, D.E., Dey, S., Shi, L.: SINR-based DoS attack on remote state estimation: a game-theoretic approach. IEEE Trans. Control Netw. Syst. **4**(3), 632–642 (2017)
4. Wu, H., Gao, Q., Tao, X., Zhang, N., Chen, D., Han, Z.: Differential game approach for attack-defense strategy analysis in internet of things networks. IEEE Internet Things J. **9**(12), 10340–10353 (2021)
5. Zargar, S.T., Joshi, J., Tipper, D.: A survey of defense mechanisms against distributed denial of service (DDoS) flooding attacks. IEEE Commun. Surv. Tutor. **15**(4), 2046–2069 (2013)
6. Xu, K., Xu, M.W., Wu, J.P.: Research on distributed denial-of-service attacks: a survey. Minimicro Syst. Shenyang **25**, 337–346 (2004)
7. Sun, C.H., Liu, B.: Survey on new solutions against distributed denial of service attacks. Acta Electonica Sinica **37**(7), 1562–1570 (2009)
8. Qin, J., Li, M., Shi, L., Yu, X.: Optimal denial-of-service attack scheduling with energy constraint over packet-dropping networks. IEEE Trans. Autom. Control **63**(6), 1648–1663 (2017)
9. Peng, L., Shi, L., Cao, X., Sun, C.: Optimal attack energy allocation against remote state estimation. IEEE Trans. Autom. Control **63**(7), 2199–2205 (2017)
10. Gohary, R.H., Huang, Y., Luo, Z.Q., Pang, J.S.: A generalized iterative water-filling algorithm for distributed power control in the presence of a jammer. IEEE Trans. Signal Process. **57**(7), 2660–2674 (2009)
11. Ding, K., Ren, X., Leong, A.S., Quevedo, D.E., Shi, L.: Remote state estimation in the presence of an active eavesdropper. IEEE Trans. Autom. Control **66**(1), 229–244 (2021)
12. Ren, X., Wu, J., Johansson, K.H., Shi, G., Shi, L.: Infinite horizon optimal transmission power control for remote state estimation over fading channels. IEEE Trans. Autom. Control **63**(1), 85–100 (2018)
13. Wu, G., Sun, J., Chen, J.: Optimal data injection attacks in cyber-physical systems. IEEE Trans. Cybern. **48**(12), 3302–3312 (2018)
14. Shiu, Y.S., Chang, S.Y., Wu, H.C., Huang, S.C.H., Chen, H.H.: Physical layer security in wireless networks: a tutorial. IEEE Wirel. Commun. **18**(2), 66–74 (2011)
15. Osborne, M.J., Rubinstein, A.: A Course in Game Theory. MIT Press, Cambridge, MA, USA (1994)
16. Schwartz, H.M.: Multi-agent Machine Learning: A Reinforcement Approach. Wiley, Hoboken, NJ, USA (2014)
17. Nash, J.: Non-cooperative games. Ann. Math. **54**(2), 286–295 (1951)
18. Littman, M.L.: Markov games as a framework for multi-agent reinforcement learning. In: Proceedings of Machine Learning, pp. 157–163. (1994)

Chapter 8
Game-Based FDI Attack Strategy Against CPSs via Zero-Sum Game

Abstract The security problem of cyber-physical systems (CPSs) under the injection attack is investigated in this chapter. The FDI attack strategies based on the game approach are proposed, which make the attacker consume less energy and more difficult to defend. In addition, the constraint of the limited attack resources is considered. Noted that the attacker and the defender are antagonistic in this chapter, the framework of a zero-sum game is used to describe this relation, and a finite-horizon quadratic cost function is also defined. The objective of this chapter is to find an appropriate offensive and defensive strategy to optimize this function. Different from the existing games, the upper bound on the energy consumption of each attack is considered. Pontryagin's maximum principle and saturation equivalent approach are used to solve this game problem. Thus, the optimal offensive and defensive strategy can be derived. Finally, a numerical simulation and a physical experiment with the wheeled mobile robot (WMR) validate the effectiveness of the proposed schemes.

8.1 Introduction

Game-based FDI attack strategy is a natural extension of the classic FDI attack strategy. It has gradually appeared because the development of smart sensors and artificial intelligence technologies makes the attacked systems have the ability to actively fight. Therefore, when designing the attack strategies, the game between the defender and the attacker needs to be considered. Among the works on game-based FDI attack strategy, the relationship between the defender and the attacker is described as a leader-follower game (i.e., a Stackelberg game) [1]. Due to the fact that the majority of the relationships between defenders and attackers are non-leader-followers, the use of the Stackelberg game cannot be applied directly. Recently, game-based FDI attack strategies and defense mechanisms for smart grids are investigated [2]. For this method, the attacker and the defender play a game against the measured values in the smart grid, and a more general way of non-cooperative game is studied. The game-based FDI attack strategy design needs to use the system states, the inputs or the outputs. Therefore, if this information is not available, the game-based FDI attack strategies cannot work. In this case, a common alternative choice is to use the

S. Gao et al., *Cyberattack Strategies Against Cyber-Physical Systems: From a Control Perspective*, https://doi.org/10.1007/978-981-95-9645-4_8

estimations instead of the actual values, and the purpose is to ensure the performance of the cost function in the FDI attack strategy. The primary effort has been made on the game-based FDI attack strategy design. The confrontation relationship between Kalman filter and attacker is constructed into a zero-sum game framework [3]. This scheme considers the active defense of the filter and the accurate estimations that can be obtained under malicious attacks.

In the aforementioned game-based FDI attack strategy design, the energy consumption of the attacker is not considered, which is assumed to be unlimited. However, in many actual attack events, the energy consumed by the attacker at each instant has an upper bound. The reasons for this upper bound of energy consumption are usually due to the maximum reception power of the attacked device, the threshold of the detector, the maximum use power, and the maximum transmission power of the device generating the attack signal. Therefore, the energy consumed by the attacker each time to inject the designed false data needs to be limited to an upper bound. This scenario can be constrained by a quadratic inequality, which added to the game-based FDI attack strategy design. Some primary efforts have been made in the design of attack strategies considering energy consumption. The impact of the optimal attack strategy on the estimated average performance when the energy is limited is studied [4]. Further, the effect of attacks on control performance is studied [5], that is, the attackers try to maximize the cost of LQG control under energy constraints. Unfortunately, not until very recently, the game-based FDI attack strategy design with limited energy consumption has received attention. This problem is important and challenging for the design of attack strategy against CPSs, especially for those systems with a smart sensor embedded with intelligent algorithms, which makes the system have the ability to resist attack. The design of the FDI attack strategy based on the zero-sum game is mainly discussed in this chapter, in which the energy consumed by the attacker is limited at each time. Designing the attack strategy with resource constraints is significant to the cyber security of CPSs, that is, it can make the defenders understand the behavior of the attacker and help them formulate defense strategies, which motivates this research. In addition, the attenuation effect of the information transmission also needs to be considered in practical application.

Game-based FDI attack strategies under resource constraint against CPSs are developed in this chapter. The main contributions of this chapter are summarized as follows:

(1) In order to describe the confrontation relationship between the attacker and the defender of CPSs, a zero-sum game framework is proposed. In this framework, the optimal game-based FDI attack strategy and the defense strategy of the CPSs are obtained, which makes the cost function optimal.
(2) Considering the limited energy resources of the attacker, a quadratic inequality is introduced. A novel game-based FDI attack strategy with limited attack energy resources is designed.
(3) The switching mode of the attack is considered in the design of the game-based FDI attack strategy, and the optimal switching sequence is obtained. In other words, the FDI attack mode for all channels from the controller to the actuator is

a special case of the switching mode. It is worth noting that the switching mode of attack has the advantages of low-energy consumption, high flexibility, and stealthiness.

The rest of this chapter is organized as follows. The problem formulation for cyber-physical systems is shown in Sect. 8.2. Section 8.3 describes the switching data injection attack strategy with the saturation constraint and non-switching data injection attack strategy under the zero-sum game framework. In Sect. 8.4, the effectiveness of the proposed scheme is demonstrated with examples and experiment. Finally, this chapter is concluded in Sect. 8.5.

8.2 Problem Formulation

8.2.1 System Model

The physical plant model of CPS is described as the following discrete-time system,

$$\begin{aligned} x_{k+1} &= A_k x_k + B_k u_k, \\ u_k &= K_k x_k, \end{aligned} \tag{8.1}$$

where $x_k \in \mathbb{R}^n$ and $u_k \in \mathbb{R}^l$ are the process state and the control input vector at time k, $k = 0, 1, 2, \ldots, N$, and N is terminating time. The matrices $A_k \in \mathbb{R}^{n\times n}$ and $B_k \in \mathbb{R}^{n\times l}$ are given. The matrix $K_k \in \mathbb{R}^{l\times n}$ is a feedback controller gain matrix, which makes the closed-loop system stable.

By optimal control theory [6–8], the following lemmas can be obtained.

Lemma 8.1 ([6]) *The optimization problem is expressed as*

$$\begin{aligned} \min_{u_k} \;\; & J_1(x_k, u_k) = x_{N+1}^T Q_{N+1} x_{N+1} + \frac{1}{2}\sum_{k=1}^{N}(x_k^T Q_k x_k + u_k^T R_k u_k), \\ s.t. \;\; & \begin{cases} x_{k+1} = A_k x_k + B_k u_k, \\ Q_k \succeq 0,\, R_k \succ 0, \end{cases} \end{aligned}$$

where $Q_k \in \mathbb{R}^{n\times n}$ is a positive semi-definite matrix and $R_k \in \mathbb{R}^{l\times l}$ is a positive definite matrix at time k.

The optimal input is

$$u_k^* = K_k^* x_k, \tag{8.2}$$

and the optimal gain is

$$K_k^* = -(R_k + B_k^T P_{k+1} B_k)^{-1} B_k^T P_{k+1} A_k,$$

where P_k satisfies the following Algebraic Riccati Equation

$$P_k = Q_k + A_k^T P_{k+1} A_k - A_k^T P_{k+1} B_k (R_k + B_k^T P_{k+1} B_k)^{-1} B_k^T P_{k+1} A_k, \quad (8.3)$$

and the final value condition satisfies

$$P_{N+1} = Q_{N+1}.$$

Lemma 8.2 ([7, 8]) *For the following optimization problem,*

$$\begin{aligned} \min_{u_k} \quad & J_1(x_k, u_k) \\ s.t. \quad & x_{k+1} = A_k x_k + B_k u_k, \end{aligned}$$

if the requirement $Q_k \succeq 0$ is not satisfied, a necessary and sufficient condition for (8.2)–(8.3) to provide a unique solution to the affine-quadratic discrete-time optimal control problem is

$$R_k + B_k^T P_{K+1} B_k \succ 0.$$

Remark 8.1 Lemma 8.1 presents the classical theory of linear quadratic regulator (LQR) control strategy under a discrete-time framework, a theory with extensive and profound influence in the field of control engineering. Unlike Lemmas 8.1, 8.2 represents a significant extension and deepening of the LQR theoretical framework. It focuses on a more complex and practical scenario, when the weight matrix Q_k fails to satisfy the traditional LQR requirement of positive semi-definite matrix, the convexity of the optimization problem ceases to exist, transforming into a nonconvex form. Under these non-standard conditions, Lemma 8.2 provides necessary and sufficient conditions for obtaining optimal control inputs, thereby offering a methodology for addressing a broader range of practical control optimization problems.

8.2.2 Attacked Model

This subsection presents the offense-defense game framework and elaborates on the dynamic model of the attacked system and the energy consumption constraints of attack behavior. First, this chapter defines the attacker and controller as the two game parties, corresponding to Player 1 (controller) and Player 2 (attacker), whose objectives are diametrically opposed. Specifically, the controller, as the defender, focuses its core strategy on minimizing the system's weighted quadratic performance index through optimized control strategy while simultaneously reducing its own energy consumption. Conversely, the attacker, as the offensive party, strives to maximize the system's weighted quadratic performance index through carefully designed attack strategies while maintaining low attack energy consumption, thereby threatening the

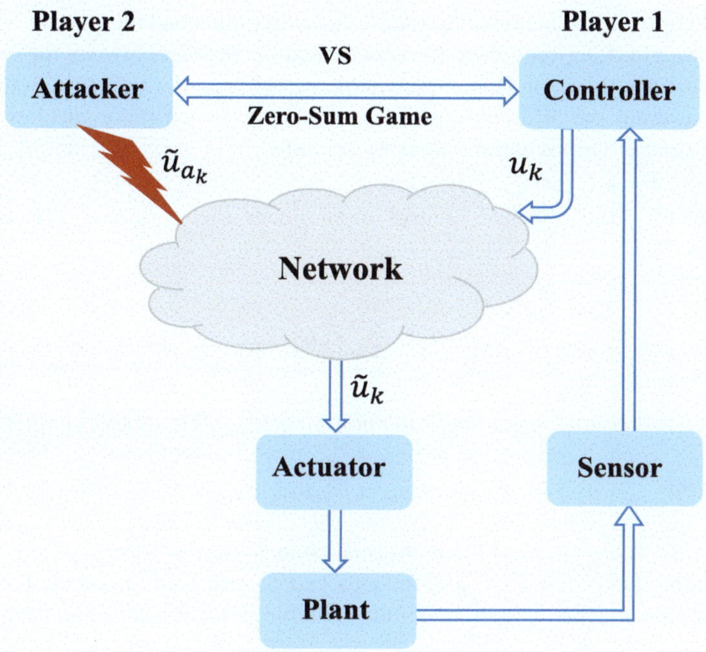

Fig. 8.1 The block diagram of the attacked system

system's stability and security. The block diagram of the attacked system is shown in Fig. 8.1.

For the attacker, the following assumptions are given.

Assumption 8.1 ([9, 10]) The attacker has complete knowledge of the system (8.1) including A_k, B_k, and K_k through eavesdropping on the input and output data for a sufficient time.

Assumption 8.2 The attacker has the ability to inject the calculated false data vector into the actuator synchronously with the control signal.

Remark 8.2 Assumption 8.1 describes the ability of the attacker to access the system information. Since the focus of this chapter is on the design of the attack strategies, the restrictions on the acquisition of system information have been appropriately relaxed. Some similar assumptions are also used in [9, 10]. Assumption 8.2 describes the ability of the attacker to invade the attacked system. In fact, due to the openness and fragility of the wireless network communication channel, the assumption of the attacker's ability to invade the system is reasonable.

The switching and non-switching false data injection attack vectors are described as $\sum_{i=1}^{N_a} \varepsilon_k^i D_{a_k}^i u_{a_k}$ and $D_{a_k} u_{a_k}$, where $u_{a_k} \in \mathbb{R}^l$ is the attack input at time k, N_a is the total number of controller to actuator channels, the switching attack matrix $D_{a_k}^i \in \mathbb{R}^{l\times l}$ is the attack weight describing the i-th communication channel between the controller and the actuator at time k and the non-switching attack matrix $D_{a_k} \in \mathbb{R}^{l\times l}$ is the attack weight describing the all communication channel between the controller and the actuator at time k. The switching input $\Phi_k(t) = [\varepsilon_k^1, \varepsilon_k^2, \ldots, \varepsilon_k^{N_a}] \in \tilde{\Phi}_k, \varepsilon_k^i \in \{0, 1\}$, and the set is $\tilde{\Phi}_k = \{(1, 0, \ldots, 0), (0, 1, \ldots, 0), \ldots, (0, 0, \ldots, 1)\}$.

Subsequently, the attacked control input under the switching FDI attack is designed as

$$\tilde{u}_k = u_k + \sum_{i=1}^{N_a} \varepsilon_k^i D_{a_k}^i u_{a_k}, \tag{8.4}$$

and the attacked control input under the non-switching FDI attack is expressed as

$$\tilde{u}_k = u_k + D_{a_k} u_{a_k}, \tag{8.5}$$

where $\tilde{u}_k \in \mathbb{R}^l$ is the attacked l-dimensional control input at time k.

Combining the system (8.1) and the attacked control input under switching FDI attack (8.4), the attacked system under the switching FDI attack can be described as

$$\tilde{x}_{k+1} = A_k \tilde{x}_k + B_k u_k + \sum_{i=1}^{N_a} \varepsilon_k^i B_{a_k}^i u_{a_k}. \tag{8.6}$$

Similarly, introducing the attacked control input under non-switching FDI attack (8.5) into the system (8.1) yields the following representation of the attacked system under non-switching FDI attacks,

$$\tilde{x}_{k+1} = A_k \tilde{x}_k + B_k u_k + B_{a_k} u_{a_k}, \tag{8.7}$$

where $\tilde{x}_k \in \mathbb{R}^n$ is the attacked n-dimensional state at time k, $B_{a_k}^i = B_k D_{a_k}^i$ and $B_{a_k} = B_k D_{a_k}$.

Considering that the attacker's energy consumption is constrained by the maximum receive power of the attacked device, the detector threshold, and the maximum operating power and maximum transmission power of the attack signal generation device, the attacker's energy consumption at time k can be expressed by the following constraints,

$$u_{a_k}^T u_{a_k} \le U_{a_0}^2, \tag{8.8}$$

where U_{a_0} is a positive definite constant, which means that there is an upper bound on the maximum energy of the attacker in each attack.

Remark 8.3 U_{a_0} is a positive constant related to the signal transmission energy E_T of the device and the threshold Ω_D of the detector. The attacker can acquire the range of detector threshold Ω_D by analyzing the alarm log of the attacked system, Ω_D satisfies $\underline{\Omega}_D \le \Omega_D \le \overline{\Omega}_D$. Therefore, U_{a_0} can be obtained by $U_{a_0} = f(E_T, \underline{\Omega}_D, \overline{\Omega}_D)$, where f is the function describing the relationship between U_{a_0}, E_T, $\underline{\Omega}_D$, and $\overline{\Omega}_D$. In this chapter, $U_{a_0}^2 = \min\{E_T, \underline{\Omega}_D, \overline{\Omega}_D\}$ is adopted. It is a natural fact that low-energy attack signals are difficult to be detected.

8.2.3 Optimal FDI Attack Problem via Zero-Sum Game

According to the objectives of the attacker and defender described in Sect. 8.2.3 and Fig. 8.1, the zero-sum game-based objective function for the attack-defense confrontation can be expressed as

$$J_2(\tilde{x}_k, u_k, u_{a_k}) = \frac{1}{2}\sum_{k=1}^{N}(\tilde{x}_k^T \tilde{Q}_k \tilde{x}_k + u_k^T R_k u_k - \gamma^2 u_{a_k}^T \tilde{R}_k u_{a_k}), \tag{8.9}$$

where $\tilde{Q}_k \in \mathbb{R}^{n\times n}$ is a positive semi-definite matrix at time k, $\tilde{R}_k \in \mathbb{R}^{p\times p}$ is a positive definite matrix at time k, and $\gamma \ge 0$ is a prescribed constant disturbance attenuation level.

In the objective function (8.9), the energy consumption terms of the controller and the attacker are denoted as $u_k^T R_k u_k$ and $u_{a_k}^T \tilde{R}_k u_{a_k}$, respectively. Since their purposes are opposite, their relationship can be described as a zero-sum game. Then, integrating the attacked system (8.6) under switching attacks, the attack energy consumption constraint (8.8), and the objective function (8.9), the optimal switching FDI attack optimization problem based on zero-sum games can be described as follows:

Problem 8.1

$$\max_{u_{a_k}} \min_{u_k} \quad J_2(\tilde{x}_k, u_k, u_{a_k})$$

$$\text{s.t.} \quad \begin{cases} \tilde{x}_{k+1} = A_k\tilde{x}_k + B_k u_k + \sum_{i=1}^{N_a} \varepsilon_k^i B_{a_k}^i u_{a_k}, \\ u_{a_k}^T u_{a_k} \le U_{a_0}^2,\ \tilde{Q}_k \succeq 0,\ R_k \succ 0,\ \tilde{R}_k \succ 0. \end{cases}$$

Likewise, based on the attacked system (8.7) and the objective function (8.9), considering the ideal scenario where attack energy consumption is unconstrained, the optimal non-switching FDI attack optimization problem based on zero-sum games can be described as

Problem 8.2

$$\max_{u_{a_k}} \min_{u_k} \quad J_2(\tilde{x}_k, u_k, u_{a_k})$$

$$\text{s.t.} \quad \begin{cases} \tilde{x}_{k+1} = A_k\tilde{x}_k + B_k u_k + B_{a_k} u_{a_k}, \\ \tilde{Q}_k \succeq 0,\ R_k \succ 0,\ \tilde{R}_k \succ 0. \end{cases}$$

Remark 8.4 From the literature [11], it is known that Problems 8.1 and 8.2 studied in this chapter satisfy the condition that the second-order partial derivatives of the Hamiltonians with respect to u_k and u_{a_k} have opposite signs. In other words, the objective function is strongly concave for the attacker and pseudoconvex for the defender. Therefore, whether it is a minimax problem or a maximin problem, a unique solution exists, namely the Nash equilibrium solution, defined as $\max_{u^*_{a_k}} \min_{u^*_k} = \min_{u^*_k} \max_{u^*_{a_k}}$.

8.3 Game-Based FDI Attack Strategies

This section systematically presents optimal FDI attack strategies within the zero-sum game framework and their algorithmic implementations, specifically containing both switching and non-switching forms.

8.3.1 Game-Based Switching FDI Attack Strategy

Before presenting the game-based switching FDI attack strategy, the following key lemmas are introduced.

Lemma 8.3 (The matrix inversion lemma [21]) *For $\vartheta \in \mathbb{R}^{n\times n}$, $\varpi \in \mathbb{R}^{n\times m}$, $\varsigma \in \mathbb{R}^{m\times n}$ and ϑ is a non-singular matrix, then $\vartheta + \varpi\varsigma$ has the inverse matrix as follows:*

$$(\vartheta + \varpi\varsigma)^{-1} = \vartheta^{-1} - \vartheta^{-1}\varpi(I + \varsigma\vartheta^{-1}\varpi)^{-1}\varsigma\vartheta^{-1},$$

where I is the identity matrix of the appropriate dimension.

Lemma 8.4 ([12, 13]) *Given matrices $D,\ L \in \mathbb{R}^{p\times q}$, for $\zeta \in \mathbb{R}^q$, suppose that $\|L\zeta\|_\infty \le 1$, then*

$$Sat[D\zeta] \in co\{V_d D\zeta + V_d^- L\zeta, d = 1, 2, \ldots, 2^p\},$$

where V_d is $p \times p$ diagonal matrix with elements either 1 or 0, $V_d^- = I_d - V_d$.

Consequently, $Sat[D\zeta]$ can be described as

$$Sat[D\zeta] = \sum_{d=1}^{2^p} \ell_d(V_d D + V_d^- L)\zeta,$$

where ℓ_d are real parameters, $0 \le \ell_d \le 1$, $\sum_{d=1}^{2^p} \ell_d = 1$ and I_d is the identity matrix of the appropriate dimension.

Lemma 8.5 ([14]) *For $\Upsilon \in \mathbb{R}^{n\times n}$, the necessary and sufficient condition for matrix Υ to be positive definite is that the matrix $(\Upsilon + \Upsilon^T)$is positive definite.*

Remark 8.5 Lemma 8.4 is mainly adopted to deal with the nonlinear saturation function $Sat[\star]$ caused by the strong constraint of attack input. Lemma 8.5 provides the basis for determining the positive definiteness of asymmetric matrices in the domain of the real numbers.

For simplicity and clarity of reading, Table 8.1, outlining each symbol's definition, is presented.

Theorem 8.1 *Under Assumptions 8.1–8.2, if $\Theta_k \succ 0$ holds, the Problem 8.1 can be solved with the optimal control input u_k^* and the optimal attack input $u_{a_k}^*$ as follows:*

$$u_k^* = -R_k^{-1} B_k^T S_{k+1} \eta_k^{-1} A_k \tilde{x}_k, \tag{8.10}$$

Table 8.1 The definition of symbols

Symbol	Definition
η_k	$I_1 + B_k R_k^{-1} B_k^T S_{k+1} - \sum_{i=1}^{N_a} \sum_{d=1}^{2^p} \varepsilon_k^i B_{a_k}^i U_{a0} \ell_k^d (\tilde{V}_k^d \tilde{D}_k + \tilde{V}_k^{d-} \tilde{L}_k)$
G_k	$I_1 + B_k R_k^{-1} B_k^T S_{k+1} - \left[\sum_{i=1}^{N_a} \sum_{d=1}^{2^p} \varepsilon_k^i B_{a_k}^i U_{a0} (\tilde{V}_k^d \tilde{D}_k + \tilde{V}_k^{d-} \tilde{L}_k)\right]$
Θ_k	$G_k + G_k^T$
Ξ_k	$\tilde{x}_k^T A_k^T (\eta_k^{-1})^T \Gamma_k \eta_k^{-1} A_k \tilde{x}_k$
Γ_k	$2S_{k+1} + S_{k+1} B_k R_k^{-1} B_k^T S_{k+1} - \gamma^2 \left[\sum_{d=1}^{2^p} U_{a0} \ell_k^d (\tilde{V}_k^d \tilde{D}_k + \tilde{V}_k^{d-} \tilde{L}_k)\right]^T \tilde{R}_k \left[\sum_{d=1}^{2^p} U_{a0} \ell_k^d (\tilde{V}_k^d \times \tilde{D}_k + \tilde{V}_k^{d-} \tilde{L}_k)\right]$
$\digamma_k$	$\gamma^{-2} (2S_{k+1} + S_{k+1} B_k R_k^{-1} B_k^T S_{k+1}) (\tilde{I}^T \tilde{R}_k \tilde{I})^{-1}$
ℓ_k^d	The real parameters, and satisfy $\ell_k^d \geq 0, \sum_{d=1}^{2^p} \ell_k^d = 1$
$\tilde{V}_k^d$	The $m \times m$ diagonal matrix with elements either 1 or 0 at the k moment
$\tilde{V}_k^{d-}$	$I_2 - \tilde{V}_k^d$
$\tilde{D}_k$	$\frac{1}{U_{a0}} \sum_{i=1}^{N_a} \varepsilon_k^i \gamma^{-2} \tilde{R}_k^{-1} B_{a_k}^{i\,T} S_{k+1}$
$\tilde{L}_k$	The given matrix and satisfies $\Vert\tilde{L}_k \tilde{x}_{k+1}\Vert_\infty \leq 1$
I_1	The identity matrix with dimension $n \times n$
I_2	The identity matrix with dimension $m \times m$
$\tilde{I}$	The unit vector in the $\sum_{i=1}^{N_a} \varepsilon_k^i \tilde{R}_k^{-1} B_{a_k}^{i\,T} S_{k+1} \eta_k^{-1} A_k \tilde{x}_k$ direction

$$u_{a_k}^* = U_{a_0} Sat\left[\frac{1}{U_{a_0}} \sum_{i=1}^{N_a} \varepsilon_k^i \gamma^{-2} \tilde{R}_k^{-1} B_{a_k}^{i^T} S_{k+1} \eta_k^{-1} A_k \tilde{x}_k\right], \tag{8.11}$$

and the optimal switching sequence is

$$i^* = \begin{cases} \arg\max\limits_{i \in \{1,2,\ldots,N_a\}} \Xi_k, & U_{a_0}^2 \leq F_k, \\ \arg\min\limits_{i \in \{1,2,\ldots,N_a\}} \Xi_k, & U_{a_0}^2 > F_k, \end{cases} \tag{8.12}$$

where S_{k+1} satisfies the following equation

$$S_k = \tilde{Q}_{k+1} + A_k^T S_{k+1} \eta_k^{-1} A_k, \tag{8.13}$$

with $S_{N+1} = 0$.

Remark 8.6 In Theorem 8.1, the condition $\Theta_k \succ 0$ needs to be satisfied. Since Θ_k contains the attack matrix $D_{a_k}^i$ which the attacker can adjust, in practical applications, the attacker only needs to adjust $D_{a_k}^i$ to make the condition satisfied.

Proof The Hamiltonian function is defined as

$$\begin{aligned} H_1(\tilde{x}_k, u_k, u_{a_k}, \lambda_{k+1}, \varepsilon_k^i) =& \frac{1}{2}(\tilde{x}_k^T \tilde{Q}_k \tilde{x}_k + u_k^T R_k u_k - \gamma^2 u_{a_k}^T \tilde{R}_k u_{a_k}) + \lambda_{k+1}^T \\ & \times \left(A_k \tilde{x}_k + B_k u_k + \sum_{i=1}^{N_a} \varepsilon_k^i B_{a_k}^i u_{a_k}\right), \end{aligned} \tag{8.14}$$

with the constraint $u_{a_k}^T u_{a_k} \leq U_{a_0}^2$.

Since the Hamiltonian function H_1 is constrained only with respect to u_{a_k}, according to [15], the following steps are taken to obtain the optimal attack $u_{a_k}^*$ under this constraint.

Step 1. When H_1 is derivable. The maximum principle condition is applied to the Hamiltonian function H_1 as

$$\frac{\partial H_1}{\partial u_{a_k}} = -\gamma^2 \tilde{R}_k u_{a_k} + B_{a_k}^T \lambda_{k+1} = 0.$$

Thus, one has

$$u_{a_k}^* = \sum_{i=1}^{N_a} \gamma^{-2} \varepsilon_k^i \tilde{R}_k^{-1} B_{a_k}^{i^T} \lambda_{k+1}.$$

Step 2. When H_1 is underivable. Pontryagin's maximum principle is applied to obtain a solution by directly maximizing H_1.

$$\begin{aligned} \max_{u_{a_k}} \quad & H_1(\tilde{x}_k^*, u_k^*, u_{a_k}, \lambda_{k+1}^*, \varepsilon_k^{i^*}) \\ & u_{a_k}^T u_{a_k} \le U_{a_0}^2. \end{aligned} \tag{8.15}$$

It follows from (8.15) that

$$H_1(\tilde{x}_k^*, u_k^*, u_{a_k}, \lambda_{k+1}^*, \varepsilon_k^{i^*}) \le H_1(\tilde{x}_k^*, u_k^*, u_{a_k}^*, \lambda_{k+1}^*, \varepsilon_k^{i^*}). \tag{8.16}$$

Substituting (8.14) into (8.16), one has

$$\begin{aligned} &\frac{1}{2}(\tilde{x}_k^{*^T} \tilde{Q}_k \tilde{x}_k^* + u_k^{*^T} R_k u_k^* - \gamma^2 u_{a_k}^T \tilde{R}_k u_{a_k}) + \lambda_{k+1}^{*^T} \left(A_k \tilde{x}_k^* + B_k u_k^* + \sum_{i=1}^{N_a} \varepsilon_k^{i^*} B_{a_k}^{i^*} u_{a_k} \right) \\ \le &\frac{1}{2}(\tilde{x}_k^{*^T} \tilde{Q}_k \tilde{x}_k^* + u_k^{*^T} R_k u_k^* - \gamma^2 u_{a_k}^{*^T} \tilde{R}_k u_{a_k}^*) + \lambda_{k+1}^{*^T} \left(A_k \tilde{x}_k^* + B_k u_k^* + \sum_{i=1}^{N_a} \varepsilon_k^{i^*} B_{a_k}^{i^*} u_{a_k}^* \right), \end{aligned}$$

which yields

$$u_{a_k}^{*^T} \left(\frac{1}{2}\gamma^2 \tilde{R}_k u_{a_k}^* - \sum_{i=1}^{N_a} \varepsilon_k^{i^*} B_{a_k}^{i^{*^T}} \lambda_{k+1}^* \right) \le u_{a_k}^T \left(\frac{1}{2}\gamma^2 \tilde{R}_k u_{a_k} - \sum_{i=1}^{N_a} \varepsilon_k^{i^*} B_{a_k}^{i^{*^T}} \lambda_{k+1}^* \right).$$

Therefore, the optimal attack $u_{a_k}^*$ needs to minimize the scalar product

$$\min_{u_{a_k}} \left\{ \langle u_{a_k}, \frac{1}{2}\gamma^2 \tilde{R}_k u_{a_k} - \sum_{i=1}^{N_a} \varepsilon_k^{i^*} B_{a_k}^{i^{*^T}} \lambda_{k+1}^* \rangle \right\}. \tag{8.17}$$

Through the knowledge of mathematical vectors, the angle between the vector u_{a_k} and the vector $(\frac{1}{2}\gamma^2 \tilde{R}_k u_{a_k} - \sum_{i=1}^{N_a} \varepsilon_k^{i^*} B_{a_k}^{i^{*^T}} \lambda_{k+1}^*)$ is 180°, i.e., they should be parallel and in the opposite direction. The corresponding geometric explanation is shown as Fig. 8.2.

Noted that the optimal attack $u_{a_k}^*$ satisfies (8.8). The optimal attack $u_{a_k}^*$ is

$$u_{a_k}^* = U_{a_0} \frac{\sum_{i=1}^{N_a} \varepsilon_k^{i^*} B_{a_k}^{i^{*^T}} \lambda_{k+1}^*}{\| \sum_{i=1}^{N_a} \varepsilon_k^{i^*} B_{a_k}^{i^{*^T}} \lambda_{k+1}^* \|} = U_{a_0} \frac{\sum_{i=1}^{N_a} \gamma^{-2} \varepsilon_k^{i^*} \tilde{R}_k^{-1} B_{a_k}^{i^{*^T}} \lambda_{k+1}^*}{\| \sum_{i=1}^{N_a} \gamma^{-2} \varepsilon_k^{i^*} \tilde{R}_k^{-1} B_{a_k}^{i^{*^T}} \lambda_{k+1}^* \|},$$

where $\frac{\sum_{i=1}^{N_a} \gamma^{-2} \varepsilon_k^{i^*} \tilde{R}_k^{-1} B_{a_k}^{i^{*^T}} \lambda_{k+1}^*}{\| \sum_{i=1}^{N_a} \gamma^{-2} \varepsilon_k^{i^*} \tilde{R}_k^{-1} B_{a_k}^{i^{*^T}} \lambda_{k+1}^* \|}$ represents the unit vector in the $\sum_{i=1}^{N_a} \gamma^{-2} \varepsilon_k^{i^*} \tilde{R}_k^{-1} B_{a_k}^{i^{*^T}}$ λ_{k+1}^* direction.

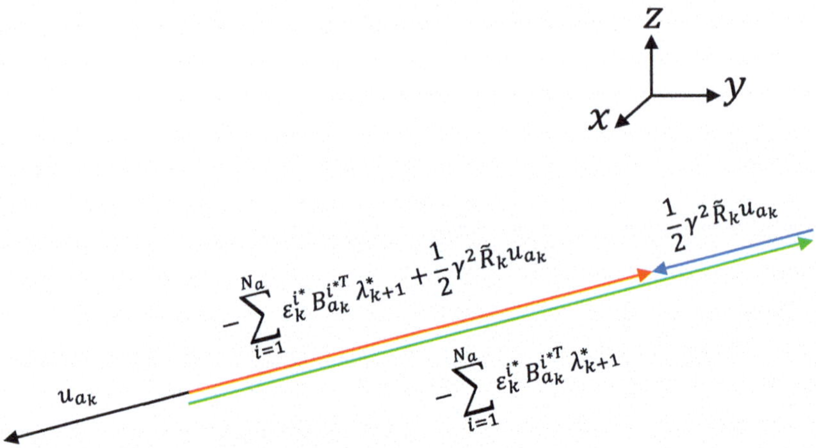

Fig. 8.2 The geometric interpretation of (8.17)

Combined with **Step 1** and **Step 2**, the optimal attack $u_{a_k}^*$ is expressed as

$$u_{a_k}^* = U_{a_0} Sat\left[\frac{1}{U_{a_0}}\sum_{i=1}^{N_a}\varepsilon_k^i\gamma^{-2}\tilde{R}_k^{-1}B_{a_k}^{i^T}\lambda_{k+1}\right]. \tag{8.18}$$

From the fact that Hamiltonian function H_1 is unconstrained with respect to $\tilde{x}_k$, u_k, λ_k, and ε_k^i, through the control theory, one has

$$\frac{\partial H_1}{\partial u_k} = R_k u_k + B_k^T\lambda_{k+1} = 0.$$

Thus, the optimal control input u_k^* is described as

$$u_k^* = -R_k^{-1}B_k^T\lambda_{k+1}. \tag{8.19}$$

By the co-state equation, one can obtain

$$\lambda_k = \frac{\partial H_1}{\partial \tilde{x}_k} = \tilde{Q}_k\tilde{x}_k + A_k^T\lambda_{k+1}. \tag{8.20}$$

And the transversality condition is given as

$$\lambda_{N+1} = 0. \tag{8.21}$$

Through the optimal theory [6–8], one can assume that

$$\lambda_k = S_k\tilde{x}_k, \tag{8.22}$$

where S_k is a positive definite matrix.

Therefore, the followings can be obtained.

$$\lambda_{k+1} = S_{k+1}\tilde{x}_{k+1},$$

$$u_k^* = -R_k^{-1}B_k^T S_{k+1}\tilde{x}_{k+1},$$

$$u_{a_k}^* = U_{a_0} Sat\left[\frac{1}{U_{a_0}}\sum_{i=1}^{N_a}\varepsilon_k^i \gamma^{-2}\tilde{R}_k^{-1}B_{a_k}^{i^T}S_{k+1}\tilde{x}_{k+1}\right], \tag{8.23}$$

$$\begin{aligned}\tilde{x}_{k+1} =& A_k\tilde{x}_k - B_k R_k^{-1}B_k^T S_{k+1}\tilde{x}_{k+1} + \sum_{i=1}^{N_a}\varepsilon_k^i B_{a_k}^i U_{a_0} Sat\Bigg[\frac{1}{U_{a_0}}\sum_{i=1}^{N_a}\varepsilon_k^i\gamma^{-2}\tilde{R}_k^{-1} \\ &\times B_{a_k}^{i^T}S_{k+1}\tilde{x}_{k+1}\Bigg].\end{aligned} \tag{8.24}$$

From (8.23), (8.24) and Lemma 8.4, one has

$$u_{a_k}^* = \sum_{d=1}^{2^p} U_{a_0}\ell_k^d(\tilde{V}_k^d\tilde{D}_k + \tilde{V}_k^{d^-}\tilde{L}_k)\tilde{x}_{k+1},$$

thus,

$$\begin{aligned}\tilde{x}_{k+1} =& A_k\tilde{x}_k - B_k R_k^{-1}B_k^T S_{k+1}\tilde{x}_{k+1} + \sum_{i=1}^{N_a}\sum_{d=1}^{2^p}\varepsilon_k^i B_{a_k}^i U_{a_0}\ell_k^d(\tilde{V}_k^d\tilde{D}_k \\ &+ \tilde{V}_k^{d^-}\tilde{L}_k)\tilde{x}_{k+1},\end{aligned} \tag{8.25}$$

where ℓ_k^d, $\tilde{V}_k^d$, $\tilde{V}_k^{d^-}$, $\tilde{D}_k$ and $\tilde{L}_k$ are described in Table 8.1.

Through (8.25), $\tilde{x}_{k+1}$ can be rewritten as

$$\tilde{x}_{k+1} = \eta_k^{-1}A_k\tilde{x}_k, \tag{8.26}$$

where η_k is described in Table 8.1.

Then λ_{k+1} can be expressed as

$$\lambda_{k+1} = S_{k+1}\eta_k^{-1}A_k\tilde{x}_k. \tag{8.27}$$

Substituting (8.27) into (8.19) and (8.18), one can obtain (8.10) and (8.11). Then substituting (8.22) and (8.27) into (8.20), one can get (8.13), and the final value condition is obtained through joints (8.21) and (8.22).

It can be known from the Lemma 8.4 that ℓ_k^d satisfies

$$0 \leq \ell_k^d \leq 1, \sum_{d=1}^{2^p} \ell_k^d = 1.$$

Thus,

$$\begin{aligned} \eta_k &= I_1 + B_k R_k^{-1} B_k^T S_{k+1} - \sum_{i=1}^{N_a} \sum_{d=1}^{2^p} \varepsilon_k^i B_{a_k}^i U_{a_0} \ell_k^d (\tilde{V}_k^d \tilde{D}_k + \tilde{V}_k^{d^-} \tilde{L}_k) \\ &\succ I_1 + B_k R_k^{-1} B_k^T S_{k+1} - \sum_{i=1}^{N_a} \sum_{d=1}^{2^p} \varepsilon_k^i B_{a_k}^i U_{a_0} (\tilde{V}_k^d \tilde{D}_k + \tilde{V}_k^{d^-} \tilde{L}_k). \end{aligned} \tag{8.28}$$

By using Lemma 8.2, the optimal solution for the Problem 1 is unique if and only if

$$I_1 + B_k R_k^{-1} B_k^T S_{k+1} - \left[\sum_{i=1}^{N_a} \sum_{d=1}^{2^p} \varepsilon_k^i B_{a_k}^i U_{a_0} (\tilde{V}_k^d \tilde{D}_k + \tilde{V}_k^{d^-} \tilde{L}_k) \right] \succ 0. \tag{8.29}$$

Since the positive definiteness of matrices is mainly for symmetric matrices, combined with Lemma 8.5, (8.29) is equivalent to

$$\begin{aligned} &\left\{ I_1 + B_k R_k^{-1} B_k^T S_{k+1} - \left[\sum_{i=1}^{N_a} \sum_{d=1}^{2^p} \varepsilon_k^i B_{a_k}^i U_{a_0} (\tilde{V}_k^d \tilde{D}_k + \tilde{V}_k^{d^-} \tilde{L}_k) \right] \right\} + \left\{ I + B_k R_k^{-1} B_k^T \right. \\ &\left. \times S_{k+1} - \left[\sum_{i=1}^{N_a} \sum_{d=1}^{2^p} \varepsilon_k^i B_{a_k}^i U_{a_0} (\tilde{V}_k^d \tilde{D}_k + \tilde{V}_k^{d^-} \tilde{L}_k) \right] \right\}^T \succ 0, \end{aligned}$$

that is, the condition $\Theta_k \succ 0$ needs to be satisfied, which means that G_k is a positive definite matrix, combined with (8.28), and so is η_k. Therefore, it satisfies the condition for the existence of a unique solution mentioned in Lemma 8.2.

The following proof process is to obtain the optimal switching conditions. Substituting (8.18) and (8.19) into (8.14), the Hamiltonian function is rewritten as

$$H_1 = \frac{1}{2}\Bigg\{\tilde{x}_k^T \tilde{Q}_k \tilde{x}_k + 2\lambda_{k+1}^T A_k \tilde{x}_k - \lambda_{k+1}^T B_k R_k^{-1} B_k^T \lambda_{k+1} + 2\sum_{i=1}^{N_a}\sum_{d=1}^{2^p} \varepsilon_k^i B_{a_k}^i U_{a_0} \ell_k^d (\tilde{V}_k^d \tilde{D}_k + \tilde{V}_k^{d^-} \tilde{L}_k)\eta_k^{-1} A_k \tilde{x}_k - \gamma^{-2}\Bigg[\sum_{d=1}^{2^p} U_{a_0} \ell_k^d (\tilde{V}_k^d \tilde{D}_k + \tilde{V}_k^{d^-} \tilde{L}_k)\eta_k^{-1} A_k \tilde{x}_k\Bigg]^T \tilde{R}_k \sum_{d=1}^{2^p} U_{a_0} \times \ell_k^d (\tilde{V}_k^d \tilde{D}_k + \tilde{V}_k^{d^-} \tilde{L}_k)\eta_k^{-1} A_k \tilde{x}_k\Bigg\}.$$

There are terms related to $\varepsilon_k = [\varepsilon_k^1, \varepsilon_k^2, \ldots, \varepsilon_k^{N_a}]^T$ in H_1 include ε_k^i and λ_{k+1}, maximizing H_1 by ε_k can be simplified to maximize

$$\begin{aligned}\Xi_k =& 2\lambda_{k+1}^T A_k \tilde{x}_k - \lambda_{k+1}^T B_k R_k^{-1} B_k^T \lambda_{k+1} + 2\sum_{i=1}^{N_a}\sum_{d=1}^{2^p} \varepsilon_k^i B_{a_k}^i U_{a_0} \ell_k^d (\tilde{V}_k^d \tilde{D}_k + \tilde{V}_k^{d^-} \tilde{L}_k) \\ &\times \eta_k^{-1} A_k \tilde{x}_k - \gamma^2\Bigg[\sum_{d=1}^{2^p} U_{a_0} \ell_k^d (\tilde{V}_k^d \tilde{D}_k + \tilde{V}_k^{d^-} \tilde{L}_k)\eta_k^{-1} A_k \tilde{x}_k\Bigg]^T \tilde{R}_k \sum_{d=1}^{2^p} U_{a_0} \ell_k^d \\ &\times (\tilde{V}_k^d \tilde{D}_k + \tilde{V}_k^{d^-} \tilde{L}_k)\eta_k^{-1} A_k \tilde{x}_k.\end{aligned} \tag{8.30}$$

Substituting (8.27) into (8.30), Ξ_k can be simplified as

$$\Xi_k = \tilde{x}_k^T A_k^T (\eta_k^{-1})^T \Gamma_k \eta_k^{-1} A_k \tilde{x}_k, \tag{8.31}$$

where Γ_k is expressed as

$$\begin{aligned}\Gamma_k =& 2S_{k+1} + S_{k+1} B_k R_k^{-1} B_k^T S_{k+1} - \gamma^2\Bigg[\sum_{d=1}^{2^p} U_{a_0} \ell_k^d (\tilde{V}_k^d \tilde{D}_k + \tilde{V}_k^{d^-} \tilde{L}_k)\Bigg]^T \\ &\times \tilde{R}_k\Bigg[\sum_{d=1}^{2^p} U_{a_0} \ell_k^d (\tilde{V}_k^d \tilde{D}_k + \tilde{V}_k^{d^-} \tilde{L}_k)\Bigg].\end{aligned} \tag{8.32}$$

Through (8.29) and the properties of the positive definite matrix, the inverse of a positive definite matrix is also a positive definite matrix. Thus,

$$R_k^{-1} \succ 0.$$

From the definition of the positive definite matrix, one can get

$$B_k R_k^{-1} B_k^T \succeq 0, \quad \forall B_k.$$

Therefore

$$S_{k+1} B_k R_k^{-1} B_k^T S_{k+1} \succeq 0.$$

Then, from the expression of the saturation function, one has

$$\sum_{d=1}^{2^p} U_{a_0}\ell_k^d(\tilde{V}_k^d \tilde{D}_k + \tilde{V}_k^{d^-}\tilde{L}_k) \leq U_{a_0}\tilde{I},$$

where $\tilde{I}$ is described as

$$\tilde{I} = \frac{\sum_{i=1}^{N_a} \varepsilon_k^i \tilde{R}_k^{-1} B_{a_k}^{i^T} S_{k+1}\eta_k^{-1} A_k \tilde{x}_k}{\|\sum_{i=1}^{N_a} \varepsilon_k^i \tilde{R}_k^{-1} B_{a_k}^{i^T} S_{k+1}\eta_k^{-1} A_k \tilde{x}_k\|},$$

and it represents the unit vector in the direction of $\sum_{i=1}^{N_a} \varepsilon_k^i \tilde{R}_k^{-1} B_{a_k}^{i^T} S_{k+1}\eta_k^{-1} A_k \tilde{x}_k$.

Then, (8.32) satisfies the following

$$\Gamma_k \geq 2S_{k+1} + S_{k+1}B_k R_k^{-1} B_k^T S_{k+1} - \gamma^2 U_{a_0}^2 \tilde{I}^T \tilde{R}_k \tilde{I}.$$

Thus,

$$\begin{cases} \Xi_k \geq 0, & U_{a_0}^2 \leq F_k, \\ \Xi_k < 0, & U_{a_0}^2 > F_k, \end{cases}$$

where F_k is described as Table 8.1.

Since ε_k maximizing Ξ_k is equivalent to maximize Ξ_k through the determination of i, maximizing H_3 is equivalent to

$$\max\ H_1 = \begin{cases} \max\limits_{i\in\{1,2,\ldots,N_a\}} \Xi_k, & U_{a_0}^2 \leq F_k, \\ \min\limits_{i\in\{1,2,\ldots,N_a\}} \Xi_k, & U_{a_0}^2 > F_k. \end{cases}$$

Therefore the optimal switching can be obtained as (8.12).

The algorithm of the switching data injection attack via zero-sum game is given in Algorithm 9.

Remark 8.7 In Theorem 8.1 and Algorithm 9, the condition $\Theta_k \succ 0$ can be judged by the eigenvalue of Θ_k, $D_{a_k}^i(\alpha_k^i)$ and $\Theta_k(\alpha_k^i)$ to denote the matrices containing the variable α_k^i. The optimal switching attack input can be implemented by online real-time calculation of the switching sequence i^*, which increases the flexibility and concealment of attacks.

8.3.2 Game-Based FDI Attack Strategy

The following presents the optimal FDI attack strategy and algorithm under the non-switching form based on zero-sum games. It is worth noting that the FDI attack under the non-switching form here is a special case of the switching form, and resources are unrestricted.

Theorem 8.2 *Under Assumptions 8.1–8.2, if* $\Psi_k = \gamma^2 \tilde{R}_k - B_{a_k}^T M_{k+1} B_{a_k} \succ 0$ *holds, the Problem 8.2 can be solved with the optimal control input* u_k^* *and the optimal attack input* $u_{a_k}^*$ *as follows:*

$$u_k^* = -R_k^{-1} B_k^T M_{k+1} \Lambda_k^{-1} A_k \tilde{x}_k \tag{8.33}$$

Algorithm 9: Game-Based Switching FDI Attack Strategy

Initialize: Set suitable parameters and matrices, γ, N, N_a, $\tilde{Q}_k$, R_k, $\tilde{R}_k$, $D_{a_k}^i$, $\tilde{V}_k^d$, $\tilde{L}_k$, U_{a0}, $x(0)$, preset eigenvalue vector β_k^i and $\Xi_k = S_{N+1} \leftarrow \mathbf{0}^{n \times n}$, $\varepsilon_k^i \leftarrow 0$, $k = 1, 2, \cdots, N$, $i = 1, 2, \cdots, N_a$.
Compute the matrix $\tilde{V}_k^{d-}$: $\tilde{V}_k^{d-} \leftarrow I - \tilde{V}_k^d$.
for $k = N$ *to* 1 **do**
  for $i = 1$ *to* N_a **do**
    Step 1: Compute the matrices η_k, S_k, F_k, Γ_k, $\tilde{D}_k$ and Θ_k, respectively.
    if $\text{eig}(\Theta_k) > 0$ **then**
      Continue;
    else
      Step 2: $D_{a_k}^i = D_{a_k}^i(\alpha_k^i)$, and compute $\Theta_k(\alpha_k^i)$.
      Step 3: Adopt the newton-raphson method [16] to solve
      $\mathcal{F}(\alpha_k^i) = \text{eig}(\Theta_k(\alpha_k^i)) - \beta_k^i = 0$, then the solution α_k^{i*} can be obtained.
      Step 4: $D_{a_k}^i = D_{a_k}^i(\alpha_k^{i*})$, and repeat **Step 1**.
while $k \le N$ **do**
  for $i = 1$ *to* N_a **do**
    Compute Ξ_k using (8.31)
    if $F^i \ge U_{a0}^2$ **then**
      $i^{optimal} = \arg\max \Xi_k$;
    else
      $i^{optimal} = \arg\min \Xi_k$;
  $\varepsilon_k^{i^{optimal}} \leftarrow 1$, and compute u_k and u_{a_k} using (8.10) and (8.11), respectively.
  $\tilde{u}_k = u_k + \sum_{i=1}^{N_a} \varepsilon_k^i D_{a_k}^i u_{a_k}$.
  $\tilde{x}_{k+1} = A_k \tilde{x}_k + B_k u_k + \sum_{i=1}^{N_a} \varepsilon_k^i B_{a_k}^i u_{a_k}$.

$$u^*_{a_k} = \gamma^{-2}\tilde{R}_k^{-1}B_{a_k}^T M_{k+1}\Lambda_k^{-1}A_k\tilde{x}_k \tag{8.34}$$

where

$$\Lambda_k = I + B_k R_k^{-1} B_k^T M_{k+1} - \gamma^{-2} B_{a_k}\tilde{R}_k^{-1} B_{a_k}^T M_{k+1} \tag{8.35}$$

and M_{k+1} *satisfies the following equation*

$$M_k = \tilde{Q}_{k+1} + A_k^T M_{k+1}\Lambda_k^{-1} A_k \tag{8.36}$$

and $M_{N+1} = 0$.

The algorithm of the false data injection attack via zero-sum game is given in Algorithm 10.

Algorithm 10: Game-Based Non-Switching FDI Attack Strategy

Initialize: $\tilde{Q}_k$, R_k, $\tilde{R}_k$, D_{a_k}, γ, U_{a0}, $x(0)$, preset eigenvalue vector β_k and $M_{N+1} \longleftarrow \mathbf{0}^{n\times n}$, $k = 1, 2, \cdots, N$.
for $k = N$ *to* 1 **do**
  Step 1: Compute the matrices Ψ_k, Λ_k, M_k, respectively.
  if $\mathrm{eig}(\Psi_k) > 0$ **then**
    Continue;
  else
    Step 2: $D_{a_k} = D_{a_k}(\alpha_k)$, and compute $\Psi_k(\alpha_k)$.
    Step 3: Adopt the newton-raphson method [16] to solve $\mathcal{F}(\alpha_k) = \mathrm{eig}(\Psi_k(\alpha_k)) - \beta_k = 0$, then the solution α_k^* can be obtained.
    Step 4: $D_{a_k} = D_{a_k}(\alpha_k^*)$, and repeat **Step 1**.
while $k \le N$ **do**
  Step 5: Compute u_k and u_{a_k} using (8.33) and (8.34), respectively.
  Step 6: $\tilde{u}_k = u_k + D_{a_k}u_{a_k}$.
  Step 7: $\tilde{x}_{k+1} = A_k\tilde{x}_k + B_k u_k + B_{a_k}u_{a_k}$.

Remark 8.8 In Theorem 8.2 and Algorithm 10, the constraint $\Psi_k \succ 0$ can be verified using the eigenvalues of Ψ_k, similar to Theorem 8.1. This constraint is satisfied by adjusting the attack matrix D_{a_k}, with the specific adjustment process detailed in Algorithm 10. Furthermore, in this subsection, the attacker's energy consumption is unconstrained. Therefore, the attack-defense game strategy proposed in this chapter aligns with the zero-sum game strategy in game theory. This further demonstrates that the optimal switching injection attack strategy based on the zero-sum game under resource-constrained conditions, as presented in this chapter, constitutes a general extension of the previous game-based attack-defense strategies.

8.4 Simulation and Experiment

In this section, the numerical example and the networked Qbot-2e, a two-wheeled mobile robot produced by Quanser, are given to validate the proposed strategies.

8.4.1 *Numerical Simulation*

In this subsection, a discrete time-varying kinematic model of vehicle trajectory tracking error is considered, and the detailed derivation of the vehicle tracking error system is shown as follows. The kinematic equation of the vehicle is

$$\begin{cases} \dot{x}(t) = v(t)\cos(\varphi(t)), \\ \dot{y}(t) = v(t)\sin(\varphi(t)), \\ \dot{\varphi}(t) = \omega(t), \end{cases} \tag{8.37}$$

where $(x(t), y(t))$, $\varphi(t)$, $v(t)$, and $\omega(t)$ represent the actual lateral and longitudinal positions, yaw angle, linear velocity, and angular velocity of the vehicle in the continuous-time inertial coordinate system, respectively. Let $\tilde{\xi}(k) = \begin{bmatrix} x(t) \\ y(t) \\ \varphi(t) \end{bmatrix}$, $\tilde{U}(t) = \begin{bmatrix} v(t) \\ \omega(t) \end{bmatrix}$, then (8.37) can be expressed as

$$\dot{\tilde{\xi}}(t) = f(\tilde{\xi}(t), \tilde{U}(t)) = \begin{bmatrix} v(t)\cos(\varphi(t)) \\ v(t)\sin(\varphi(t)) \\ \omega(t) \end{bmatrix}. \tag{8.38}$$

The desired trajectory equation is

$$\dot{\tilde{\xi}}_r(t) = f(\tilde{\xi}_r(t), \tilde{U}_r(t)), \tag{8.39}$$

where $x_r(t)$ and $y_r(t)$ are the desired lateral and longitudinal positions in the continuous-time inertial coordinate system, respectively. $\varphi_r(t)$ denotes the desired yaw angle. $v_r(t)$ and $\omega_r(t)$ stand for the desired linear velocity, and angular velocity of the vehicle in the continuous-time inertial coordinate system, respectively. $\tilde{\xi}_r(k) = \begin{bmatrix} x_r(t) \\ y_r(t) \\ \varphi_r(t) \end{bmatrix}$ and $\tilde{U}_r(t) = \begin{bmatrix} v_r(t) \\ \omega_r(t) \end{bmatrix}$ are the desired state and control input.

The linearization of (8.38) near $(\tilde{\xi}_r(t), \tilde{U}_r(t))$ can be obtained

$$\dot{\tilde{\xi}}(t) = f(\tilde{\xi}_r(t), \tilde{U}_r(t)) + f_{\tilde{\xi}(t)} \cdot (\tilde{\xi}(t) - \tilde{\xi}_r(t)) + f_{\tilde{U}(t)} \cdot (\tilde{U}(t) - \tilde{U}_r(t)), \tag{8.40}$$

where $f_{\tilde{\xi}(t)} = \frac{\partial f(\tilde{\xi}(t),\tilde{U}(t))}{\partial \tilde{\xi}(t)}$ and $f_{\tilde{U}(t)} = \frac{\partial f(\tilde{\xi}(t),\tilde{U}(t))}{\partial \tilde{U}(t)}$.

Combining (8.39) and (8.40), we can obtain

$$\dot{\tilde{\xi}}(t) - \dot{\tilde{\xi}}_r(t) = \begin{bmatrix} 0 & 0 & -v(t)\sin(\varphi(t)) \\ 0 & 0 & v(t)\cos(\varphi(t)) \\ 0 & 0 & 0 \end{bmatrix} (\tilde{\xi}(t) - \tilde{\xi}_r(t)) + \begin{bmatrix} \cos(\varphi(t)) & 0 \\ \sin(\varphi(t)) & 0 \\ 0 & 1 \end{bmatrix} (\tilde{U}(t) - \tilde{U}_r(t)).$$

Then, define the state and input of the error system as $\xi(t) = \begin{bmatrix} x(t) - x_r(t) \\ y(t) - y_r(t) \\ \varphi(t) - \varphi_r(t) \end{bmatrix}$ and $\mathbf{U(t)} = \begin{bmatrix} v(t) - v_r(t) \\ \omega(t) - \omega_r(t) \end{bmatrix}$, we can get

$$\dot{\xi}(t) = \begin{bmatrix} 0 & 0 & -v(t)\sin(\varphi(t)) \\ 0 & 0 & v(t)\cos(\varphi(t)) \\ 0 & 0 & 0 \end{bmatrix} \xi(t) + \begin{bmatrix} \cos(\varphi(t)) & 0 \\ \sin(\varphi(t)) & 0 \\ 0 & 1 \end{bmatrix} U(t). \tag{8.41}$$

The discrete time-varying kinematic model of vehicle trajectory tracking error can be obtained through the discretization of (8.41), that is

$$\begin{aligned} \xi(k+1) &= (I + f_{\tilde{\xi}(k)}\tau)\xi(k) + f_{\tilde{U}(k)}\tau U(k) \\ &= A_k\xi(k) + B_k U(k), \end{aligned} \tag{8.42}$$

where $\mathbf{A}_k = \begin{bmatrix} 1 & 0 & -\tau v(k)\sin(\varphi(k)) \\ 0 & 1 & \tau v(k)\cos(\varphi(k)) \\ 0 & 0 & 1 \end{bmatrix}$, $\mathbf{B}_k = \begin{bmatrix} \tau\cos(\varphi(k)) & 0 \\ \tau\sin(\varphi(k)) & 0 \\ 0 & \tau \end{bmatrix}$, $\tau = 0.01$ is the sampling time, $(x(k), y(k))$, $\varphi(k)$, $v(k)$, and $\omega(k)$ represent the actual lateral and longitudinal positions, yaw angle, linear velocity, and angular velocity of the vehicle in the discrete-time inertial coordinate system, respectively. $(x_r(k), y_r(k))$, $\varphi_r(k)$, $v_r(k)$, and $\omega_r(k)$ represent the desired lateral and longitudinal positions, yaw angle, linear velocity, and angular velocity of the vehicle in the discrete-time inertial coordinate system, respectively. $\tilde{\xi}(k)$ and $\tilde{U}(k)$ are the state and control input of the discrete-time vehicle trajectory tracking error system.

The weighting matrices are defined as $\tilde{Q} = \text{diag}(200, 200, 200)$, $R = \text{diag}(2, 2)$ and $\tilde{R} = \text{diag}(5, 5)$. And the signal interference attenuation factor $\gamma = 0.95$. The constraint of attack input is $U_{a_0} = 0.8$. For the desired lateral and longitudinal positions and yaw angle, they are defined as $x_r(k) = 6 * (k-1) * \tau$, $y_r(k) = 6 * (k-1) * \tau$, $\varphi_r(k) = \pi/4$. The desired speed and angular velocity of the vehicle are defined as $v_r(k) = 0.6\,m/s$, $\omega_r(k) = 0rad/s$.

The switching attack direction matrices are $D_{a_k}^1 = [1 * \sin(0.05 * \pi * k), 0; 0, 0]$ and $D_{a_k}^2 = [0, 0; 0, 9 * \sin(0.005 * \pi * k)]$, and the non-switching attack direction matrix is $D_{a_k} = [1 * \sin(0.05 * \pi * k), 0; 0, 9 * \sin(0.005 * \pi * k)]$. Then the designed defense mechanism and attack policy are employed for 1000 times of sim-

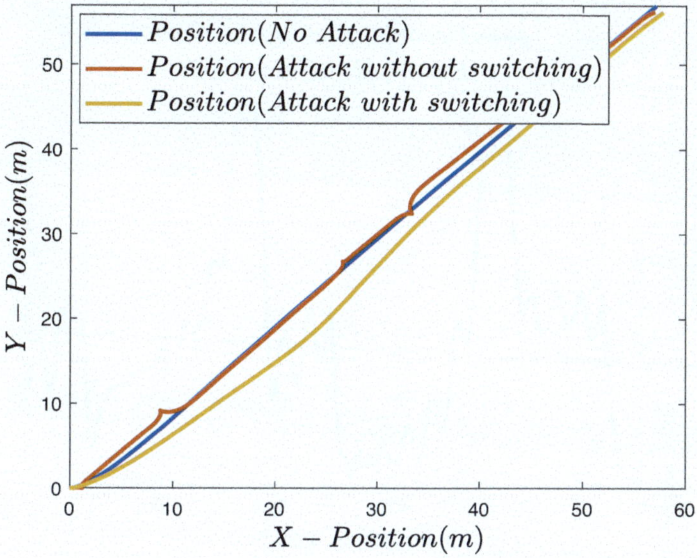

Fig. 8.3 The trajectory of vehicle without attack, with non-switching attack and with switching attack

ulation, and the simulation results are shown in Figs. 8.3, 8.4, 8.5, 8.6, 8.7, 8.8, 8.9, 8.10, 8.11, and 8.12.

It can be seen from Fig. 8.3 that the designed optimal control strategy can make the system perform well without divergence. Meanwhile, it also shows that the designed optimal switching attack strategy can make the system state have the largest error, and the performance suffers the greatest damage under the objective function. The optimal control strategy and the constrained switching attack policy designed are shown in Figs. 8.5 and 8.6, the optimal control strategy and the non-switching attack policy designed are shown in Figs. 8.7 and 8.8. It is clearly to see from Figs. 8.10 and 8.11 that the energy consumption of the attacker who uses the switching attack satisfies the condition that the maximum energy consumption is less than $U_{a_0}^2$ at each time. Combined with the position error effect in Fig. 8.4 and compared with the energy consumption of non-switching attack in Figs. 8.9, it can be seen that the constrained switching attack can have greater damage effect, and the energy consumption is far less than that of non-switching attack. It is worth noting that the energy consumption of non-switching attack is very high in a few times, which makes its position error better than that of switching attack, but the overall effect of switching attack is better than that of non-switching attack.

Furthermore, the optimal switching condition $i(k)$ with 1000 times of simulation can be seen in Fig. 8.12. During each switching attack, the dwell time of the corresponding controller actuator channel is the sampling time, and the next switching occurs at the end of the sampling time.

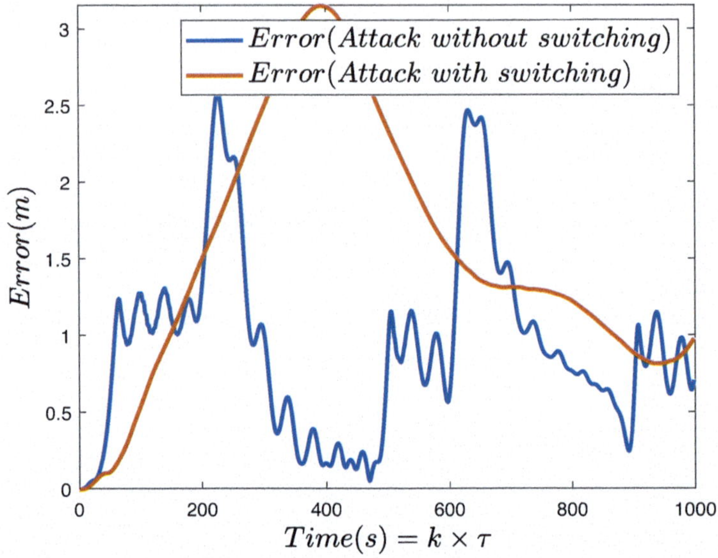

Fig. 8.4 The position error of vehicle under non-switching attacks and switching attacks

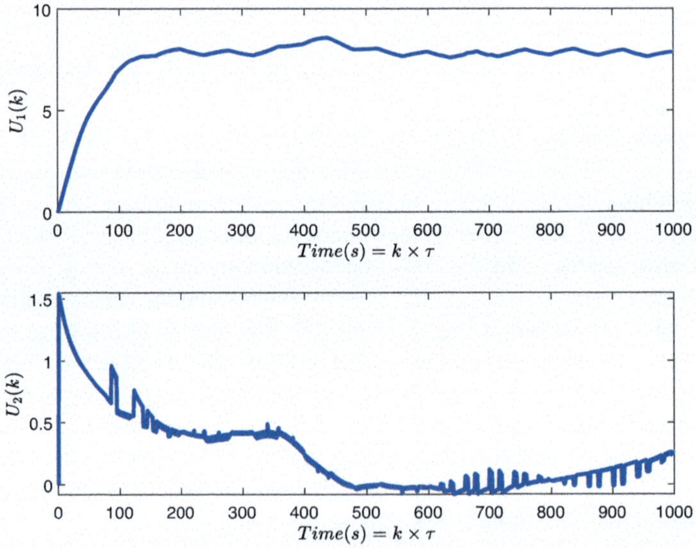

Fig. 8.5 The control input under switching attack

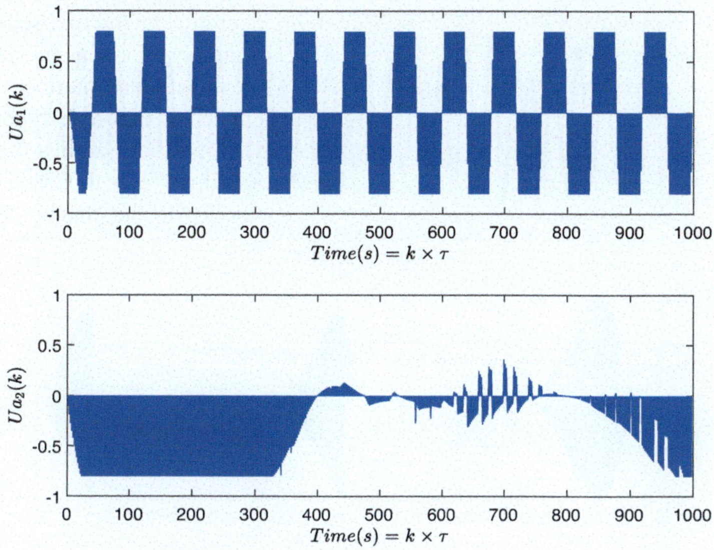

Fig. 8.6 The attack input under switching attack

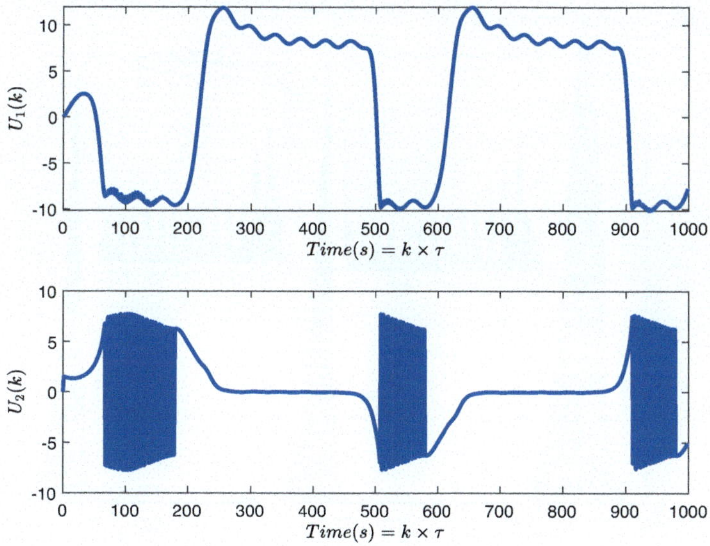

Fig. 8.7 The control input under non-switching attacks

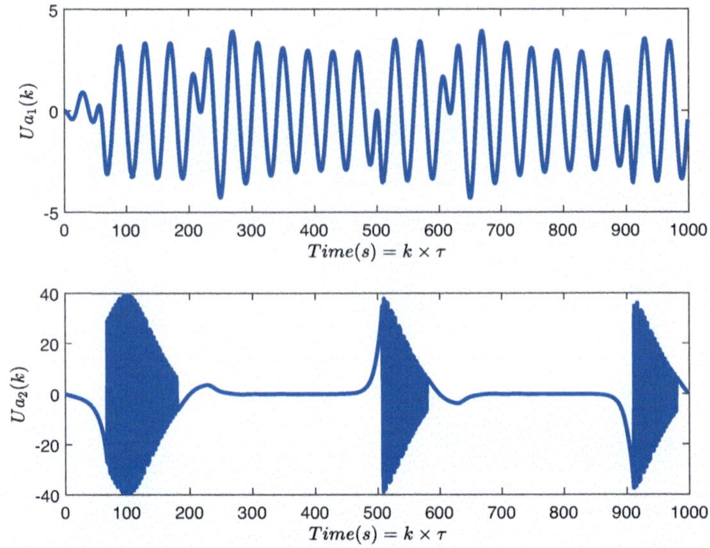

Fig. 8.8 The attack input under non-switching attacks

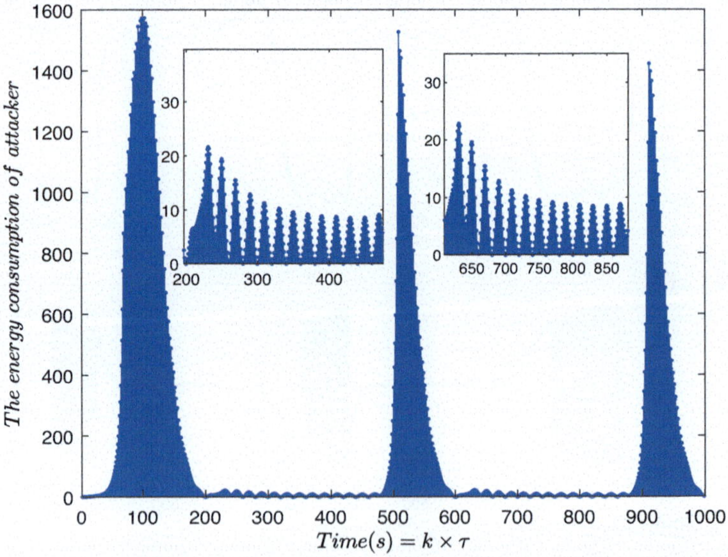

Fig. 8.9 The energy consumption of the attacker uses non-switching attacks

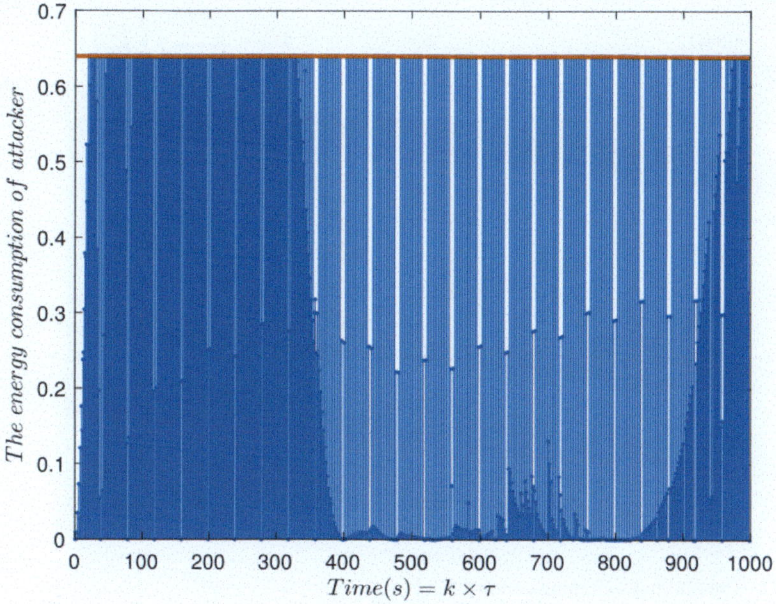

Fig. 8.10 The energy consumption of the attacker uses switching attacks

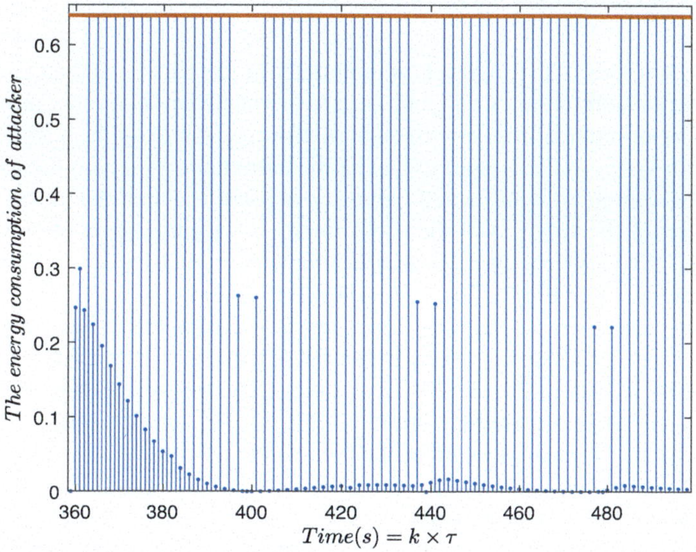

Fig. 8.11 The energy consumption of the attacker uses switching attacks within [360, 500] times

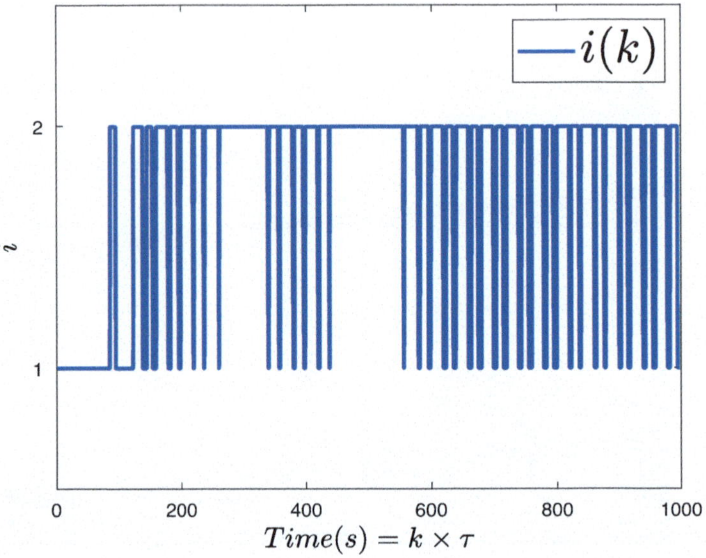

Fig. 8.12 The optimal switching condition $i(k)$

8.4.2 Experiment

In this part, we verify the proposed optimal defense mechanism and attack strategy based on the game framework on Qbot-2e, a two-wheeled mobile robot produced by Quanser. The structure of the experimental platform is shown in Fig. 8.13a, in which the host with Wi-Fi router is used to handle the communication with the robot.

A motion capture system named Optitrack is used to obtain the global position and direction of the robot. And the robot interacts with the Quirc program in the host through Wi-Fi, where Quarc is a rapid control prototype software developed by Quanser, it's shown in Fig. 8.13b, for MATLAB Simulink.

The main parameters related to QBot-2e are shown in Table 8.2. The related matrices γ, U_{a_0}, $x_r(k)$, $y_r(k)$, and $\varphi_r(k)$ are the same as Sect. 8.4.1, the weighting matrices are defined as $\tilde{Q} = \text{diag}(20, 20, 20)$, $R = \text{diag}(2, 2)$, and $\tilde{R} = \text{diag}(5, 5)$, and the switching attack direction matrices are $D^1_{a_k} = [1, 0; 0, 0]$ and $D^2_{a_k} = [0, 0; 0, 9]$. The desired speed and angular velocity of the vehicle are defined as $v_r(k) = 0.6\,m/s < v_{max}$, $\omega_r(k) = 0rad/s$.

The experimental results are shown in Figs. 8.14, 8.15, 8.16 and 8.17. Compared with the Figs. 8.14 and 8.15, the trajectory direction of the robot applied in the design scheme tends to be parallel in the later stage, the green arrow indicates the direction of the unattended trajectory, and the red arrow indicates the direction of the trajectory of the attack, but there is a large error. It can be seen from Figs. 8.16 and 8.17 that the designed scheme can make the system state have the largest error, and the performance suffers the greatest damage under the objective function.

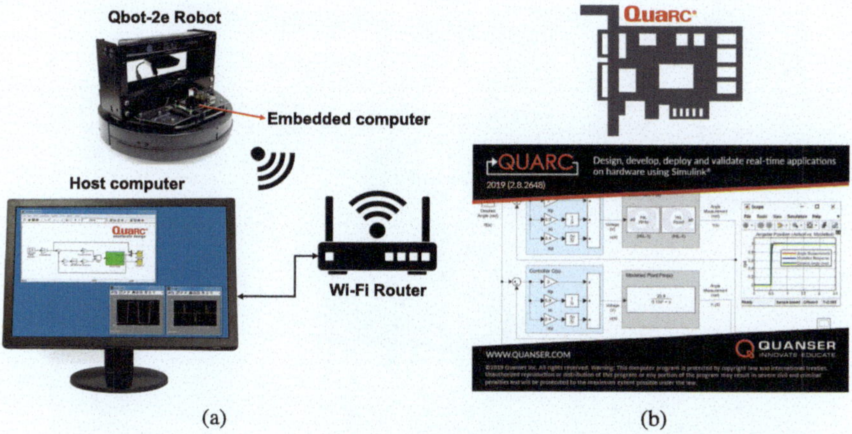

Fig. 8.13 **a** The system under non-switching attacks. **b** The total energy of the attacker uses non-switching attacks

Table 8.2 QBot-2e specifications

Symbol	Description	Value
D	Diameter of the QBot-2e	0.35 m
d	Distance between the left and right wheels	0.235 m
h	Height of the QBot-2e (with Kinect mounted)	0.27 m
v_{max}	Maximum speed of the QBot-2e	0.7 m/s
M	Total mass of the QBot-2e	3.82 kg

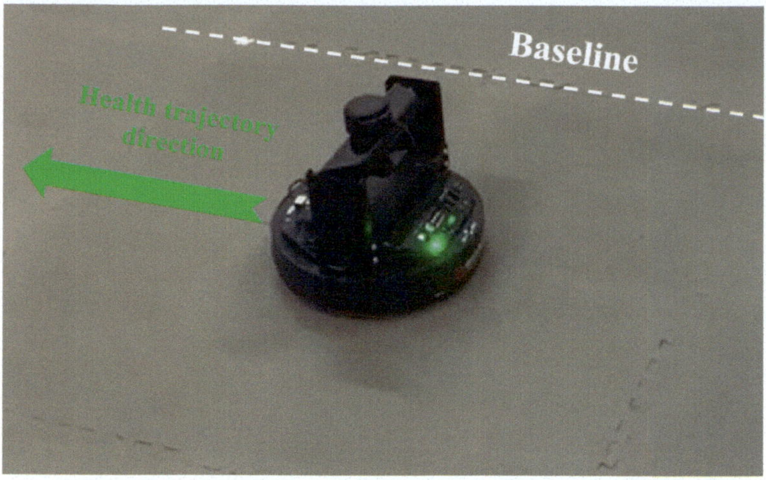

Fig. 8.14 The motion trajectory direction of Qbot-2e under LQR control without attack

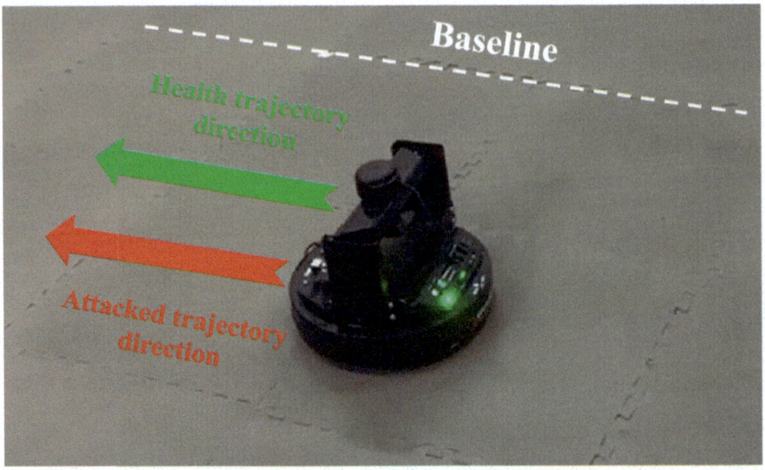

Fig. 8.15 The motion trajectory direction of Qbot-2e under LQR control with designed scheme

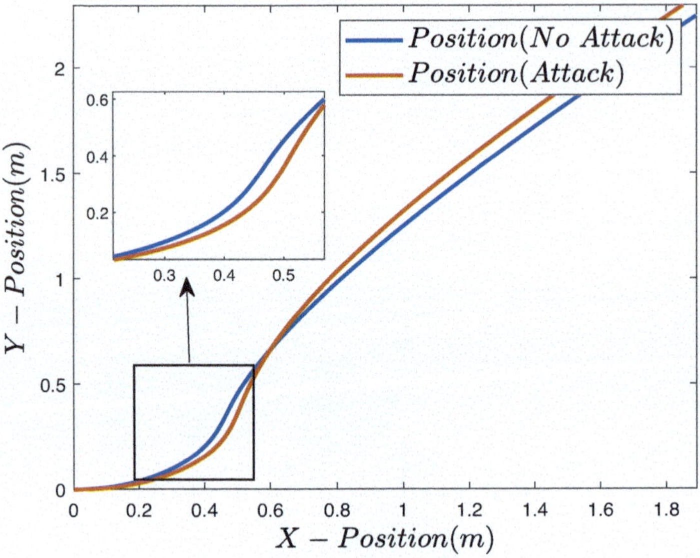

Fig. 8.16 The trajectory of Qbot-2e without attack and with switching attack

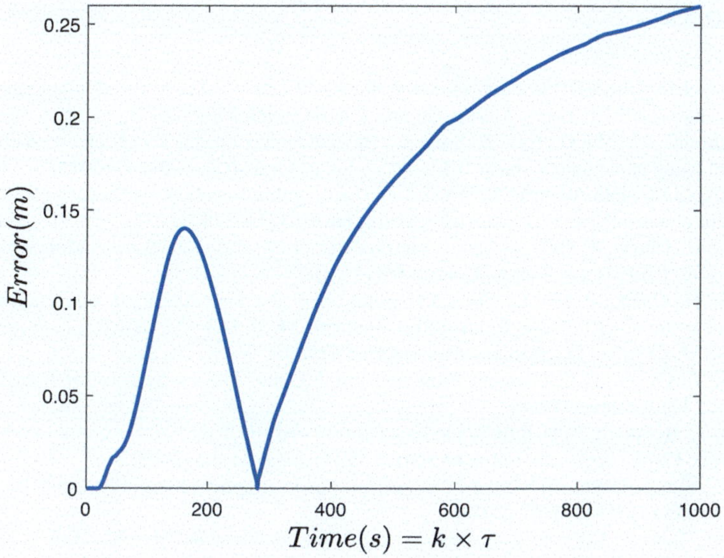

Fig. 8.17 The position error of Qbot-2e under switching attacks

8.5 Conclusion

The problem of zero-sum game-based switching data injection attack with saturation constraint for cyber-physical systems has been solved in this chapter. The physical plant is described as a linear time-varying discrete-time model, and the non-cooperative relationship between the controller and the attacker is transformed into a constrained zero-sum game problem. Under the framework of zero-sum game, optimal control mechanism and switching attack strategy are obtained to achieve the desired cost and system performance. By using the mathematical vector methods and Pontryagin's maximum principle, the optimal switching attack strategy and the game algebraic Riccati equation are derived. Finally, the effectiveness of the proposed method is verified by numerical simulation and physical experiment. In the future research, we will introduce the multi-player game framework to study the non-cooperative relationship between the systems and the attackers and the cooperation and non-cooperative coexistence between the multiple controllers.

References

1. Li, Y., Shi, D., Chen, T.: False data injection attacks on networked control systems: a Stackelberg game analysis. IEEE Trans. Autom. Control **63**(10), 3503–3509 (2018)
2. Esmalifalak, M., Shi, G., Han, Z., Song, L.: Bad data injection attack and defense in electricity market using game theory study. IEEE Trans. Smart Grid. **4**(1), 160–169 (2013)
3. Wu, C., Li, X., Pan, W., Liu, J., Wu, L.: Zero-sum game-based optimal secure control under actuator attacks. IEEE Trans. Autom. Control **66**(8), 3773–3780 (2021)
4. Zhang, H., Cheng, P., Shi, L., Chen, J.: Optimal denial-of-service attack scheduling with energy constraint. IEEE Trans. Autom. Control **60**(11), 3023–3028 (2015)
5. Zhang, H., Cheng, P., Shi, L., Chen, J.: Optimal DoS attack scheduling in wireless networked control system. IEEE Trans. Control Syst. Technol. **24**(3), 843–852 (2016)
6. Lewis, F.L., Vrabie, D., Syrmos, V.L.: Optimal Control. Wiley (2012)
7. Başar, T., Olsder, G.J.: Dynamic Noncooperative Game Theory. Society for Industrial and Applied Mathematics (1998)
8. Başar, T., Bernhard, P.: H-Infinity Optimal Control and Related Minimax Design Problems: A Dynamic Game Approach. Springer Science & Business Media (2008)
9. Wu, G., Sun, J., Chen, J.: Optimal data injection attacks in cyber-physical systems. IEEE Trans. Cybern. **48**(12), 3302–3312 (2018)
10. Wu, G., Sun, J.: Optimal data integrity attack on actuators in cyber-physical systems. In: Proceedings of the 2016 American Control Conference, pp. 1160–1164 (2016)
11. Zhang, H., Wang, Q.: Optimal Control Theory and Application. Higher Education Publication, Beijing (2006)
12. Hu, T., Lin, Z.: Control Systems with Actuator Saturation: Analysis and Design. Springer Science & Business Media (2001)
13. Zhang, X., Li, S., Li, B.: Robust stabilization of fractional-order linear systems subject to input saturation with the fractional order $0 < \alpha < 1$. In: Proceedings of the 2016 Chinese Control and Decision Conference, pp. 1618–1622 (2016)
14. Johnson, C.R.: Positive definite matrices. The Am. Math. Mon. **77**(3), 259–264 (1970)
15. Rusnak, I., Meir, L.: Optimal guidance for high-order and acceleration constrained missile. J. Guid., Control, Dyn. **14**(3), 589–596 (1991)
16. Verbeke, J., Cools, R.: The newton-raphson method. Int. J. Math. Educ. Sci. Technol. **26**(2), 177–193 (1995)

The manufacturer's authorised representative in the EU is Springer Nature Customer Service Centre GmbH, Europaplatz 3, 69115 Heidelberg, Germany. If you have any concerns regarding our products, please contact ProductSafety@springernature.com

Printed and bound by CPI Group (UK) Ltd, Croydon, CR0 4YY
07/10/2026
02229350-0001